STEVE
   BROTHERS
102 1/2 S.W. 6th
COLLEGE PLACE, WA

THE ECONOMY OF NATURE

A Textbook in Basic Ecology

ROBERT E. RICKLEFS

University of Pennsylvania

CHIRON PRESS

INCORPORATED

1816 S.W. Hawthorne Terrace
Portland, Oregon 97201

Library of Congress Catalog Card Number 75-18263

ISBN 0-913462-04-7

Table of Contents

Preface   v

Chapter

1   Introduction   1

2   Life and the Physical Environment   8

3   Aquatic and Terrestrial Environments   23

4   Soil Formation   36

5   Variation in the Environment   50

6   The Diversity of Biological Communities   72

7   Primary Production   110

8   Energy Flow in the Community   128

9   Nutrient Cycling   156

10   Environment and the Distribution of Organisms   180

11   Homeostatic Responses of Organisms   198

12   Evolutionary Responses   219

13   Population Growth and Regulation   235

14   Competition   266

15   Predation   299

16   Extinction   333

17   The Community as a Unit of Ecology   350

18 Community Development   377

19 Community Stability   400

Conversion Factors   419

Glossary   422

Selected Readings and Text References   430

Illustration Credits, Acknowledgments, and
     References   441

Index   449

Preface

The recent proliferation of books in ecology still leaves a gap between long texts that are more or less comprehensive and short books, usually paperbacks, that are inadequate even for beginning courses lasting a quarter or a semester. I have written *The Economy of Nature* to provide a broad, integrated treatment of ecological principles in a book of moderate length.

I have emphasized the dynamics of populations, communities, and ecosystems while keeping sight of the organism as the basic unit of the biological community. I have tried to balance ideas and principles on the one hand with examples of structure and functioning of natural systems on the other. I believe such balance makes clear the complementary roles of theory and observation in the development of science. I have also tried to convey the diversity of biological communities and the remarkable manifestation of basic principles under different environmental conditions.

I have deliberately avoided the problem of man's ecological crisis. One could not hope to do justice to a topic of such importance and complexity in a chapter or two tacked on to the end of a text whose subject is the basic principles of ecology. Where that problem can be understood or solved, students will readily discern applications of the principles discovered in the study of natural systems. I have omitted descriptions of mathematical and statistical techniques in ecology because they do not so much aid our understanding of principles as they provide useful tools for the professional ecologist.

The Economy of Nature is meant to be a basic exposition of ecology, not a source book for advanced students and professional ecologists. I have accordingly omitted literature references and, wherever possible, scientific names in the text so as to give force to the narrative itself. References, selected readings, and source books are listed by chapter at the back of the book.

Philadelphia   R.E.R.

Introduction

Ecology is the study of plants and animals, as individuals and together in populations and biological communities, in relation to their environments — the physical, chemical, and biological characteristics of their surroundings. Ecology has recently been expanded in concept by the awakening perception in many of us that man, like all other creatures, also dwells in an environment with which he must come to terms to ensure his survival. To be sure, man can, like no other animal, modify his surroundings according to his own design. Yet blight, plague, and pollution constantly remind us of mankind's fallibility. If we are going to come to terms with Nature, it will have to be on her terms for the most part. These conditions of Nature, the bounds beyond which man cannot step, are the subject of this book. It is to these basic laws, obeyed by all other organisms, that man must eventually bow.

Ecology has become enough of a popular catchall to include the likes of sanitary engineering, regional planning, paper recycling, and organic gardening. Most of these endeavors are merely attempts, however necessary, to soften the blow of Nature's verdict on our flagrant violations of her laws — our unwillingness to play the game by the old and tried rules — and to delay the sentencing a little while. But as surely as there are no superficial remedies, so the case against man's environmental policy does not lie in superficial evidence: not in the sewage dumped into rivers, or the pesticides sprayed on crops, or the guns and harpoons of the hunters, or the exhausts of our cars, or suburban sprawl. It has grown out of man's failure to heed basic economic principles of nature.

We are all aware, consciously or not, of our environmental predicament. To emphasize its symptoms here would only add to the confusion, indignation, frustration, gloom, despair, or perhaps even apathy in reaction to the unfolding catastrophe we view each day through newspapers, television, and direct contact with our surroundings.

Technological, economic, and political remedies are outside the scope of this book. My aim in writing is to show how natural assemblages of animals and plants are put together and how these assemblages function. I should hope that man's general place in this scheme — how, like other organisms, he is a part of the natural world and how his activities influence the natural world — will be obvious despite my failure to consider his specific case at length in the pages that follow.

The Realm of Ecology: History

The word ecology is derived from the Greek *oikos* meaning house, man's immediate surroundings. The origin of the word in the middle of the last century is obscure, but its general usage can be traced to the definition given by the influential German biologist, Ernst Haeckel, in 1870. "By ecology," he wrote, "we mean the body of knowledge concerning the economy of nature — the investigation of the total relations of the animal both to its organic and to its inorganic environment; including above all, its friendly and inimical relation with those animals and plants with which it comes directly or indirectly into contact — in a word, ecology is the study of all the complex interrelations referred to by Darwin as the conditions of the struggle for existence."

The period in which Haeckel and Darwin worked was a period of exploration. Naturalists were just beginning to discover the bewildering variety of plants and animals and their peculiar ways of life. Charles Darwin's theory of evolution by natural selection had placed the organism in the context of the environment: form and behavior were adapted to the particular environment in which the organism dwelled. Ecology first flourished as the study of natural histories of organisms, the life stories of animals and plants: where and when they were found, what they ate, what ate them, and how they responded to changes in their surroundings. This narrow view of ecology gave way, towards the end of the nineteenth century, to a broader perception of the interrelationships of all plants and animals. Whereas *autecology* related the organism to its surroundings, *synecology*, as this broader view has been called, recognized that assemblages of plants and animals had characteristic properties of structure and function shaped by the environment. Does a grassland not differ from a forest, and can these differences not be related to temperature, rainfall, and soils?

Synecology, and its inherent treatment of all animals and plants together as a biological unit, the *community*, achieved its extreme expression in conceiving the community as a superorganism. The parallels between community and organism are obvious. Each is made up of distinctive subunits. The organism has its liver, muscle, and heart; the community its green plants, predators, and decomposers. A forest community grows on cleared land, progressing through field and shrub stages of succession to its mature form, just as the organism progresses through its developmental stages.

Ecologists have always viewed organisms and communities in the context of their physical environments. Evolutionary adaptations and developmental responses enable plants and animals to respond to variation in the environment. Biological structure and function are molded by their physical surroundings. As recently as the 1940's, ecologists began to realize that the biological community and its environment could be considered together as a single unit. The physical and biological worlds form a larger system — the *ecosystem* — within which the material substances of life are continually passed back and forth between the earth, air, and water on one hand and plants and animals on the other.

A gradually enlarging view has not been the only contributor to the maturity of ecology. Since the first part of this century, ecology has been a meeting ground for ideas from genetics, physiology, mathematics, agriculture, and animal husbandry. Indeed, for many years the infusion of ideas and approaches from other areas of endeavor had diverted the central movement of ecology to many seemingly divergent paths. One led to population biology, another to physiological ecology, a third to the study of community energetics, and so on. We now seem to be on the threshold of a period of coalescence, a recognition of the ties between the separate disciplines of ecology and an achievement of unity.

The Realm of Ecology: Organization

I cannot imagine the form of the ultimate unification of ecology as a science, nor, indeed, be certain whether it shall ever take a form as well ordered as the structure given by theorems to mathematics and by laws to physics. There is a principle that prevents a scientific discipline from gelling prematurely into a mold shaped by misconception and ignorance. I shall, however, try to provide a basic organization, or plan, of the realm of ecology. I do not mean to describe all the conceptual, often semantic, pigeonholes into which ecologists stuff their observations of nature. To keep these distinctions to a minimum is to hasten progress toward understanding nature which, except for rather restricted use by pigeons, has no pigeonholes. Organizations, plans, schemes of classification, are like the crampon, alpenstock, and piton that help the mountaineer on his way to the summit. We take what help we can find, and this book is in part a catalogue, an encyclopedia, of devices, useful now, but to be put aside along the way as the terrain changes and the path to understanding grows more gentle.

Ecology is a three-dimensional construction of horizontal layers stacked on top of each other, representing a hierarchy of biological

organization from the individual through the population and community to the ecosystem, and of vertical sections cutting through all layers,
representing form, function, development, regulation, and adaptation.
If we follow the community layer across its sections, we find a form
section with the numbers and relative abundances of species; a function
section with the interaction between predator and prey populations and
the mutually depressing influences of competitors; a development section with plant succession, as in the reversion of cleared land to forest; a
regulation section with the elusive property of inherent community
stability hiding in a corner; and an adaptation section with the evolution
of antipredator adaptations. Taking a particular stack of sections, that
representing function, for example, we find energy flow and nutrient
cycles in the ecosystem layer; predator–prey interactions and competition between species in the community layer; birth, death, immigration,
and emigration in the population layer; physiology and behavior of
individuals in the organism layer.

Each layer of ecological organization has unique properties of structure and function. Each section in each layer of the construction represents a unique constellation of observed phenomena, perceived patterns, and abstracted concepts. Yet all are presumably governed by a
fundamental set of natural laws. Such laws are the quest of ecology the
science.

The Need for Ecology

One may wonder where this view of ecology as a science fits into the far
vaster construction that is human civilization. How can abstract concepts deduced from observations of unspoiled nature fit next to the
realities of corn blight and eutrophication? Ecology is, of course, of dual
nature. There is the desire for knowledge for its own sake, a kind of
perversity limited, as far as I know, to the human species. But an active
science searching out patterns and explanations must be placed with the
arts and letters in any definition of civilization. And there is the application of knowledge and understanding to solutions of environmental
problems.

The two natures of ecology go hand in hand, because the basic
principles found in the study of natural communities ought to pertain to
disturbed communities as well. The physical and chemical principles of
mineral solubility, chemical reaction, surface tension, and change of
state are all pertinent to the management of soils. When desert soils are
irrigated for crops, for example, water dissolves salts from the soil
particles. Rather than being washed out of the surface layers of the soil,
the salts are deposited at the surface as water is drawn upward by rapid

evaporation from the ground under the hot desert sun. That irrigation would eventually turn thousands of acres of desert into salt flats unfit for growing crops could have been predicted from general principles of chemistry and physics.

The mathematics of population ecology tells us that populations decline when removal of individuals by predators exceeds recruitment through reproduction. Applied to game species, principles of demography show that harvest rates are greatest when hunting effort is adjusted to maintain game populations at levels that are low enough to ensure sufficient food for vigorous reproduction but high enough to provide a high sustained yield. Effective game management is achieved by regulating the hunting season, the daily bag limits, the type of weapon used, and even the price of licenses. Wildlife biologists follow ecological principles when they flood areas to encourage waterfowl and burn shrublands to encourage the new plant growth on which deer thrive. If game management is one's concern, basic ecological principles point the way to sound practice.

Ecology is the mistress of much unheeded advice. Plant geneticists have been trying for years to breed *the* perfect strain of wheat, resistant to all pests and diseases. Even while man applies artificial selection to obtain a breed with superior characteristics, he also applies selection to beetles and viruses and fungi, and so obtains the superior characteristic (for the pest, at least) of their being able to eat and infect man's superior wheat. This kind of evolutionary cat-and-mouse game could, in a less demanding world, lead to a harmless stalemate, while providing employment for many plant geneticists. But nature is not so forgiving. Pushing a strain of wheat towards greater pest resistance might reduce its ability to compete with weeds or resist drought. The adaptations of an organism together are like a ball of clay. It can be pinched out in one place only if it is pinched in somewhere else. There is only so much clay in the ball.

Herbicides and irrigation can balance losses in competitive ability and drought tolerance incurred for the sake of gains in pest resistance, but the costs of chemicals and of irrigation systems may more than offset the gains. I am reminded of the attempt to introduce cotton into Costa Rica. In its drier regions, this tropical country offers an ideal climate and in the beginning cotton crops flourished. As with cotton crops everywhere, insect damage can cause great economic loss. At first, insecticides kept the insect damage under control, but as pest populations evolved resistance to the chemicals, greater and greater quantities of poisons had to be applied. The cotton boom in Costa Rica ended when the cost of pesticides ate up the profits from growing cotton. All that remain are rusting tractors and weed-covered fields. That cotton could have been profitable in Costa Rica without pecticides is doubtful.

That the economic and ecologic cost of the pesticides would be unbearable is, in the wisdom of the hindsight, certain.

I am also reminded of a complex case study of the direct and indirect efforts of man's disturbance to natural communities that points to the need for fundamental understanding of ecosystem function. Clear Lake, California, is a beautiful body of water twenty miles long, once providing idyllic vacations and well-known for its fishing and water skiing. In fact, the only drawback to this earthly paradise was the summer swarming of small gnatlike flies, known as midges, around the edges of the lake. The midges, which passed their larval stage in bottom sediments of shallow parts of the lake, posed no threat to health and did not bite. But their great numbers, congregating around the lights of houses in the evening, were a nuisance.

Suitable control programs eluded the efforts of researchers until the end of World War II, when pesticides such as DDD and DDT became widely available. In 1949, DDD was applied to the entire lake at a level of less than two-hundredths of a part per million of insecticide in the water. This measure was so successful that few midges were found near the lake shore over the next two or three years. After 1951, however, the number of midges began to increase. DDD was again applied in 1954, again resulting in substantial killing off of the bothersome insects. Later that year, residents began to find dead waterbirds washed up on the shore of the lake. No one connected these deaths with the midge control program.

The midge population recovered more rapidly after the second application of DDD than after the first. A third treatment, tried in 1957, was less successful, and biologists began to suspect that the midges had become resistant to the insecticide. That year, large numbers of water birds, mostly western grebes, were found dead along the shore. This time, two dying birds were sent to the Bureau of Chemistry of the California Department of Agriculture. Analyses showed that the midge control program had been a death warrant for Clear Lake. The concentration of DDD residues in the fat of the dead grebes was almost one hundred thousand times greater than the concentration of the original application of insecticide to the lake. It was later discovered that fish had also assimilated and concentrated the poison to levels that made many species unsafe to eat.

The effects of the midge control program had been to breed a DDD-resistant strain of midge that is still a local nuisance, to eliminate the western grebe from the lake and reduce populations of other water birds, and to place in jeopardy the lake's most important asset, its fishery. It is ironic that human activities precipitated the pest problem to begin with: nutrients draining into the lake from fertilized croplands and in the form of raw sewage had enriched the bottom sediments of the lake to favor the growth of midge larvae.

The ultimate results of the midge control program came as a complete surprise. Its designers ignored basic principles of ecology. They did not count on the ability of the midge population to evolve resistance to a toxic insecticide, nor did they foresee that residues of DDD would accumulate in the bodies of animals and be passed from prey to predator up the food chain, concentrated at each step until they became lethal. Unforeseen consequences have plagued nearly all our efforts to bring under control the complicated system of checks and balances that maintains the stability of the natural world. As our activities continue to exert an increasing impact on the environment, control of environmental problems will become more difficult and complex, and recognition and application of basic ecological principles will become all the more crucial to survival of life, including our own.

2

Life and the Physical Environment

We often contrast the living and the nonliving as opposites — biological versus physical and chemical, animate versus inanimate, organic versus inorganic, active versus passive, biotic versus abiotic. While these two great realms of the natural world are almost always readily distinguished and separable, they do not exist one apart from the other. The dependence of life upon the physical world is obvious. The impact of living beings on the physical world is more subtle, but this impact is equally important to the continued existence of life on Earth. Soils, the atmosphere, lakes and oceans, and many sediments turned to stone by geological forces owe their characteristics in part to the activities of plants and animals.

The Uniqueness of Life

Many properties held in common by all forms of life set organisms apart from stones and other inanimate objects. Motion and reproduction are the two most obvious of these properties, even though many plants may be said to move very little indeed, and one might describe the growth of crystals as a kind of reproduction. Exceptional cases which appear to fall on the wrong side of the great fence separating the biotic and the abiotic are less worrisome than they seem at first, for motion is a superficial expression of a more fundamental property of life, namely the ability to perform work directed toward a predetermined goal, and reproduction represents, above all, the emancipation of biological structure and function from the direct influence of physical laws. One might compare the genetic material passed from generation to generation by the act of reproduction to language in its abstraction and specificity.

Organisms are like internal combustion engines transforming energy to perform useful work. Whether the organism directs this work toward pursuing prey, producing seeds, keeping warm, or maintaining such basic body functions as breathing, blood circulation, and salt balance, it perpetually strives to maintain itself *out* of equilibrium with the physical forces of gravity, heat flow, diffusion, and chemical reaction. In a sense, this is the secret of life. A boulder rolling down a steep slope releases energy during its descent but no useful work is performed. The source of energy, gravity in this case, is external and as soon as the boulder comes to rest in the valley below, it is once more brought into equilibrium with the forces in its physical environment.

A bird in flight must constantly expend energy to maintain itself aloft against the pull of gravity. The bird's source of energy is internal, being the food which it has assimilated into its body, and the work performed serves a purpose useful to the bird in its pursuit of prey, escape from predators, or migration. To be able to act against external physical forces is the one common property of all living forms, the source of animation that distinguishes the living from the nonliving. Bird flight may be a supreme realization of animation, but plants just as surely perform work to counter physical forces when they absorb soil minerals into their roots or synthesize the highly complex carbohydrates and proteins that provide their structure.

Physical forces in the environment could not be held at bay without the expenditure of energy to perform work. The ultimate source of energy for life is external, the light from the Sun. Plants have evolved special pigments, among them chlorophyll, that absorb light energy. That energy is then converted to food energy in the form of sugars manufactured from simple inorganic compounds — carbon dioxide and water. The energy trapping process is called *photosynthesis*, literally a putting together with light. Energy contained in the chemical bonds of the manufactured sugars, and thence proteins and fats, may then be used by the plant, or by animals that either eat plants or eat other animals that eat plants, and so on, to perform the work required of an animate existence.

The Interaction of Life and the Physical World

Life is totally dependent on the physical world. On one hand, organisms receive their nourishment from the physical world, and on the other, the distributions of plants and animals are limited by their tolerance of the physical environment. The heat and dryness of deserts prevent the occurrence of most life forms, just as the bitter cold of polar regions prevents the establishment of all but the most hardy organisms. Form and function are also brought under the yoke of the physical world. The viscosity and density of water require that fish be streamlined according to rigid hydrodynamic rules if they are to be swift. The concentration of oxygen in the atmosphere, at 21 per cent, places upper bounds on the metabolic rates of organisms. Similarly, the ability of plants and animals to dissipate body heat — accomplished by the purely physical means of evaporative cooling, conduction, and radiation of heat from the body surface to the surroundings — limits their rate of activity and their safe exposure to direct sunlight.

The activities of organisms also affect the physical world, sometimes in a profound manner. The oxygen that we take for granted with every breath was produced largely by green plants during photosynthesis. Before green plants evolved in primitive seas, the atmosphere of the Earth was composed mostly of methane (CH_4), ammonia (NH_3), water vapor (H_2O), and hydrogen (H_2). As early aquatic plants began to utilize sunlight as a source of energy, they began to liberate oxygen, some of which escaped from the oceans and accumulated in the atmosphere. Over the past two billion years, the span of life on Earth, most of the hydrogen once contained in the Earth's primitive atmosphere has escaped into space. The carbon contained in atmospheric methane and the nitrogen in ammonia have been assimilated by plants, and their place in the atmosphere has been partly taken by oxygen released during photosynthesis. PRIMARY PRODUCTION

Plants play an equally influential role in the development of soil properties. Plant roots find their way into tiny crevices and pulverize rock as they grow and expand. Bacteria and fungi hasten the weathering of rock by chemical means. Fungi secrete acids to dissolve nutritive minerals out of unaltered rock, thereby weakening the crystalline structure of the rock and speeding its decomposition. Rotting plant detritus also releases acids that do the work of chemical decay, while fragments of detritus alter the physical structure of the soil. Animals, by burrowing, trampling, and defecating, also play a part in the development of soil.

The role of plants and animals in maintaining soil characteristics is shown most dramatically when communities are disturbed. The development of the Dust Bowl in the midwestern part of the United States during the 1930's provides a vivid example. The Dust Bowl area is normally dry and windy, but the root systems of the natural vegetation, mostly perennial grasses, were extensive enough to hold the soil in place. When the area was plowed and the prairies converted to agriculture, the perennial grasses were replaced by annual crops, whose root systems were less extensive and which were ploughed up each year. A series of dry years reduced crop growth and turned the soil surface into fine dust. The result, shown in Figure 2–1, is legendary.

Plants also influence movement of water. Rain does not accumulate where it falls. If it did, New York State would be under 200 feet of water within a lifetime. Some water flows over the soil surface or through the underlying earth to enter rivers, lakes, and, eventually, the ocean. The remainder escapes by evaporation from the ground surface and from leaves. The leaf area in an eastern deciduous forest is, on the average, about four times the ground surface area; that is, there are about four acres of leaf surface per acre of forest floor. Plants are thus the major pathway of evaporation. When a forest is cut down, most of the water that normally would have evaporated from the leaves flows instead over the ground surface into rivers. The consequences of clear-cutting without provision for extensive replanting are flooding, increased erosion and the silt deposition that accompanies it, and removal of mineral nutrients from the denuded soil.

Evaporation of water from plant leaves tends to retain water in a locality, for much of the water vapor quickly condenses and falls as rain nearby. In some areas, particularly in the tropics, the presence of forest vegetation increases local precipitation. Extensive clearing of forests for agriculture has caused significant drying trends in some local climates.

The Ecosystem Concept

The interdependence of the physical and biological realms is the basis of the ecosystem concept in ecology. In spite of the ecosystem's being the largest and, in many ways, the most fundamental unit of ecology, the term itself was not used until 1935, when it was coined by the English botanist A. G. Tansley. The ecosystem, he wrote, includes ". . . not only the organism–complex, but also the whole complex of physical factors forming what we call the environment of the biome — the habitat factors in the widest sense. Though the organism may claim our primary interest, when we are trying to think fundamentally we cannot separate them from their special environment, with which they form a physical system."

The biotic and abiotic parts of the ecosystem are linked by a constant exchange of material — nutrient cycles — driven by energy from the Sun. The basic pattern of energy and material flux in the ecosystem is shown in Figure 2–2. Plants manufacture organic compounds, utilizing energy obtained from sunlight and nutrients from soil and water. The plants use these compounds as a source of building materials for their tissues and as a source of energy for their maintenance functions. To release stored chemical energy, plants break apart organic compounds into their original inorganic constituents — carbon dioxide, water, nitrates, phosphates, and so on — thus completing the nutrient cycle.

Plants manufacture their own "food" from raw materials. Hence they are referred to as *autotrophs,* literally self-nourishers. Animals obtain their energy in ready-made food by eating plants or other animals. Animals are therefore referred to as *heterotrophs,* meaning nourished from others. The specialization of living forms as food-producers and

FIGURE 2–1 The Dust Bowl area of the midwestern United States. Wind erosion begins when soils are ploughed but there is insufficient rain for crop growth. At upper left, a winter wheat crop failure in Finney County, Kansas, has resulted in soil blowing (March 1954). At lower left, wind blown soil particles — a dust storm — approaches Springfield, Colorado, in May 1937, during the height of the Dust Bowl tragedy. Dust storms completely destroyed some farming areas, as in Bacca County, Colorado, above.

food-consumers creates an energetic structure, called the *trophic structure,* in biological communities, through which energy flows and nutrients cycle. The food chain from grass to caterpillar to sparrow to snake to hawk depicts the path of organic materials and the energy and nutrient minerals they contain. Each link in the food chain, each trophic level in the community, dissipates most of the food energy it consumes as heat, motion, and, in the case of luminescent organisms, light. None of these energy forms is useful to other organisms. Hence, with each step in the food chain, the total amount of usable energy that passes through to the next higher trophic level becomes smaller. It is no wonder, then, that all the grass in Africa heaped into a big pile would dwarf a similar assemblage of grasshoppers, gazelles, zebras, wildebeests, rhinoceroses, and all other animals that eat grass. So much would the piles of plants and herbivores overwhelm us that we would probably not even notice the pitiful heap of lions, cheetahs, and hyenas nearby.

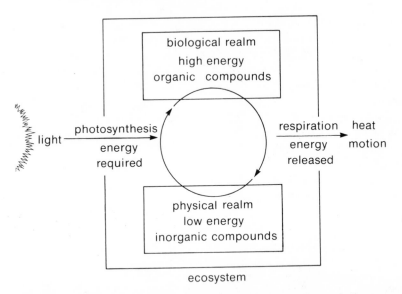

FIGURE 2–2   Schematic diagram of the flow of energy through the ecosystem and the cycling of chemical nutrients within the ecosystem. The biological and physical realms, represented by organic and inorganic compounds, together comprise the ecosystem.

Ecosystem structure and function summarize the activities of all the organisms in the biological community — their interactions with the physical environment and with each other. We must never lose sight of the fact that the lives of organisms are played out for their own benefit, and not to fulfill some role in the ecosystem as an actor fills a role in a drama. The properties of the ecosystem result from the self-serving activities of the plants and animals they contain, and this is where understanding of the ecosystem structure and function must be sought. Before considering the properties of the physical environment and over-all organic productivity, energy flow, and nutrient cycles in ecosystems, we ought to stop briefly to consider the lives of a few kinds of organisms, intriguing by themselves and indicative of the forces that shape the structure and functioning of ecosystems.

The Giant Red Velvet Mite

The Mohave Desert is a forsaken land of searing summer heat, bitter winter cold, and year-long drought. Little rain falls because clouds are intercepted by the mountains that rise between the desert and the coast of southern California. Except for struggling desert plants, little life is in evidence during most of the year. But the desert's stillness is occasionally broken during the milder days of the winter by swarms of insects and other creatures that appear on the surface or flying above it for a few hours and then disappear as mysteriously as they came. One of the more conspicuous of these creatures is the giant red velvet mite (Figure 2–3) whose scientific name, *Dinothrombium pandorae,* paints a vivid picture of this close relative of spiders. The generic name, *Dinothrombium,* is derived from the Greek *deinos,* meaning terrible, and *thrombos,* a lump or clot. This particular species is named after the mythological woman Pandora, who was sent by Zeus to bring evil as a counterbalance to the gift of fire Prometheus, in disobeying Zeus, had made to the human race. For her dowry Pandora carried a box containing all the ills humans might suffer. When her fatal flaw, curiosity, compelled Pandora to open the box, the ills escaped and have plagued the human race ever since.

Several years ago, biologists Lloyd Tevis and Irwin Newell became interested in the behavior of the mite and began a study of its activity in relation to the physical conditions of its environment. They found that the mites spend most of the year in their burrows, which are dug only where the grains of sand measure less than one-half millimeter in diameter. The particular environmental conditions that are favorable for the emergence of the giant red velvet mite occur infrequently in the Mohave Desert. During four years of observation, adult mites appeared above ground only ten times, always during the months of December, January, or February, when temperatures reach a level low enough for the mite to tolerate the desert's surface. On the basis of their observations, Tevis and Newell could predict that emergence would occur on the first sunny day after a rain of more than three-tenths of an inch, provided that air temperatures were moderate. An individual mite appeared only once each year.

FIGURE 2–3   Adult giant red velvet mite, on the surface of the ground (left) and in a vertical burrow (right, a cutaway view of the underside of the mite).

On the day of a major emergence, the mites come out of their burrows between 9:00 and 10:00 A.M., and by late morning many hundreds of mites scurry across the desert sands in all directions. At midday, between 11:30 and 12:30, the mites dig back into the sand where they wait until the following year before emerging again. The mites do not leave their burrows to terrorize unsuspecting desert travellers; during their two- or three-hour stay above ground each year, mites perform two important functions: feeding and mating. On the same day the mites emerge, large swarms of termites appear, flying over the desert sands, their emergence presumably triggered by the same physical conditions that cause the mites to leave their burrows. It is upon these termites that the mites feed. Flying termites cannot, of course, be caught by the flightless mites; each mite must locate its prey after the termites drop to the ground and shed their wings but before they burrow into the sand. All this happens very quickly, giving the mites about an hour to find their prey.

Because the mites are solitary, they must mate as well as feed during their brief sojourn above ground each year. The courtship pattern of the giant red velvet mite resembles that of spiders and their relatives. The males walk nervously around and over a feeding female, tapping and stroking her, and cover the sand around her with loosely spun webs. Males court *feeding* females for two reasons. First, the females have ravenous appetites (they haven't eaten for a year, after all) and are probably just as likely to devour a male mite as a termite. Second, and probably more important, females can produce eggs only after they have had a meal. Thus, by mating with a feeding female, the male guarantees that his reproductive efforts will not be wasted.

At about midday, after feeding and mating have taken place, the mites congregate in troughs on the windward sides of sand dunes, where surface temperatures and the size of the sand particles are just right, and burrow into the sand. The mites continue to dig their new burrows on the first day until the coolness of the late winter afternoon slows their activity. Burrowing continues on subsequent days, when the sand becomes warm enough, until the burrows are complete. During the rest of the year, the adult mite spends its time moving up and down in its burrow to follow the movement of the suitable temperature zone as the surface of the sand is heated and cooled each day.

Eggs are laid early in spring. They soon hatch, and young mites crawl to the surface of the desert to search for a host to attach themselves to, usually, a grasshopper. While they are growing, the young mites remain attached to their host, obtaining their nourishment from its body fluids. When they are full grown, the mites drop off their unwilling hosts, and seek suitable spots to dig their own burrows in the sand, thus renewing the life cycle of the giant red velvet mite.

The Oropendola and the Cowbird

The tropical surroundings of the chestnut-headed oropendola in Panama are a far cry from the rigorous environment of the giant red velvet mite. In the tropics, air temperature varies little during the year and abundant rainfall maintains the lush tropical vegetation. Life abounds, and diverse animals and plants are intricately interwoven into a rich fabric of biological interactions. In the harsh environments of the desert and polar regions such interactions are noticeably simplified. In the tropics we are not so aware of adaptations to the physical environment, rather we are impressed by adaptations to biological environments.

The situation we are going to examine involves two birds — the chestnut-headed oropendola and its brood parasite the giant cowbird (Figure 2-4). As we shall see, two insects — a fly and a wasp — also play an integral part in the interaction between the two birds. The biological sleuthing that uncovered this story was performed by Neal Smith, a staff biologist at the Smithsonian Tropical Research Institute in the Panama Canal Zone. Smith's insight, inventiveness, and perseverance proved a good match for the complexity of the oropendola–cowbird relationship.

Brood parasites (cowbirds and cuckoos are familiar ones) have been known for a long time. They are so named because the female lays her eggs in the nest of another species, the host, and the young are raised by the foster parents. Naturalists used to believe that the presence of a brood parasite in a nest always reduced the survival of host young because brood parasite young compete for food brought in by the host parents. As one would expect, many potential hosts can detect the presence of parasite eggs in their nest and eject them. To counter this defense, many brood parasite species have evolved an elaborate egg mimicry to fool the host species into accepting the alien egg as one of its own.

Smith was aware that nesting colonies of the chestnut-headed oropendola, usually consisting of ten to a hundred nests, were often parasitized by the giant cowbird. He observed a curious phenomenon. Examining the eggshells that had been thrown out of the nests by the females after the young had hatched, Smith noted that under some colonies the eggshells of the cowbirds were distinctly different from those of the oropendolas, whereas in other colonies eggshells of the two species so closely resembled each other that they could be distinguished only on the basis of shell thickness. Thus it appeared that in some colonies the cowbirds had evolved to mimic the eggs of their host while in others they had not. These promising observations were to form the basis of a detailed study of brood parasitism.

FIGURE 2–4   Female chestnut-headed oropendola at her nest (above) and giant cowbird (below). The oropendola egg (left) is compared to a mimicking cowbird egg (center) and a nonmimetic cowbird egg (right).

Smith's first problem was to devise methods of studying bird nests suspended from the outlying limbs of large trees, 20 to 60 feet above his head. One generally avoids climbing trees in the tropics because of all the other creatures that might be climbing the same tree. And even if one could reach the top of a tree without an unpleasant encounter, climbing out on the small branches from which the nests are treacherously suspended would be challenging even for a circus acrobat. To remove the nests, examine and manipulate their contents, and then replace the nests in their original position seemed impossible. The nest is constructed in the form of a long, pensile bag, made of interwoven grasses and small vines. But it should be feasible, Smith reasoned, to cut or rip the nests down with the aid of long poles and then reattach them on the original site. Most ecologists would have difficulty imagining themselves standing atop a 15-foot ladder, balancing 48 feet of extendable aluminum poles with instruments at the tip controlled from below by long ropes. But with this apparatus, Smith performed the delicate task of lowering a nest of fragile eggs 50 feet to the ground, then examining the contents and replacing it — all of this carried out at night, under the duress of incessant mosquito attack, so as not to provoke the female oropendolas to abandon their nests.

Using a formidable, Rube Goldberg array of pincers and flaps connected by lever and pulley, Smith was able to replace the nests in their original positions with a sticky variety of contact tape or contact glue. In the most highly developed version of nest replacement, he stapled rat snaptraps to the end of the nest. These clung tenaciously to any tree limb they were pressed against.

Having mastered the technique of working with nests, Smith set out to determine whether the presence or absence of egg mimicry in the cowbirds elicited different oropendola behavior toward foreign eggs. A number of objects — including mimicking cowbird eggs, nonmimicking cowbird eggs, other kinds of eggs, and a variety of other objects only remotely, if at all, resembling eggs — were put into nests in both kinds of oropendola colonies, those that tolerated nonmimicking eggs and those that did not. Smith found, as he had originally suspected, that in oropendola colonies where the cowbird eggs closely mimicked those of their hosts, the oropendolas removed virtually everything from their nests except their own eggs and very closely matching cowbird eggs. These oropendolas were discriminators and tried, often in vain, to discover and eject the cowbird eggs. On the other hand, in oropendola colonies where cowbirds were poor egg mimics, the oropendolas were willing to accept all sorts of foreign objects. Smith had thus identified two types of oropendola colonies, one in which the oropendolas discriminated against cowbirds and tried to eject everything from their nests but their own eggs, the other in which oropendolas were nondiscriminators and accepted notably different eggs and other materials as well.

In what other ways did the two colonies differ? In the nondiscriminator colonies, were the cowbirds actually beneficial to the oropendolas? Why else would their eggs be tolerated by the oropendolas? While making an extensive survey of the oropendolas in the Panama Canal Zone and nearby Panama, Smith found that in nondiscriminator colonies, young oropendolas were often infested with the larvae of a species of bot fly. The parasites sometimes killed the nestling oropendolas and frequently so weakened them that their chances of surviving to adulthood were slim. In discriminator colonies these parasites were rarely present. Here was a major difference between the two types of colonies. Was it possible that the role of the cowbird in the colonies was linked to bot fly parasitism?

Smith examined oropendola young in nondiscriminator colonies (susceptible to bot fly parasitism) and discovered that the incidence of bot flies was higher in nests that did not contain cowbirds than in nests which did, as shown by the following data:

| | Number of nestling oropendolas in nests | |
	With cowbirds	Without cowbirds
With bot fly parasites	57	382
With no parasites	619	42

Further observations plainly showed that the nestling cowbirds would snap at anything small, including adult bot flies, that moved within the nest and, furthermore, that they would remove bot fly larvae from the skin of the nestling oropendolas. This behavior on the part of the young cowbird benefitted the oropendola and accounted for the acceptance of the brood parasites in nondiscriminator colonies.

Cowbird young are well suited for grooming their nest mates. They hatch five to seven days before the oropendola young and develop precociously. Their eyes are open within 48 hours after hatching, whereas the eyes of oropendola nestlings open six to nine days after hatching. Also, the cowbird young are born with a thick covering of down, absent in the oropendola young, which presumably deters bot flies from laying their eggs on the skin of the young cowbirds. By the time the oropendolas hatch, the cowbirds are sufficiently developed to groom the oropendola young. In discriminator colonies, which are not troubled by bot fly parasitism, cowbird young perform no such useful function for the oropendolas, and because they compete with the oropendola young for food, they are detrimental to the productivity of the colony.

The role of the cowbird in this story is now evident, but we have not yet determined why bot flies are present in some colonies and absent from others. Smith noted that all the discriminator colonies, and none of the nondiscriminator colonies, were built near the nests of wasps or bees, which swarm in large numbers around their nests and virtually fill the air throughout the oropendola colony. Wasps and bees presumably prevent the bot flies from entering the colonies. Occasionally Smith found the detached wings of bot flies beneath the nests of discriminator colonies. By hanging up rolls of fly paper in the two types of colonies, Smith found that adult bot flies rarely enter the area around discriminator nests. But the protection is not perfect. Nests on the periphery of the discriminator colonies, and thus at some distance from the wasp and bee nests, are occasionally parasitized by bot flies. Because the protective influence of the wasp extends over a limited distance, discriminator colonies tend to be much more compact than nondiscriminator colonies, with nests placed close together around the wasp nests at the center.

For the oropendola, the presence or absence of wasps in the vicinity of the colony completely alters the role of the cowbird as a factor of the environment. Accordingly, the behavior of the oropendolas towards the cowbirds and their eggs also varies between the two types of colonies, and in turn affects the environment that molds the behavior of the cowbird. In discriminator colonies, adult oropendolas not only eject the eggs of cowbirds from their nests if they can distinguish them, but they also chase adult cowbirds out of the colony. The oropendolas in nondiscriminating colonies are indifferent to cowbirds. Behavior of the cowbird in the two types of colonies reflects this difference. In discriminator colonies, female cowbirds are cautious and always enter the colony singly. Behavioral adaptation has gone so far that the cowbirds mimic the behavior of the discriminator oropendolas; they often gather the stems of small vines and act as if they are beginning to build a nest, a most uncharacteristic behavior of brood parasites. On the other hand, cowbirds that parasitize nondiscriminator colonies are often gregarious and enter the colonies in small groups. They behave aggressively towards the oropendolas and sometimes even chase them from their nests.

It is not sufficient for a successful egg mimic merely to produce an egg that is indistinguishable from host eggs. It must also take care to lay the egg at the proper time. A female oropendola normally lays her two eggs on consecutive days. She is very sensitive to the appearance of new eggs in the nest before or after her laying period, and will frequently desert her nest if it contains more than three eggs. Even when confronted by perfect egg mimicry, a discriminating oropendola can be fooled only if a single egg is laid by the cowbird soon after the first oropendola egg has been laid. If the cowbird lays its egg a day too early or too late, it reveals its presence and the oropendola will abandon the

TABLE 2-1   Biological attributes of discriminator and nondiscriminator colonies of the chestnut-headed oropendola.

| | Colony type | |
	Discriminator	*Nondiscriminator*
Wasp nests	Present	Absent
Possibility of bot fly parasitism	Slight or absent	Heavy
Effect of cowbird on oropendola	Disadvantageous	Advantageous
Foreign objects in nest	Rejected	Accepted
Cowbird eggs	Mimetic	Nonmimetic
Cowbird eggs per nest	One	Several
Cowbird behavior in colony	Timid	Aggressive
Colony structure	Compact	Open
Nesting season	Late	Early

nest and start another. In nondiscriminator colonies, however, neither the number of cowbird eggs nor their appearance matters to the oropendola. Commonly, a female cowbird lays two to five eggs over several days in the nest of a single oropendola. Smith refers to these birds as dumpers.

The interaction between the oropendola and the cowbird shows how the environment determines the adaptations of the organism and how two or more kinds of organisms can be major factors in each other's environments (Table 2-1). The oropendola, the cowbird, the bot fly, and the wasp are all mutually important to each other and affect the evolution of one another. This relationship differs sharply from the relationship of an organism to the physical environment, which neither evolves nor responds by adaptation to changes in the biotic environment. Whereas the physical environment is passive, the biotic environment is responsive. It continually readjusts to evolutionary changes in any one of its living components. We might expect evolution in a biotically dominated environment to differ from evolution in a physically dominated environment. Different populations, like the oropendola and the cowbird, may coevolve with respect to each other into a mutually beneficial relationship that could not be achieved between a population and its physical environment.

3

Aquatic and Terrestrial Environments

Life arose in the sea. Conditions in shallow coastal waters were ideal for the development and diversification of the first groups of plants and animals. Temperature and salinity were relatively constant; sunlight, dissolved gases, and minerals were abundant. Water itself is buoyant and supports both delicate structures and massive bodies with equal ease.

The difficulty of the first step in colonizing the land is intimated by the gap of several hundred million years between the time life began to flourish in the sea and the appearance of life on land. Yet in spite of the harshness of terrestrial environments, life has attained a high degree of organic productivity and diversity on the land.

Perhaps we should not distinguish environments as being primarily aquatic or terrestrial, for the sea is as surely underlain by land as the terrestrial environment is drenched in an ocean of air. To appreciate fully the distinction between aquatic and terrestrial environments, we should contrast the properties of water and air rather than those of water and earth.

The qualities of water which overwhelmingly determine the form and functioning of aquatic organisms are its density (about 800 times that of air) and its ability to dissolve gases and minerals. Water provides a complete medium for life; most marine organisms are independent of the land beneath them except for those that use it as a site for attachment in shallow water or a place to burrow. In contrast, terrestrial life is narrowly confined to the interface between the atmosphere and the land, each of which makes essential contributions to the environment of life. Air provides oxygen for respiration and carbon dioxide for photosynthesis, while soil is the source of water and minerals.

Density and Viscosity of Water and Air

Water provides support for organisms, but most are slightly more dense than salt or fresh water. Aquatic plants and animals have evolved a wide variety of structures that prevent or retard sinking. Fish have swim bladders, small enclosures within the body that are filled with gas to equalize the density of the body and the surrounding water. Many large kelps, a type of seaweed found in shallow coastal water, have analogous gas-filled organs. They are attached to the bottom by holdfasts and gas-filled bulbs float their leaves to the sunlit, oxygen-rich surface waters. Microscopic unicellar algae (*phytoplankton*) float in vast numbers in the surface waters of lakes and oceans. These plant cells contain minute drops of oils which, being less dense than water, compensate for the natural tendency of the cells to sink. Tiny marine animals often have long, filamentous appendages that retard sinking, just as a parachute slows the fall of a body through air (Figure 3–1). The wings of maple seeds, the spider's silk thread, and the tufts on dandelion and milkweed seeds, provide a similar function and increase the dispersal range of land species.

Fast-moving aquatic organisms must be streamlined to reduce the drag encountered in moving through a medium as viscous as water. Mackerel and other schooling fish of the open ocean closely approach the physicist's body of ideal proportions (Figure 3–2). Air offers far less resistance to movement, having less than $\frac{1}{50}$ the viscosity of water.

Because water is more buoyant than air, gravity does not limit the maximum size of organisms to the extent that it does on land. Whales may attain more than 100 feet in length and weigh more than 100 tons, dwarfing the largest land animals. (Large elephants weigh only seven

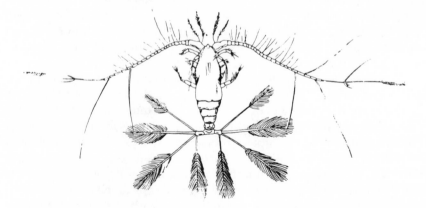

FIGURE 3–1   Filamentous and feathery projections from the body of a tropical marine planktonic crustacean (overall length about 1.2 millimeters).

FIGURE 3–2   The streamlined shapes of young mackerel reduce the drag of water on the body and allow the fish to move rapidly with minimum energy expenditure.

tons.) That water provides excellent support against gravity is illustrated by the skeletons of sharks, which are composed of flexible cartilage rather than bone, and therefore would offer little support on land. Even the air-breathing whales suffocate quickly when accidently stranded on a beach because their great weight deflates their lungs. By contrast, terrestrial organisms have rigid structures to keep their bodies upright against the pull of gravity. The bony internal skeletons of vertebrates, the chitinous exoskeleton of insects, the rigid cellulose walls of plant cells all provide a similar function: support. Rigid structures occur in aquatic organisms more for protection (the shells of molluscs) or to provide attachment sites for muscles (the shells of crabs and bony skeletons of fish) than to support the weight of the body.

Light

Light strikes the surface of the oceans and the land with equal intensity. On land, most light is absorbed or reflected by the leaves of plants. In fact, light is often less than abundant in terrestrial environments and is competed for by plants. Why do trees lift their leaves so far above the ground, if not to reach above the leaves of their neighbors to obtain more light? Where lack of water prevents vegetation from covering all the habitat, incident light is reflected back into space or is absorbed and converted to heat, warming the ground surface.

The transparency of a glass of pure water is deceptive. Water absorbs and scatters light strongly enough to severely limit the depth of the sunlit zone of the sea. Because photosynthesis requires light, the depth to which plants are found in the oceans is limited by the penetration of light into a fairly narrow zone close to the surface, in which photosynthesis exceeds plant respiration, called the *euphotic* zone. The lower limit of the euphotic zone, where photosynthesis just balances the rate of respiration, is called the *compensation point*. If algae in the phytoplankton sink below the compensation point or are carried below it by downward water movements, and not soon returned to the surface by upwelling currents, they will die.

In some exceptionally clear marine and lake waters, particularly in tropical seas, the compensation point may be 100 meters below the surface, but this is a rare condition. In productive waters with dense phytoplankton, or in turbid waters with suspended silt particles, the euphotic zone may be as shallow as one meter. In some exceptionally polluted rivers, little light extends beyond a few centimeters.

Because plants require light, large *benthic* algae (forms attached to the bottom) occur only near the edges of continents where the depth of the water does not exceed 100 meters. In the vast open reaches of the oceans, as well as in shallower coastal water, one-celled floating plants compose the phytoplankton of the euphotic zone. The small, floating animals (*zooplankton*) that prey upon the phytoplankton are also found primarily in this region, where their food is most abundant. But animal life is not restricted to the uppermost layers of water. Even the deepest parts of the ocean, under several miles of water, harbor a diverse fauna fed by the constant rain of dead organisms that sink from the sunlit regions above.

Oxygen

Nearly all organisms, including green plants, require oxygen for *respiration* (the biochemical release of energy from organic compounds). Although oxygen abounds in the atmosphere, comprising about one-fifth its weight, oxygen does not dissolve readily in water. Even for terrestrial organisms, whose bodies are mostly water, the procurement of oxygen, and its distribution throughout the body, is a major problem. The manner in which different organisms have solved this problem demonstrates the important influence exerted by physical properties of the environment on the form and functioning of organisms (Table 3–1).

Small aquatic organisms obtain oxygen by *diffusion* from the surrounding water into their tissues. Terrestrial plants also rely on diffusion for gas exchange with the atmosphere. Diffusion is a physical process by which molecules tend to move from regions of high concentration to regions of low concentration until their distribution is uniform. If the concentration of oxygen in an organism's tissues is lower than in the surrounding medium, oxygen diffuses *into* the body. Because animals utilize oxygen in respiratory metabolism, the concentration of oxygen in the body is kept low and continually tends to diffuse into the organism's tissues. Diffusion cannot, however, supply adequate oxygen to tissues over distances greater than one millimeter. This problem is circumvented in large organisms by circulatory systems that move body fluids from the organism's surface — its skin, gills, or lungs — to tissues deep within its body.

Although circulation greatly aids the distribution of oxygen (and other substances) throughout the body, water cannot carry enough dissolved oxygen to support rapid metabolism. The solubility of oxygen in water (up to 1 per cent by volume, or about 0.0014 per cent by weight) just cannot supply sufficient oxygen to fulfill the needs of active tissues. Many groups of animals have complex proteins, such as hemoglobin, in their blood which increase oxygen carrying capacity. Oxygen readily combines with these molecules, whereby it is removed from the blood plasma, in which the abundance of oxygen is limited by its ability to

TABLE 3–1   Some problems encountered in supplying oxygen to the tissues of large organisms, and their solutions.

Problem	Solution	Biological occurrence
Not critical in small or inactive organisms	Oxygen obtained by simple diffusion through cells	Protozoa, sponges, coelenterates
Diffusion distance from surface to body core is too great in large organisms	Circulatory system pumps fluids from surface to core	Widespread pumping by body muscles in roundworms; open system without capillaries in arthropods and many molluscs; closed capillary systems in vertebrates
The solubility of oxygen in water limits oxygen transport by circulating fluids	Oxygen-binding proteins (e.g., hemoglobin) incorporated into blood	Hemoglobin is widespread in vertebrates, but sporadic in lower groups where other pigments may be found; insects notably lack blood pigments because air is carried directly to cells by a tracheal system
High concentrations of protein increase blood viscosity	Blood–respiratory proteins packaged in red blood cells	All vertebrates, some molluscs and echinoderms

dissolve in water. Hemoglobin, for example, can carry 50 times more oxygen than plasma. Large quantities of protein in blood create the additional problem of increasing blood viscosity. Vertebrates and some marine invertebrates have overcome the problem of thick blood by packaging hemoglobin in red blood cells which easily slip past each other in the blood stream.

Many of these adaptations for procuring oxygen are adjusted in response to variation in the availability of oxygen in the environment. The solubility of oxygen in water decreases when either temperature or salinity increase. Under conditions that most favor solubility — 0°C in fresh water — oxygen is less than one-fourth as concentrated as in air. Natural bodies of water never contain as much dissolved oxygen as conditions of temperature and salinity would allow. One rarely finds concentrations of oxygen greater than 6 cc per liter of water — about $\frac{1}{30}$ the concentration of oxygen in air. Stagnant waters, particularly in swamps or at the bottom of *eutrophic* lakes, where bacterial decomposition of organic matter uses up available oxygen, may completely lack oxygen. An environment without oxygen is said to be *anaerobic*.

Fish living in stagnant water and birds and mammals living at high altitudes, where air is less dense and hence oxygen is less available than at sea level, usually have higher concentrations of hemoglobin in their blood than animals living in habitats with more oxygen. The hemoglobin molecule itself becomes adapted in the presence of low oxygen levels in the environment to bind oxygen more readily, thereby increasing the ease of oxygen procurement. In humans dwelling at sea level, the normal oxygen capacity of blood is 21 per cent by volume. After volunteers in one experiment spent several weeks at 17,600 feet, the oxygen capacity of their blood increased to 25 per cent, mostly owing to increased amounts of hemoglobin, but the response fell short of the 30 per cent oxygen capacity of permanent residents of the high-altitude locality. In addition to changes in hemoglobin, adaptation of lung size, breathing rate and volume, heart size, rate and stroke volume of the heart beat, and capillary density also influence the procurement of oxygen.

The rate at which an organism can obtain dissolved oxygen from water depends in part on how rapidly it can pass water through its respiratory structures. The high viscosity of water compared to air, therefore, further restricts the availability of oxygen to aquatic organisms. Clams and some fish pump water past their gills in an unceasing stream. Other fish swim constantly to maintain an adequate flow of water. By comparison, terrestrial animals can move air in and out of their lungs rapidly.

Oxygen is evenly distributed in the atmosphere, but its concentration in water varies considerably because oxygen diffuses slowly. The abundance of oxygen in water generally decreases away from the interface of air and water, being lower at the bottom of a pond than at its surface, for example. Still water usually contains less oxygen than rapidly flowing water, whose riffles, waterfalls, and waves cause extensive mixing of water and air. Photosynthesis also contributes oxygen to water. The concentration of dissolved oxygen in most aquatic environments increases during the day, owing to the photosynthetic production of oxygen. This increase is balanced at night when plant and animal respiration consumes oxygen and none is produced by photosynthesis. Although carbon dioxide is more abundant and better distributed in water than is oxygen, its abundance varies in a daily cycle inverse to that of oxygen.

Thermal Properties of Water

Heat and temperature are often used interchangeably to refer to some property we sense in bathwater, but the terms have quite distinct meanings. *Heat* is a measure of the energy content of a substance, the total kinetic energy of its molecules. *Temperature* is a measure of the rate of motion of molecules in a substance. At a given temperature, individual molecules in different substances have similar kinetic energies, but the substances may contain different amounts of heat energy depending on their density and the relative weights of their molecules. For example, a cubic meter of water at 30°C contains about 500 times more heat than the same volume of air at 30°C, because the water contains so many more molecules.

Changes in the amount of heat energy in a substance are related to temperature change by a quantity known as the *specific heat,* the amount of heat energy that must be added to or removed from a substance to change its temperature. A small amount of heat applied to a given volume of air changes its temperature rapidly, but the same amount of heat added to an equal volume of water would change its temperature little. This is why it takes longer to bring a large pot of water to a boil than to heat up an oven, and why air temperatures fluctuate between greater extremes than the temperatures of oceans and large lakes.

The high *specific heat* of water (about 500 times that of an equal volume of air) and the high *thermal conductance* of water (rate of diffusion of heat through water: approximately 30 times that of air) result in a thermally constant and uniform environment.* Even in small lakes, surface water temperatures do not vary more than a few degrees each

---

* Specific heat is a measure of the amount of heat energy that must be added to or removed from a substance to cause a given change in temperature. Because the specific heat of water is so much greater than that of air, the temperature of water changes more slowly. Thermal conductance is a measure of the rate at which heat energy is transferred through a substance between regions of high and low temperature.

day when the temperature of the surrounding air may vary as widely as 10 to 20°C. Water similarly moderates seasonal variation in temperature. Aquatic organisms are not faced with the extremes of temperature experienced by terrestrial organisms living outside the tropics.

The thermal conductivity of water enhances the uniformity of the aquatic environment, but at the same time causes heat to be drawn rapidly from the bodies of aquatic organisms. Air is a better insulator than water, a quality that has permitted the evolution of warm-bloodedness in terrestrial mammals and birds. The limited availability of oxygen in water has further restricted the evolution of warm-bloodedness in aquatic animals. The trait is limited to air-breathing aquatic mammals and birds (whales, seals, penguins, and ducks), which can maintain high metabolic rates.

Minerals

Organisms are not made only of hydrocarbons. They require a wide variety of mineral elements to form their structure and maintain their proper function. The elements required in greatest amount, after hydrogen, oxygen, and carbon (which are incorporated into living matter by photosynthesis), are nitrogen, phosphorus, sulfur, potassium, calcium, magnesium, and iron. Their primary functions are summarized in Table 3-2. Many other nutrients are known to be required in smaller quantity.

Mineral nutrients are acquired by plants in the form of dissolved

TABLE 3–2   Major nutrients required by living organisms, and their functions.

Element*	Function
Nitrogen (N)	Structural component of proteins and nucleic acids
Phosphorus (P)	Structural component of nucleic acids, phospholipids, and bone
Potassium (K)	Major solute in animal cells
Sulfur (S)	Structural component of many proteins
Calcium (Ca)	Regulator of cell permeability; structural component of bone and material between woody plant cells
Magnesium (Mg)	Structural component of chlorophyll; involved in function of many enzymes
Iron (Fe)	Structural component of hemoglobin and many enzymes
Sodium (Na)	Major solute in extracellular fluids of animals

* Chemical symbol in parentheses.

ions, electrically charged parts of compounds which have dissociated in water. (For example, table salt, sodium chloride [NaCl], dissociates into sodium [Na$^+$] and chloride [Cl$^-$] ions upon solution.) Plants obtain nitrogen in the form of ammonia ion (NH$_4^+$) or nitrate ion (NO$_3^-$), phosphorus in the form of phosphate ion (PO$_4^=$), calcium and potassium in the form of their simple (elemental) ions (Ca^{++}, K$^+$), and so on. The solubility of these substances, which determines their availability, varies with temperature, acidity, and the presence of the other dissolved substances.

All natural water contains some dissolved minerals. Even rainwater is never pure because minerals in dust particles or droplets of ocean spray quickly dissolve in raindrops. Most lakes and rivers contain 0.01 to 0.02 per cent dissolved minerals. These eventually find their way to the sea where salts and other minerals have accumulated over the millenia to an average concentration of 3.5 per cent. In hot climates, where dissolved minerals are concentrated by the evaporation of water more rapidly than rainfall dilutes them, lakes without natural drainage outlets, like the Great Salt Lake of Utah, may contain up to 10 per cent dissolved substances.

Dissolved minerals in fresh and salt water differ in composition as well as in quantity (Table 3–3). Sea water abounds in sodium, magnesium, chloride, and sulphate ions, whereas calcium and carbonate ions predominate in fresh water.

Dissolved substances pose several problems for plants and animals. First, organisms must obtain minerals from the soil, water, or their food. Second, they must maintain concentrations of minerals in their bodies above their concentrations in the environment. For marine organisms, a third problem is the presence of many ions in greater concentrations in the water than in their body fluids. Left to their own devices, ions diffuse across cell membranes from regions of high to regions of low concentration, thereby tending to equalize their concentrations. Water

TABLE 3–3 Percentage composition of substances in dissolved minerals in two rivers, in sea water, and in the blood plasma and cells of frogs.

Substance	Delaware River	Rio Grande River	Sea water	Frog plasma	Frog cells
Sodium	6.7	14.8	30.4	35.4	1.3
Potassium	1.5	0.9	1.1	1.3	77.7
Calcium	17.5	13.7	1.2	1.2	3.1
Magnesium	4.8	3.0	3.7	0.4	5.3
Chlorine	4.2	21.7	55.2	39.0	0.8
Sulfate	17.5	30.1	7.7	—	—
Carbonate	33.0	11.6	0.4	22.7	11.7

also moves across membranes (*osmosis*) toward regions of high ion concentration, tending to dilute concentrations of dissolved minerals.

It would be quite futile for a salt-water fish, with blood containing half the concentration of mineral ions as sea water, to try to dilute the oceans with water from its body. Keeping ions out of the body is as big a problem for salt-water fish as the retention of ions is for fresh-water species. The solution to the problem lies in the ability of cell membranes to selectively pump ions and water in one direction or the other. The gills and kidneys of aquatic organisms are constantly active in maintaining the proper ion concentration in the body. The kidneys of fresh-water fish actively prevent the passage of ions by the osmosis of water through the skin into the body. Additionally, the gills of fresh-water fish actively take up ions from the surrounding water. The gills and kidneys of salt-water fish selectively excrete ions to counter the tendency of ions to diffuse into the body from the surrounding water. Marine fish also drink sea water to replenish water lost by osmosis across their skin.

Water Loss

Terrestrial environments pose a severe problem to life: the conservation of water within the body. Form and function of most terrestrial organisms are, in fact, adapted to deal with the problem of desiccation. Outer coverings of organisms — the chitin of arthropods (insects, spiders, and others), the skin of reptiles, birds, and mammals, the bark and cuticle of plants — are nearly impermeable to water. Moreover, the respiratory organs, whose surfaces must be kept moist for gas exchange to occur, have been relocated from external positions (as in the gills of fish and amphibians) to more protected internal positions — the lungs of vertebrates and the tracheae (air passages) of insects. Gas exchange in terrestrial plants is restricted to small openings (stomata) distributed over the surface of the leaf (Figure 3-3). These morphological adaptations considerably reduce the loss of water through evaporation. Terrestrial organisms lacking well-developed, water-conserving adaptations, earthworms for example, are restricted to moist soil within which air is saturated with water vapor.

Water loss and salt (ion) balance are as intimately linked to each other as are the physical processes of diffusion and osmosis. Animals that eat meat obtain salts in their food in excess of their requirements. Fresh-water carnivores can rid their bodies of salts simply by drinking copious amounts of water and flushing the salts out in a very dilute urine. If water is scarce, animals cannot afford to be so wasteful and some method must be found to concentrate the salts in the urine, so that less water is lost for a given amount of salt excreted. Concentration of salts is performed primarily by the kidneys, but many reptiles and birds,

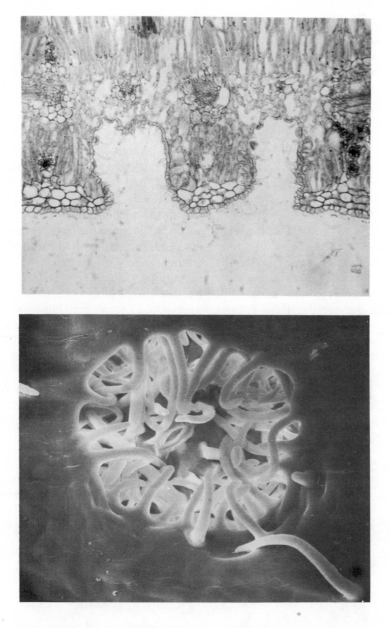

FIGURE 3–3 The leaf of oleander, a drought-resistant plant. At top, a cross section shows the locations of the stomata — openings to the leaf interior through which gas exchange takes place. The stomata lie deep within hair-filled crypts on the undersurface of the leaf. The hairs reduce air movement and trap moisture, thereby reducing water loss from the leaf. At bottom, a scanning electron micrograph of the undersurface of the leaf shows the hair-filled pit magnified about 500 times.

especially sea birds, have salt-secreting glands, similar to our tear glands, near their eyes. As one would expect, desert animals have champion kidneys. For example, whereas man can concentrate salts in his urine to only four times their level in blood plasma, the kangaroo rat's kidney can produce a urine with 18 times the salt concentration in its blood plasma. This adaptation enables the kangaroo rat to live in a desert environment without drinking any water (Figure 3–4).

Carnivores consume excess nitrogen, as well as excess salts, in their food. Nitrogen, consumed in the form of proteins, must be eliminated from the body when the proteins are metabolized. Aquatic organisms produce the simple nitrogenous by-product, ammonia (NH_3). Although ammonia is poisonous to living tissues, aquatic organisms can rapidly eliminate the toxin in their copious, dilute urine before it reaches a dangerous concentration. Terrestrial animals cannot afford to lose so much water for the sake of nitrogen excretion, and protein metabolism produces a less toxic nitrogen by-product, which can be concentrated in the blood stream and urine without dangerous side effects. In mammals this waste product is urea, which has the chemical formula $CO(NH_2)_2$. Birds and reptiles have carried adaptation to terrestrial life one step further by producing a more complex nitrogenous waste product, uric

FIGURE 3–4   Merriam's kangaroo rat is well-adapted to hot deserts in the southwestern United States. Its kidneys concentrate urine, thereby conserving water, and it ventures out of its cool burrows to feed only at night when heat stress on the surface is least.

acid ($C_5H_4N_4O_3$). Uric acid has the distinct advantage in dry environments of being excreted as a dry crystal, thus requiring very little
excretory water loss.

And so we see some of the major differences between aquatic and
terrestrial habitats and the unique problems each poses for organisms.
The characteristics of aquatic and terrestrial environments are determined primarily by the density and thermal properties of water and air,
and the relative availability of oxygen, water, and minerals. One characteristic aspect of almost all terrestrial environments, absent from aquatic
habitats, is soil. As we shall see in the following chapter, soil is a
complex combination of weathered rock and decaying organic detritus,
within which much of the energy transformation and mineral exchange
in terrestrial ecosystems takes place. Soil does not form in aquatic
habitats because wave action prevents the accumulation of loose sediments in many areas, and aquatic plants do not have leathery leaves and
woody branches which, when they decompose, contribute greatly to the
structure of terrestrial soils. Probably the closest thing to soil in the sea
are mucky, organic sediments in ocean basins. These sediments are
habitats for many burrowing detritus feeders, as are terrestrial soils.
When churned up by water movement, they also contribute dissolved
minerals to open waters. But the analogy stops there. Unlike aquatic
plants, the trees, shrubs, and herbs of terrestrial habitats obtain all their
minerals from the soil and, in turn, contribute to its structure and
composition. Also, the properties of soils are determined partly by the
underlying bedrock, whereas marine sediments merely accumulate on
top of the ocean floor. Finally, aquatic sediments are fully saturated with
water at all times and do not develop the zones characteristic of terrestrial soils.

In the next chapter, we shall take a close look at the process of soil
formation and the environmental factors which determine the structure
and composition of soils in different regions.

4

Soil Formation

We take the dirt under our feet for granted, unwisely it would seem, because most of the vital mineral exchange between the biosphere and the inorganic world occurs in the soil. Plants obtain water and nutrients from the soil. Leaves and branches die and return to the soil, where they are decomposed and their minerals released. Organisms responsible for decomposition — the myriad bacteria and fungi, the minute arthropods and worms, the termites and millipedes — abound in the surface layers of the soil where dead organic matter is freshest. Their activities contribute to the development of soil properties from above, while physical and chemical decomposition of the bedrock contribute to the soil from below.

Once formed, soils remain in a dynamic state. Some minerals are removed by ground water; others blow in as dust or are released by the decomposition of underlying rock layers. Although soil is in a constant state of flux, soils of most regions attain characteristic steady-state properties. Soil characteristics vary greatly over the world and influence the distribution of vegetation types. Five factors are largely responsible for variation in soils: climate, parent material (bedrock), vegetation, local topography, and, to some extent, age. In general, the decomposition and weathering of parent rock and the addition of organic material to the soil proceed most rapidly in warm, wet climates. As a result, the influence of parent rock on the structure and composition of soil *decreases* with increasing rainfall, temperature, and age.

In dry areas, rainfall is so sparse that chemical weathering of rock is slow, and plant production is so low that little organic detritus is added to the soil. Soils of arid regions are typically shallow; bedrock lies close to the surface. Such soils are often referred to as *lithosols* (from the Greek *lithos,* meaning stone) because of their close association with the underlying rock (Figure 4-1). Weathering may extend to the depth of only one foot; lithosols can be shallower or even absent where erosion removes

FIGURE 4–1 Profile of a lithosol from Logan County, Kansas, illustrating shallow soil depth and poor soil zonation.

weathered rock and organic detritus as rapidly as they are formed. The faces of cliffs and rocks in the upper regions of intertidal zones at the edge of the sea are extreme examples of sites where soil formation is prevented by erosion. *Alluvial* soils are another class in which the weathering process does not have a chance to work because seasonal flood waters deposit fresh layers of silt each year. At the other extreme, bedrock is most highly weathered in parts of the humid tropics where chemical alteration of the parent material may extend to depths of 20 feet or more. Most temperate soils are intermediate in depth, usually extending to a few feet.

Soil Horizons

Where a recent roadcut or excavation exposes the soil in cross section, one is often struck by the presence of distinct layers, called *horizons* (Figure 4-2). Soil horizons have been described with complex and sometimes conflicting terminology by soil classifiers. A generalized, and somewhat simplified, soil profile has three major divisions — the A, B,

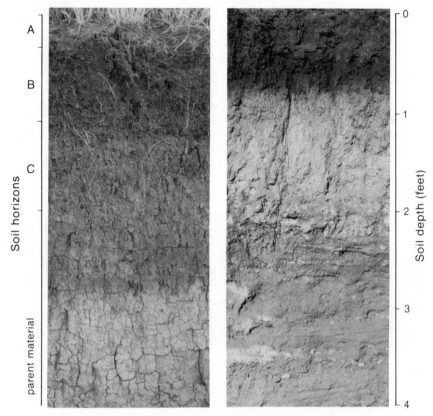

FIGURE 4–2   Soil profiles from the central United States illustrating distinct horizons. The profile at left, from eastern Colorado, is weathered to a depth of about two feet where the subsoil contacts the original parental material, consisting of loosely aggregated calcium-rich, wind-deposited sediments. The A_1 and A_2 horizons are not clearly distinguished except that the latter is somewhat lighter-colored. The B horizon contains a dark band of redeposited organic materials which were leached from the uppermost layers of the soil. The C horizon is light-colored and has been leached of much of its calcium. Some of the calcium has been redeposited at the base of the C horizon and at greater depths in the parent material. The profile at right is that of a typical prairie soil from Nebraska. Rainfall is sufficient to leach readily soluble ions completely from the soil. Hence there are no B layers of redeposition, as in the drier Colorado soil at left, and profile is more homogeneous. The A horizon is weakly subdivided into a darker upper layer and lighter lower layer. The weathered soil lies upon a parent material composed of loess, the windblown remnants of glacial activity. The depth scale, in feet, at right, applies to both profiles; the soil horizons, at left, apply only to the left-hand profile.

and C horizons — and three minor divisions of the uppermost, the A horizon. Arrayed in order descending from the surface of the soil, the horizons and their predominant characteristics are:

HUMI-FICATION

A_0 primarily dead organic litter. Most soil organisms are found in this layer.

A_1 a humus layer, consisting of partly decomposed organic material mixed with mineral soil.

A_2 a region of extensive *leaching* of mineral ions from the soil. Because minerals are dissolved by water (mobilized) in this layer, plant roots are concentrated here, where mineral ions are most readily available.

MINERALIZATION

B a region of little organic material whose chemical composition closely resembles that of the underlying rock. Mineral ions leached out of the overlying A_2 horizon are sometimes deposited here.

DONE BY DECAY OR DETRITUS FOOD CHAIN

C primarily weakly weathered material, which closely resembles the parent rock.   — *SOIL FAUNA*

The soil horizons demonstrate the decreasing influence of climate and the increasing influence of bedrock with increasing depth. Soil formation is greatly complicated, however, by the movement of soil minerals upward and downward through the soil profile. Before considering these processes in detail, we shall examine the initial weathering of the bedrock and how it influences soil characteristics.

Weathering

Decomposition of rock at the bottom of a soil profile, or on a newly exposed rock surface, is fostered by the action of both physical and chemical agents. Repeated freezing and thawing of water in crevices breaks up rock into smaller pieces and exposes new surfaces to chemical action. Initial chemical alteration of the rock occurs when water dissolves some of the more soluble minerals, particularly sodium (Na^+) and calcium (Ca^{++}) and their associated chloride (Cl^-) and sulfate ($SO_4^=$) ions. Other minerals, particularly the oxides of titanium, aluminum, iron, and silicon, are resistant to leaching under most conditions.

The weathering of granite exemplifies some basic processes of soil formation. Granite is an igneous rock formed when the less dense molten material deep within the earth rose to the surface, cooled, and crystallized. Granite consists chiefly of three minerals: feldspar, mica, and quartz. Feldspar, which consists of oxides* of potassium,

---

* Oxides are compounds consisting of oxygen and one or more other elements.

aluminum, and silicon ($K_2O \cdot Al_2O_3 \cdot 6\ SiO_2$), weathers rapidly due to the removal of potassium (K) as potassium carbonate (K_2CO_3) in the presence of carbonic acid. (Carbonic acid [H_2CO_3], formed when carbon dioxide dissolves in water, is always present in rainwater.) The remainder of the feldspar mineral is reorganized with water to form kaolinite ($Al_2O_3 \cdot 2\ SiO_2 \cdot 2\ H_2O$), a type of clay. As a general class of materials, clays perform an extremely important function in the soil, that of providing sites for ion exchange between the soil and plants. We shall look in detail at the role of clay particles later.

The mica grains in granite are composed of oxides of potassium, magnesium (Mg), iron (Fe), aluminum (Al), and silicon (Si). When granite weathers, potassium and magnesium are removed rapidly and the remaining oxides of iron, aluminum, and silicon form clay particles. Quartz, a form of silica (SiO_2), is not soluble in acidic water and therefore remains more or less unaltered in the soil as sand grains. Changes in chemical composition of granite as it weathers from rock to soil are summarized in Figure 4–3. Calcium, magnesium, sodium, and potassium disappear quickly while aluminum oxides remain.

Removal of minerals from weathered granitic rock varies greatly with climate. Similar parent materials in localities with progressively higher temperature and greater rainfall (Massachusetts, Virginia, and British Guiana) exhibit greater loss of total rock volume and increased removal of specific minerals, particularly silica and potassium (Figure 4–4).

The decomposition of granite illustrates some of the principal chemical influences on soil formation, but weathering follows quite different courses on different types of bedrock. Pure quartz sand (silica) and pure limestone (calcium carbonate) do not produce clay readily because they lack iron and aluminum oxides; soil formation thus proceeds slowly unless other materials are mixed into the bedrock. The composition of the bedrock and its initial weathering determine the relative amounts of

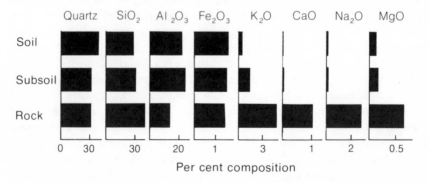

FIGURE 4–3   Percentage composition of soil layers and parent rock (granite) for each mineral in a soil profile from British Guiana.

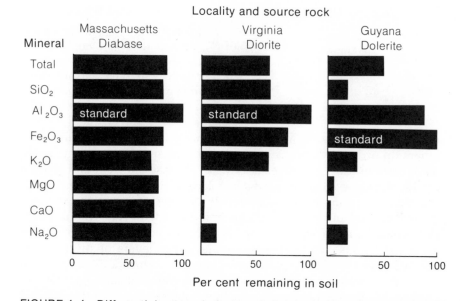

FIGURE 4–4 Differential removal of minerals from granitic igneous rocks as a result of weathering in Massachusetts, Virginia, and British Guiana. Values are compared to either aluminum or iron oxides, which are assumed to be the most stable components of the mineral soil.

clay and sand in derived soils. These qualities in turn influence the availability of mineral ions in the soil and the capacity of the soil to hold water.

The Clay–Humus Complex

Plants can obtain minerals from the soil only in the form of soluble ions. Because ions dissolve in water, they would quickly be washed out of the soil if they were not strongly attracted to stable soil particles. Clay and humus join in a close association referred to as a complex. The particles of the clay–humus complex, called *micelles,* are large enough to form a stable component of the soil, where they actively play a role in flux of mineral ions in the ecosystem. The surface of each complex particle has numerous sites with negative charges that attract positive ions such as calcium, magnesium, and potassium, and so retain them in the soil (Figure 4–5).

The role of the clay–humus complex in soil chemistry is not, however, so simple. The bonds between the mineral ions and the micelle are relatively weak, so they constantly break and reform. When a potassium ion (K^+) dissociates from a micelle, its place may be taken by any other

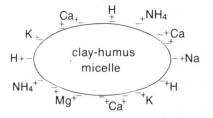

FIGURE 4–5 Schematic representation of a clay–humus complex particle (micelle) with hydrogen ions and mineral ions attracted by negative charges at its surface.

ion that is close by. Some ions cling more strongly to micelles than others. In order of decreasing tenacity, the common ions are hydrogen (H^+), calcium (Ca^{++}), magnesium (Mg^{++}), potassium (K^+), and sodium (Na^+). Hydrogen ions thus tend to displace calcium and all other ions on the micelle. If ions were not added to or removed from the soil, the relative proportions of mineral ions associated with clay–humus particles would reach a steady state. But the carbonic acid in rainwater continually adds hydrogen ions to the upper layers of the soil; the hydrogen ions readily displace other minerals, which are then washed out of the soil and into ground water. The influx of hydrogen ions in rainwater is largely responsible for the mobility of ions in the soil and the differentiation of layers in the soil profile, as we shall see below.

Soil Water

In addition to determining the clay content of the soil, the mineral composition of the bedrock determines the size and abundance of sand grains and silt particles, collectively known as the *skeleton* of the soil. The materials of the soil skeleton are chemically inert but they influence the physical structure of the soil and its water-holding capacity.

Water is very sticky. The capacity of water molecules to cling to each other and to surfaces they touch underlies the familiar phenomena of surface tension and the rise of water against gravity in capillary tubes. Water clings tightly to surfaces of the soil skeleton. Because total surface area of particles in the soil increases as particle size decreases, silty soils hold more water than coarse sands, through which water drains quickly.

Water capacity is not equivalent to water availability. Plant roots easily take up water that clings loosely to soil particles by surface tension, but water near the surface of sand and silt particles is bound tightly to the soil particles by stronger forces. Botanists measure the strength with which the cells of root hairs can absorb water from the soil in terms of equivalents of atmospheric pressure. Capillary attraction holds water in the soil with a force equivalent to a pressure of one-third atmosphere, about 5 pounds per square inch. (Sea level atmospheric pressure is 14.7 pounds per square inch.) Water attracted to soil particles with less force than one-third atmosphere (water in the middle of large interstices between soil particles, hence at great distance from their surfaces) drains out of the soil, under the pull of gravity, into the ground water in the crevices of the bedrock below. The amount of water held against gravity by forces of attraction greater than one-third atmosphere is called the *field capacity* of the soil.

A force equivalent to one-third atmosphere is sufficient to raise a column of water about 10 feet above the water surface. We know that plant roots can exert a much greater pull on water in the soil, because water rises in the tallest trees to leaves more than 300 feet above the ground. In fact, plants can exert a pull of about 15 atmospheres on soil water and therefore can take up water held in the soil by forces weaker than 15 atmospheres. Water close to the surface of soil particles is bound by forces of attraction that often exceed 30 atmospheres. Water held by forces greater than 15 atmospheres is unavailable to plants and is called the *wilting capacity* of the soil. Once plants under drought stress have taken up all the water in the soil held by forces weaker than 15 atmospheres, they can no longer obtain water and wilt, even though water remains in the soil.

In soils with small particles, the surface area of the soil skeleton is relatively large and a correspondingly large proportion of soil water is held by forces greater than 15 atmospheres. Soils with larger skeletal particles have less surface area and larger interstices between particles. A larger proportion of soil water is held loosely but these soils have a lower capacity. Availability of water to plants is greater in soils with skeletal particles intermediate in size between sand and mud (Figure 4-6). Such soils are called *loams*.

Podsolization

Under mild, temperate conditions of temperature and rainfall, sand grains and clay particles are resistant to weathering and form stable components of the soil skeleton. Under conditions of high soil acidity, however, clay particles break down and their soluble ions are leached out of the soil. This process, known as *podsolization*, reduces the ion exchange capacity, and therefore the fertility of the soil, by reducing the clay content.

Acid conditions are found primarily in cold regions where rainfall exceeds evaporation during the year and where coniferous trees dominate the forests. Under these moist conditions, water continually moves downward through the soil profile, so there is little upward transport of new clay materials from the weathered bedrock below. The slow decomposition of acidic plant litter under conifer forests produces organic acids, which increase the acidity of the soil and promote leaching of clay particles.

Podsolization advances furthest under spruce and fir forests in the

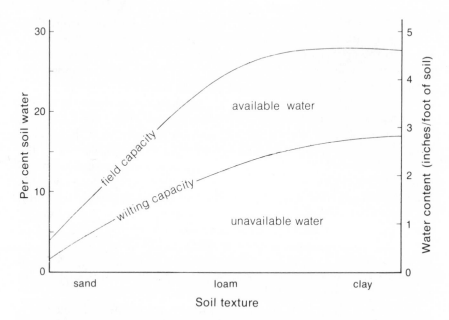

FIGURE 4–6   The water content of soils as a function of the particle size of the soil skeleton material. The difference between the field capacity and the wilting capacity of the soil represents the water available to plants.

northern United States and across a wide belt of southern and north-western Canada. A typical profile of a podsolized soil (Figure 4–7) has striking bands corresponding to regions of leaching (*eluviated* horizons) and redeposition (*illuviated* horizons). The topmost layers of the profile (A_0 and A_1) are dark and rich in organic matter. These are underlain by a light-colored A_2 horizon, which has been leached of most of its clay content. As a result, the A_2 consists mainly of sandy skeletal material that holds neither water nor nutrients well. One usually finds a dark-colored band of deposition immediately under the eluviated A_2 horizon. This is the uppermost layer of the B horizon, often distinguished as B_1, where iron and aluminum oxides are redeposited, giving the layer a dark color. Other, more mobile minerals may accumulate to some extent in lower parts of the B horizon, which then grades almost imperceptibly into a C horizon and the parent material.

Laterization

Whereas acidic conditions foster the removal of clay materials (iron and aluminum oxides) from the soil, a basic or alkaline soil reaction facilitates the removal of silica (SiO_2). Leaching of silica, called *laterization*, occurs extensively in tropical regions. Soil scientists do not fully understand

FIGURE 4–7   Profile of a podsolized soil in Plymouth County, Massachu-setts. The light-colored, eluviated A_2 horizon and the dark-colored, illuviated B_1 horizon immediately below it form distinct bands. Note the general ab-sence of roots in the A_2 horizon compared to the lower B_1 horizon.

why tropical soils tend to have low acidity (high alkalinity). We do know, however, that plant detritus decays rapidly in the tropics, owing to high temperatures and moisture. Humic acids, therefore, do not persist in tropical soils as long as they do in cooler regions. Furthermore, decomposition of plant litter is accomplished primarily by bacteria that produce no acid. The plant litter of cool forests is decomposed in part by fungi which produce acids to aid the chemical breakdown of organic detritus.

Laterization has an effect on the soil profile opposite to that of podsolization. Removal of silica from the top layers of the soil increases the proportion of iron and aluminum oxides in these layers. These oxides give tropical soils their characteristic red color. If laterization proceeds far enough, all the silica disappears from the soil, including the silica in clay particles, leaving behind a material called *laterite,* which is more like concrete than soil.

Laterization does not normally alter soil completely in undisturbed tropical forests. Organic humus particles, which accumulate in the upper layers of the soil, maintain a soil structure suitable for root growth. Strongly laterized layers may, however, form deeper in the soil profile.

Disturbance of tropical forests can have disastrous effects where soil is prone to laterization. Removal of trees for agriculture or housing exposes the soil to the drying effects of the Sun. Evaporation of water from the ground surface frequently reverses the usual downward flow of water through the soil profile. Iron and aluminum oxides are then brought to the surface where they cement soil particles into a substance so hard that it is at best suitable for masonry. A completely laterized soil is nearly impervious to water and thus promotes surface runoff and erosion. Such soils are, of course, useless for agriculture, and their hardness and low water content slows the regeneration of natural vegetation.

Calcification

In dry regions, where evaporation exceeds rainfall, water does not percolate completely through the soil. Minerals, such as calcium, dissolved in the topmost layers of the soil profile, are often redeposited in these same layers when water evaporates from the soil, or they may be transported downward to the lower limit of water penetration. The results of this process, called *calcification,* can be seen in the left-hand soil profile in Figure 4-2, in which a narrow, diffuse band of calcium carbonate (light-colored) has been redeposited about two feet below the soil surface. This horizon marks the lower limit of water percolation, immediately below which one finds relatively unweathered parent material.

The presence of ground water close to the surface greatly modifies soil profiles in dry regions. Minerals dissolved in the ground water are drawn to the surface by evaporation and the upward pull of capillary movement. Evaporating water then leaves the minerals behind at the surface, sometimes forming thick crusts called *caliches* (Figure 4-8). Caliche deposits, which are excessively alkaline, inhibit plant growth.

In desert basins where ground water is close enough to the surface to be drawn upwards by evaporation, caliche deposits form the "dry lakes" that are widespread in the Mohave Desert and Great Basin of the western United States. Standing water on the surface in such regions is usually so full of dissolved minerals that it is undrinkable. (For many early pioneers who crossed the deserts, the choice between dying of thirst or alkali poisoning must have been difficult.)

Irrigation *can* make a desert bloom. Dry soils become highly fertile when they are irrigated because of the high concentration of mineral ions in the upper layers of the profile. But the richness of agricultural returns is often cut short by speeded calcification of the soil. Irrigation water is ordinarily obtained from rivers which, in dry regions, are usually loaded with silt and dissolved minerals. Most water added to the soil by irrigation eventually evaporates, leaving behind the minerals it carried near the soil surface. The ultimate result is similar to what

FIGURE 4–8 An alkaline area devoid of plants in Chouteau County, Montana, where rising ground water has deposited a crust of calcium carbonate.

happens when water naturally enters the soil profile from underlying ground water. Calcium accumulates rapidly and the soil soon becomes too alkaline for agriculture.

The Role of Vegetation in Soil Formation

The initial weathering of bedrock and the secondary alteration of the soil profile by podsolization, laterization, and calcification primarily influence the inorganic composition of the soil. Yet many important characteristics of the soil, including its humus content and the availability of nitrogen and phosphorus, are determined largely by the vegetation.

Soil changes that can follow removal of vegetation in the tropics show dramatically the role of vegetation in maintaining a steady state in the soil. Denudation quickly alters the movement of water through the soil and rapidly changes patterns of leaching and deposition.

Vegetation exerts its most dramatic influence on the development of soils where the underlying parent material is freshly exposed. Primary soil development occurs where geologic agents remove layers of existing soil or add sediments over the top of existing soil horizons. Since the recession of the glaciers from the Great Lakes 10,000 to 12,000 years ago, the surface level of the Great Lakes has periodically lowered, leaving behind a chronological series of sand dunes at the southern end of Lake Michigan. The value of these dunes to the study of ecological processes was first recognized by the pioneering plant ecologist, Henry C. Cowles, who, in 1899, described changes in vegetation observed on progressively older dunes. A newly formed dune consists largely of sand (silica). Water percolates rapidly through the dune and, because clays are absent, quickly leaches out any mineral nutrients. The dune environment excludes all but the most hardy plants. Marram grass (genus *Ammophila*) colonizes the sand at an early stage by sending out rhizomes (horizontal roots) from plants growing in better soil at the edge of the dunes (Figure 4-9).

Once grasses become established, they stabilize the dune and begin to add organic detritus to the dune surface. By building up the humus content of the sand, dune grasses encourage true soil development. Grasses and shrubs dominate the first century of plant succession on dunes. These are followed by the establishment of pine and its rapid replacement by black oak at an age of 150 to 200 years.

Changes brought about by vegetation, and the ultimate attainment of a steady state in the soil, were described in a paper by Jerry Olson, published in 1958. Olson found that the humus added to the soil provided sites for mineral ion exchange, just as the clay–humus complex does in clay-based soils. Silt and clay particles are eventually added to the soil by wind deposition. The ion exchange capacity of the soil

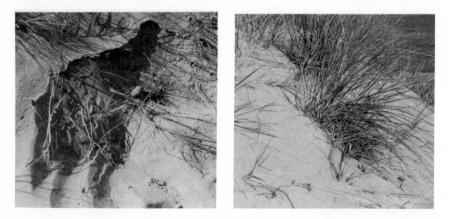

FIGURE 4–9   Marram grass growing on dunes in Indiana Dunes State Park, Indiana. At left, plants are seen extending out over fresh sand. At right, sand has been removed to expose the underground rhizomes by which the grass spreads.

increases rapidly for 500 to 1,000 years after dune stabilization, then levels off. Litter continues to accumulate on the forest floor and humic acids eventually make the soil acidic. Hydrogen ions replace other nutrients until they occupy almost half the ion exchange sites in the soil. As a result, soil fertility declines slowly, then levels off about 4,000 years after dune stabilization.

Crocker and Major, in 1955, described soil development on areas bared by receding glaciers at Glacier Bay, Alaska. The retreat of the edge of the glacier had been recorded for a century, thus Crocker and Major knew the exact age of each of their study sites. Unlike the Lake Michigan sand dunes, sediments left behind by receding glaciers contained abundant calcium and clay. Vegetation established itself rapidly and decaying plant detritus changed the hydrogen ion concentration of the soil from slightly alkaline to slightly acid in 20 years. Each species of plant influenced soil acidity differently, however. Alder thickets acidified the soil rapidly, but willow and cottonwood did so slowly. In any case, increasing soil acidity accelerated the removal of calcium, while accumulating detritus steadily added to the inorganic nitrogen content of the soil. These changes in turn influenced the suitability of the habitat to different species of plants and fostered further changes in vegetation.

We have seen how variation in climate and geology affect soil properties. Geographical and temporal variation in all aspects of the physical world broadly influence the structure and functioning of the entire ecosystem, determining not only physical and chemical properties of the soils but also levels of organic production, paths of energy flow and nutrient cycling, and the adaptations of plants and animals that give each habitat its characteristic appearance. In the next chapter, we shall examine major geographical and seasonal patterns in aquatic and terrestrial environments and look at how variation in topography creates an almost endless variety of local climates.

5

Variation in the Environment

Everyday experience reveals variation in the natural world. Climate and the appearance of biological communities are closely linked. As one travels from east to west across the United States, the gradual change in climate is paralleled by change in vegetation. Tall forests of broad-leaved trees along the east coast are replaced by grasslands in the dried midwestern states and by desert shrublands in the arid Great Basin. Amidst the topographic diversity of the western mountains, one travels through altitudinal zones of vegetation from hot desert scrub to cool montane forests of aspen and spruce. Travelling from northern Canada to Panama, one encounters even more striking change.

Major patterns of climate on the Earth's surface depend on the relation of the Earth to the Sun, placement of the continents and oceans, and the circulation of the wind and seas. But local variations in environment caused by geology and topography result in the diversification of communities within regions of uniform climate. Physical properties of the environment are further modified and diversified by vegetation and the activities of animals.

Global Climate Patterns

The Earth's climate tends to be cold and dry towards the poles and hot and wet towards the Equator, but this oversimplification has as many exceptions as a weather report. Nonetheless, climate does exhibit broadly defined patterns.

Global variation in climate is determined largely by the position of the Sun relative to the surface of the Earth. The Sun exerts its greatest warming effect on the atmosphere, oceans, and land when it is directly overhead. The Sun's warmth is diminished when it lies close to the horizon and its rays strike the surface at an oblique angle. Not only does a beam of sunlight spread over a greater area when the Sun is low, it also travels a longer path through the atmosphere where much of its light energy is either reflected or absorbed by the atmosphere and re-radiated into space as heat. The Sun's highest position each day varies from directly overhead in the tropics to near the horizon in arctic regions; hence the warming effect of the Sun increases from the poles to the Equator. This uneven distribution of the Sun's energy over the surface of the Earth creates the major geographical patterns in temperature, rainfall, and wind.

Warming air expands, becomes less dense, and thus tends to rise. Its ability to hold water vapor increases and evaporation is accelerated. The rate of evaporation from a water surface nearly doubles with each 10°C rise in temperature. The Sun heats the atmosphere most intensely in the tropics. The warmed surface air picks up water vapor and begins to rise. As the moisture-laden air gains altitude and cools, its water vapor condenses into thick clouds that drench the tropical landscape with rain. Daily cycles of heating and cooling cause most tropical rain to fall during the afternoon and evening; in temperate areas, as well, summer thunder showers, resulting from strong vertical currents of warm, moist air, occur most often late in the day. Warm tropical air can hold much more water than temperate or arctic air. Hence annual precipitation is greatest in tropical regions (Figure 5–1). The tropics are not so wet because more water occurs in tropical latitudes than elsewhere; rather water is cycled more rapidly through the atmosphere. Cycles of evaporation and precipitation are driven by the Sun; the source of energy, not the quantity of water, primarily determines latitudinal patterns in rainfall. The distribution of continental land masses exerts a secondary effect on rainfall patterns. Rainfall is more plentiful in the Southern Hemisphere because oceans and lakes

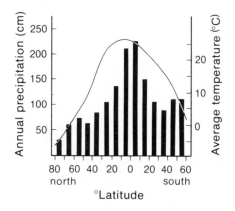

FIGURE 5–1 Average annual precipitation (vertical bars) and temperature (solid line) for 10° latitudinal belts within continental land masses. The figure represents averages for many localities, which obscures the great variation within each latitudinal belt.

cover a greater proportion of its surface (81 per cent, compared to 61 per cent of the Northern Hemisphere). Water evaporates more readily from exposed surfaces of water than from soil and vegetation.

Doldrums and Trade Winds

Winds are driven by energy from the Sun, just like the cycling of water in the atmosphere. Indeed, the two cannot be separated and wind patterns exert a strong influence on precipitation. The mass of warm air that rises in the tropics eventually spreads out to the north and south in the upper layers of the atmosphere. It is replaced from below by surface-level air from subtropical latitudes (Figure 5–2). About 30° north and south of the solar equator,* the tropical air mass that rose under the warming sun has cooled enough that it begins to sink back to the surface. It is relatively dry because condensation has removed much of its water, which fell as rain over the tropical regions where the air current originated. Its capacity to evaporate and hold water increases further as it sinks and warms. As the air strikes the Earth's surface in subtropical latitudes and spreads to the north and south it draws moisture from the land, creating zones of arid climate.

Convection currents in the atmosphere driven by the Sun's energy redistribute heat and moisture about the surface of the Earth. The region of rising air in the tropics — the doldrums — is one of high rainfall. Conversely, descending air robs the land of water, which is transported elsewhere by wind currents. The tradewinds, blowing steadily toward the Equator from the dry horse latitudes, carry moisture picked up along the way into the tropics.

Just as warm tropical air rises, cold air masses over the north and south polar regions descend and flow along the surface toward lower latitudes. When cold air meets a warmer air mass blowing poleward across temperate latitudes, the warm moist air rises up over the denser polar air and cools, bringing precipitation.

Continents and Ocean Currents

Precipitation is distributed over the surface of the Earth in such a way that most wet regions occur close to the Equator, and the major deserts occupy a belt centered about 30° latitude north and south of the Equator

* The *solar equator* is the parallel of latitude that lies directly beneath the Sun. The position of the solar equator varies seasonally from 23° North Latitude on June 21 (the summer solstice) to 23° South Latitude on December 21 (the winter solstice). The solar equator coincides with the Earth's geographical equator at the equinoxes (March 21 and September 21).

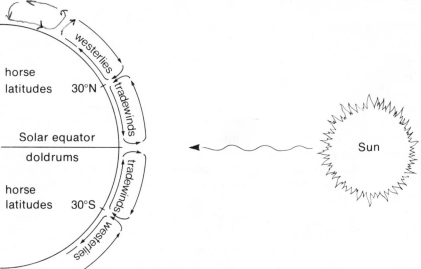

FIGURE 5–2   Simplified diagram of convection currents driven by the Sun's energy in the atmosphere of tropical and subtropical latitudes.

(Figure 5–3). Great names in deserts — the Arabian, Sahara, Kalahari, and Namib of Africa, the Atacama of South America, the Mohave and Sonoran of North America, and the Australian — all belong to regions within these belts.

Exceptions to this pattern are caused by major land masses. Mountains force air upward, causing it to cool and lose its moisture as precipitation on the windward side of a mountain range. As the air descends the leeward slopes of the mountains and travels across the

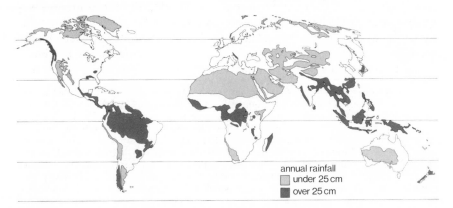

FIGURE 5–3   Distribution of the major deserts (regions with less than 10 inches [25 centimeters] annual precipitation) and wet areas (having more than 80 inches [150 centimeters] annual precipitation).

lowlands beyond, it picks up moisture and creates arid environments called *rain shadows*. The Great Basin deserts of the western United States and the Gobi Desert of Asia lie in the rain shadows of extensive mountain ranges.

The interior of a continent is usually drier than its coasts simply because the interior is further removed from the major site of water evaporation at the surface of the ocean. Furthermore, coastal (*maritime*) climates are less variable than interior (*continental*) climates, because the tremendous heat capacity of water compared to land ameliorates coastal temperatures. For example, the difference between the hottest and coolest mean monthly temperature near the Pacific coast of the United States at Portland, Oregon, is 28°F (16°C).* Farther inland, this range increases to 33°F (18°C) at Spokane, Washington; 47°F (26°C) at Helena, Montana; 60°F (33°C) at Bismarck, North Dakota.

Ocean currents also play a major role in transferring heat over the surface of the Earth. In large ocean basins, cold water tends to move toward the tropics along the western coasts of the continents, and warm water tends to move towards temperate latitudes along the eastern coasts of continents (Figure 5–4). The cold Humboldt Current moving north along the coasts of Chile and Peru is partly responsible for the presence of deserts along the west coast of South America right to the Equator, though these regions also lie within the rain shadow of the Andes Mountains. Conversely, the warm Gulf Stream, emanating from the Gulf of Mexico, carries a mild climate far to the north into western Europe.

The Changing Seasons

Although we may characterize a region's climate as hot or cold, and wet or dry, regular cycles of change are as important aspects of climate as long term averages of temperature and rainfall. Periodic cycles in climate are based upon cyclical astronomical events: the rotation of the Earth upon its axis causes daily periodicity in the environment; the revolution

---

* Although we are accustomed to using Fahrenheit (F) temperatures in the United States, the Celsius (C) scale of temperature (frequently referred to as the Centigrade scale), is used throughout most of the world and almost exclusively in scientific work. The Celsius scale is much more convenient to ecologists than the Fahrenheit scale because it forms the basis of the commonly used heat scale of calories. The Celsius scale divides the temperature range into 100 degrees between the freezing point of water (zero degrees Celsius, or 0°C) and the boiling point (100°C). On the Fahrenheit scale, the corresponding temperatures are 32°F and 212°F, delimiting a 180 degree range between the freezing and boiling points of water. Each degree Celsius is therefore equal to 1.8 degrees Fahrenheit. Convenient conversion formulae are $C = \frac{5}{9} (F - 32)$ and conversely $F = \frac{9}{5}C + 32$. Some familiar benchmarks on the Celsius scale are: 0°C (32°F), freezing point of water; 10°C (50°F), a cool day; 20°C (68°F), a mild day; 30°C (86°F), a warm day; 40°C (104°F), a very hot day. A very cold day (0°F) is about $-18°C$. Room temperatures usually range between 20 and 25°C; body temperature of man is 37°C.

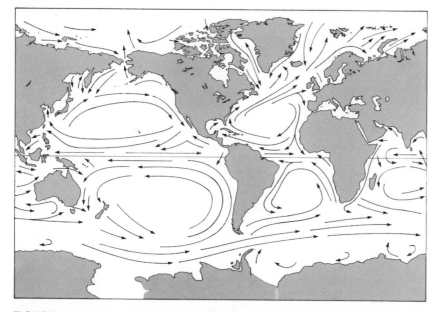

FIGURE 5–4   The major ocean currents. Water movement generally proceeds clockwise in the Northern Hemisphere and counterclockwise in the
Southern Hemisphere.

of the Moon around the Earth determines the periodicity of the tides; the
revolution of the Earth around the Sun brings seasonal change.

   The Earth's Equator is tilted slightly with respect to the path the
Earth follows in its orbit around the Sun. As a result, the Northern
Hemisphere receives more solar energy than the Southern Hemisphere
during the northern summer, less during the northern winter. The
seasonal change in temperature increases with distance from the
Equator. In boreal regions of the Northern Hemisphere, mean monthly
temperatures may vary as much as 30°C (54°F), with extremes of more
than 50°C (90°F) annually, whereas the mean temperatures of the warm-
est and coldest months in the tropics differ by as little as two or three
degrees.

   Latitudinal patterns in rainfall seasonality are complicated by belts
of wet and dry climate which move north and south with the changing
seasons. Annual variation in rainfall is greatest in broad latitudinal belts
lying about 20° north and south of the Equator (Figure 5–5). As the
seasons change, these regions are alternately crossed by the solar
equator, bringing heavy rains, and the subtropical high pressure belts,
bringing clear skies.

   Panama, at 9°N Latitude, lies within the wet tropics, but even there
the seasonal movement of the solar equator profoundly influences the
climate. The major tropical belt of high rainfall remains south of Panama

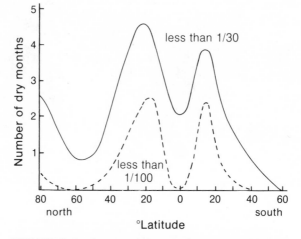

FIGURE 5–5   Seasonal distribution of rainfall as a function of latitude. The length and severity of the dry season are indicated by the number of months with rainfall less than a certain fraction ($\frac{1}{30}$ and $\frac{1}{100}$) of the yearly total. If rainfall were evenly distributed throughout the year, $\frac{1}{12}$ of the total would fall during each month.

during most of our winter, but it lies directly over Panama during the summer. Hence the winter is dry and windy and the summer humid and rainy (Figure 5-6). Panama's climate is wetter on the northern (Caribbean) side of the Isthmus, the direction of prevailing winds, than on the southern (Pacific) side. This rain shadow effect is more pronounced in nearby western Costa Rica, where a high mountain range intercepts moisture coming from the Caribbean side of the Isthmus. The Pacific lowlands are so *dry* during the winter months that most trees lose their leaves. The tinder-dry forest and bare branches contrast sharply with the wet, lush, more typically tropical state of the forest during the wet season (Figure 5-7).

Seasonal Cycles in Aquatic Environments

The steady winds, which blow in a southwesterly direction during the Panamanian dry season, create strong *upwelling* currents in the Pacific Ocean along the southern coast of the Isthmus. Warm surface waters are blown away from the coast and cooler water moves upward from deeper regions to replace it. As a result, the annual range of water temperature on the Pacific coast is about three times wider than that of the Caribbean coast.

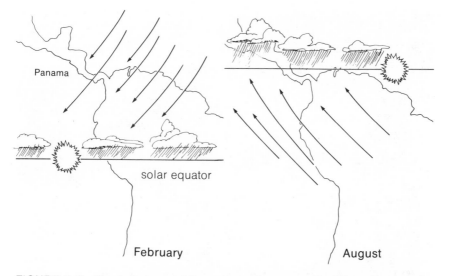

FIGURE 5–6   The influence of the seasonal movement of the solar equator on the climate of Panama. The northeast tradewinds bring dry weather during the winter and create rain shadows along the Pacific coast of Central America. The tradewinds also cause cold upwelling currents along the Pacific coast by blowing surface waters away from the shore.

The seas are warmed by the Sun just as the continents and the atmosphere are, but their great mass of water acts like a heat sink to dampen fluctuations in temperature. Small temperate-zone lakes are more sensitive than oceans to the changing seasons. Temperature cycles are an important force in the nutrient budgets of lakes because changes in temperature gradients from surface to bottom cause vertical mixing of the water twice each year, during the spring and the fall. Seasonal change in temperature profiles is illustrated in Figure 5–8. Because water is most dense at 4°C, ice covered lakes are coldest just beneath the surface, warming slightly (up to 4°C) toward the bottom. In early spring, the Sun warms the surface of the water gradually. Until surface temperature reaches 4°C, surface water tends to sink into the cooler layers immediately below. This minor vertical mixing creates a uniform temperature distribution throughout the water column. Without thermal layering to impede mixing, surface winds cause deep vertical movement of water in early spring (*spring turnover*), bringing nutrients from regions of decomposition in the bottom sediments to the surface.

As the Sun rises higher each day and the air above the lake warms, surface water warms up faster than deeper water, creating a sharp zone of temperature change, called the *thermocline,* across which water does not mix. The warm surface water literally floats on the cooler water below. The depth of the thermocline varies with local wind patterns,

FIGURE 5–7   Kiawe forest on the island of Maui, Hawaii during the peak of the dry season. Note the complete absence of leaves at this time. The grasses of the forest understory are tinder-dry.

and with the depth and turbidity of the lake. Thermoclines occur anywhere between 5 and 20 meters below the surface; lakes shallower than 5 meters usually lack stratification. Walden Pond, in Concord, Massachusetts, develops a sharp thermocline between 6 and 10 meters depth. In August, water temperatures within the thermocline drop from 25 to 5°C.

The thermocline demarcates an upper layer of warm water (the *epilimnion*) and a deep layer of cold water (the *hypolimnion*). Most of the primary production of the lake occurs in the epilimnion where sunlight is abundant. Photosynthesis supplements mixing of oxygen at the lake surface to keep the epilimnion well aerated and thus suitable for animal life, but plants often deplete dissolved mineral nutrients, thereby curtailing their own productivity. The hypolimnion is cut off from the surface of the lake by sharp temperature *stratification,* and being frequently below the euphotic zone of photosynthesis, animals and bacteria deplete the hypolimnion of oxygen, creating anaerobic conditions.

During the fall, surface layers of the lake cool more rapidly than deeper layers and, becoming heavier than the underlying water, begin to sink. This vertical mixing (*fall overturn*), persists into late fall, until the temperature at the lake surface drops below 4°C and winter stratification ensues. Fall overturn produces greater vertical mixing of water than spring overturn because temperature differences in the lake are greater during summer stratification than during winter stratification. Fall over-turn speeds the movement of oxygen to deep waters and rushes nu-trients to the surface. Where the hypolimnion becomes fairly warm in midsummer, deep vertical mixing may take place in late summer when temperatures are still warm enough for plant growth. Infusion of nu-trients into surface waters at this time often causes a later burst of phytoplankton population increase — the *fall bloom*.

Integrated Description of Climate

We find it easier to dissect climate into its component properties of temperature, humidity, precipitation, wind, and solar radiation, than to appreciate at once all the implications of these factors for the ecosystem. We must, however, understand *interactions* among climate factors be-cause these factors are clearly interdependent in their effect on life. For example, seasonal rainfall promotes plant growth more strongly during

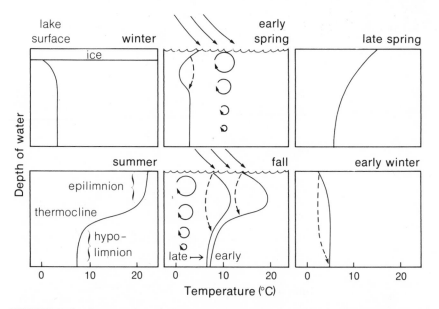

FIGURE 5–8 Seasonal cycle in the temperature profile of a temperate-zone lake. Overturn, indicated by dashed lines, is more extensive in fall than in spring. Thermoclines occur at depths of 5 to 25 meters, depending on the depth and turbidity of the lake and on local winds.

warm months than during cold months. Wind movement and solar radiation interact with temperature to determine thermal stress; temperature and humidity together influence water balance.

The *climograph* was developed to portray seasonal changes in temperature and rainfall simultaneously. The climograph has a rainfall scale (horizontal axis) and a temperature scale (vertical axis); each month is plotted on the graph according to its average temperature and rainfall. Seasonal progression of climate is portrayed on the climograph by following the points for each month of the year in succession (Figure 5–9). The horizontal spread of months represents variation in rainfall and the vertical spread, variation in temperature.

The climograph permits a visual comparison of climates at different localities. We note immediately that the seasons in Panama bring marked variation in rainfall but little change in temperature, whereas the reverse characterizes New York City. Moving east to west from Cincinnati, Ohio to Winnemucca, Nevada, climate becomes more arid

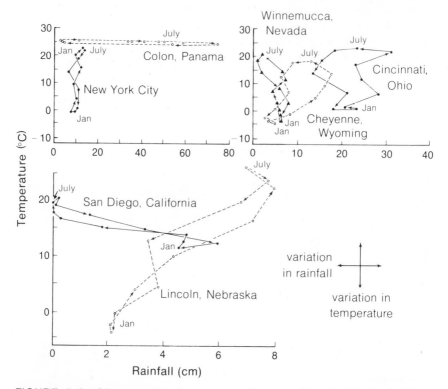

FIGURE 5–9 Climographs of representative localities in North America. Each point represents the mean temperature and rainfall for one month. Lines connecting the months at each locality indicate seasonal change in climate.

but temperatures remain within the same range. The change in vegetation from deciduous broad-leaved forest in Ohio, to short grass prairie in Wyoming, and to desert shrubs in Nevada is thus related to the water relations of different plant forms rather than temperature tolerance. Although San Diego's weather in January resembles that of Lincoln, Nebraska in April, the overall climates differ as much as their vegetation. San Diego's hot, dry summers and cool, moist winters favor slow-growing, drought-resistant shrubs (chaparral), while Lincoln, Nebraska is (or rather was) surrounded by tall grass prairie. Summer rainfall in the Great Plains states supports greater plant productivity than the winter rainfall of the west coast because water is abundant on the prairies during the warm summer months, the best growing season of the year. The winters, however, are too cold and dry for shrubby vegetation.

Although the climograph is useful for comparing localities, it fails to combine the effects of temperature and rainfall in any biologically meaningful way and it does not show the cumulative effects of weather upon environmental conditions. For example, during dry seasons evaporation and *transpiration* (evaporation of water from leaves) both remove water from the soil. If rainfall is insufficient to balance evaporation and transpiration loss, the water deficit in the soil steadily increases, perhaps for months at a time. In other words, soil water reflects last month's rainfall, as well as more recent input to the soil.

In 1948, geographer C. W. Thornthwaite devised a method for utilizing climate to estimate the availability of water in the soil. He compared the rate at which water is drawn from the soil by plants and by direct evaporation to the rate at which it is restored by precipitation. The sum of evaporation and transpiration is the total *evapotranspiration* of the habitat. In natural environments, evapotranspiration is at times limited by the availability of water in the soil. Potential evapotranspiration represents the amount of water which would be drawn from the soil if soil moisture were not limiting. If precipitation input to the soil (rainfall minus surface runoff) exceeds potential evapotranspiration at all seasons, the soil will remain saturated with water throughout the year as at Brevard, North Carolina (Figure 5–10). At Bar Harbor, Maine, precipitation falls below potential evapotranspiration during the warm summer months and the soil is depleted of water during late summer and early fall. Canton, Mississippi receives rainfall similar to that of Bar Harbor but the hotter climate of Mississippi increases the potential evaporation of water from her soils. Serious water deficits are incurred during the summer months. Manhattan, Kansas, receives much less rain than Canton, Mississippi, but because the rainfall is concentrated during the summer periods of maximum potential evapotranspiration, soil water deficits are no more serious than in Mississippi. On the other hand, Grand Junction, Colorado, has a dry climate where soils are

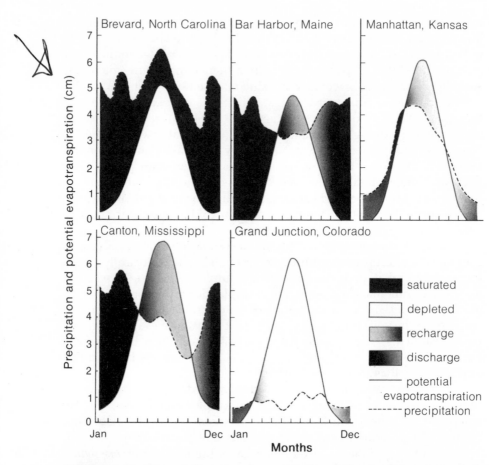

FIGURE 5–10 The relationship of precipitation and potential evapotranspiration to changes in the availability of soil moisture. When evapotranspiration exceeds precipitation, water is withdrawn from the soil until the deficiency exceeds 10 centimeters (4 inches), the average amount of moisture that soils can hold.

depleted of water most of the year and rarely become saturated. Plant productivity is correspondingly low.

Because potential evapotranspiration increases with temperature, temperature and water stress go hand in hand. Thus boreal regions receiving 10 to 20 inches of precipitation each year have a more favorable water budget for plant production than tropical regions with similar levels of precipitation.

Thornthwaite's analysis may lack the detail to predict local variation in soil moisture and plant production, but his graphs indicate the relative length and severity of seasonal drought. By keeping a running seasonal balance of gains and losses of water, one can appreciate the cumulative effects of climate on soil moisture.

Topographic Influences

The topography and geology of a region can create variation in the environment within regions of uniform climate. In hilly areas, the slope of the land and its exposure to the Sun determine the temperature and moisture content of the soil. Soils on steep slopes are well-drained, often creating conditions of moisture-stress for plants when the soils of nearby lowlands are saturated by water. Stream bottomlands and seasonally dry river beds in arid regions often support well-developed forest communities, in sharp contrast to the surrounding desert vegetation. Plant communities on shady and sunny sides of mountains and valleys frequently differ in accordance with the temperature and moisture regimes of each exposure. South-facing slopes are exposed to the direct heating effect of the Sun, which limits vegetation to shrubby, drought-resistant (*xeric*) forms. The corresponding north-facing slopes are relatively cool and wet and harbor moisture-requiring (*mesic*) vegetation (Figure 5-11). The distinction between north- and south-facing exposures largely disappears in the tropics because the Sun passes directly overhead during the middle of the day.

Air temperature decreases with altitude, by about 6°C for each 1,000-meter increase in elevation. Even in the tropics, if one could climb high enough one would eventually find freezing temperatures and

FIGURE 5–11 The effect of exposure on vegetation of a series of mountain ridges near Aspen, Colorado. The north-facing (left-facing) slopes are cool and moist, permitting the development of spruce forest. Shrubby, drought-resistant vegetation grows on the south-facing slopes.

perpetual snow. If the temperature at sea level was 30°C, freezing temperatures would be reached at about 5,000 meters (16,000 feet). This, indeed, is the approximate altitude of the snowline in the Andes of central Peru (Figure 5–12).

A 6°C drop in temperature corresponds, in temperature latitudes, to an 800-kilometer (500-mile) increase in latitude. In many respects, the climate and vegetation of high altitudes resembles that of sea level localities at higher latitudes. Despite their similarities, however, one would be mistaken to equate alpine communities particularly in the tropics and arctic communities. Alpine environments are usually less seasonal than their low-elevation counterparts at higher latitudes, even though average temperature and annual rainfall may be similar. Temperatures in tropical montane environments remain nearly constant over the year, and the occurrence of frost-free conditions at high altitudes allows many tropical plants and animals to live in the cold environments found there.

In the mountains of the southwestern United States, changes in plant communities with elevation create more or less distinct belts of vegetation, referred to as *life zones* by the early naturalist C. H. Merriam. Merriam's scheme of classification included five broad zones, which he named, from south to north (or low to high elevation): Lower Sonoran, Upper Sonoran, Transition, Canadian (or Hudsonian), and Arctic-Alpine.

FIGURE 5–12   Even in the tropics one may find communities dominated by cold temperatures. Biological communities stop abruptly at the snowline at about 16,000-feet elevation in central Peru.

Hudsonian Zone Elevation 8,500 feet

Alpine Zone Elevation 11,000 feet

Upper Sonoran Zone Elevation 5,000 feet

Transition Zone Elevation 6,500 feet

Lower Sonoran Zone Elevation 3,000 feet

Upper Sonoran Zone Elevation 4,000 feet

FIGURE 5-13 Vegetation types corresponding to different elevation in the mountains of southeastern Arizona. Lower Sonoran vegetation is dominated by saguaro cactus, small desert trees such as paloverde and mesquite, and numerous annual and perennial shrubs and small succulent cacti. Agave, ocotillo, and grasses are conspicuous elements of the Upper Sonoran Zone, with oaks appearing toward its upper edge. Large trees predominate at higher elevations — ponderosa pine in the Transition Zone, and spruce and fir in the Hudsonian Zone. These gradually give way to bushes, willows, herbs, and lichens in the Alpine Zone above the treeline.

At low elevations in the southwest, one encounters a cactus and desert shrub association characteristic of the Sonoran Desert of northern Mexico and southern Arizona (Figure 5–13). In the woodlands along stream beds, plants and animals have a distinctly tropical flavor. Many hummingbirds and flycatchers, ring-tailed cats, jaguars, and peccaries make their only temperate zone appearances in this area. At 2,500 meters higher, in the Alpine Zone, we find a landscape resembling the tundra of northern Canada and Alaska. By climbing 2,500 meters, we experience changes in climate and vegetation which would require a journey to the north of 2,000 kilometers, or more, at sea level.

Geological Influences

Local variation in the bedrock underlying a region promotes the differentiation of soil types and enhances biotic heterogeneity. In some mountainous regions of the United States, outcrops of serpentine (a kind of igneous rock) produce soils with so much magnesium that the vegetation characteristic of surrounding soil types cannot grow. Serpentine *barrens,* as they are called, are usually dominated by a sparse covering of grasses and herbs, many of which are distinct endemics (species found nowhere else) that have evolved a high tolerance for magnesium (Figure 5–14). Depending on the composition of the bedrock and the rate of weathering, granite, shale, and sandstone can also produce a barrens type of vegetation. The extensive pine barrens of southern New Jersey occur on a large outcrop of sand, which produces a dry, acid, infertile soil capable of supporting no more than knee-high pygmy forests of pines.

Cave-ins in limestone regions of the Appalachian Mountains and the melting of large blocks of ice left behind by retreating glaciers in New England and the Great Lakes region, have created local depressions in the landscape — sink holes and kettle holes. Because these depressions are more or less continually wet, they support a vegetation cover that differs from the surrounding land. *Bogs* and *glades* represent the most extensive development of these wet conditions. The water trapped within a bog is very still and affords little exchange of oxygen and mineral ions with the surrounding environment. Decomposers deplete the water of its oxygen, producing anaerobic, acid conditions which bring a halt to the decomposition of organic matter. Plant detritus accumulates. Few inorganic nutrients find their way into the bog, except in rainfall. These infertile, acid conditions that characterize bogs and glades are unsuitable for most kinds of plants. Specialized bog species, like sphagnum, Labrador tea, cranberry, and pitcher plants thrive. Only a few trees, mostly red and black spruce, invade the edges of the bog (Figure 5–15). These examples demonstrate how topography and geology create diversity within regions of uniform climate.

FIGURE 5–14 A small serpentine barren in eastern Pennsylvania. The soils surrounding the barren support oak–hickory–beech forest.

Variation in the Aquatic Environment

Although aquatic habitats lack many of the influences that shape terrestrial environments, oceans, lakes, and rivers are surprisingly diverse. Water depth is an obvious factor, but the physical substrate — whether rock, sand, or mud — and exposure to current and wave action also enhance the diversity of underwater environments.

The edge of the sea is alternately covered by water and exposed to the air by a twice-daily cycle of tides (Figure 5–16). Plants and animals with different tolerance of terrestrial conditions form distinct bands within the *intertidal* zone. The uppermost reaches of the shore are only infrequently splashed by waves and are inhabited by such creatures as the isopod, *Ligia* (a relative of the sowbug of forests and gardens). *Ligia* is so adapted to the near-terrestrial conditions in a splash zone, that it will not tolerate submersion in sea water. A little lower, near the high tide mark, one finds barnacles and periwinkles. The seaweeds *Ulva* and *Fucus* begin to appear a foot or so lower. Along the west coast of the United States, mussel beds with associated species of gooseneck barnacles, limpets, chitons, snails, starfish, and worms occupy a narrow band in the middle intertidal region. The seaweed, *Laminaria,* forms a dark green mantle over even richer and more diverse communities at the lower edge of the intertidal zone, where sea anemones, sea urchins, nudibranches, tunicates, sponges, and myriad crabs and amphipods abound. The conditions of alternate exposure to air and submersion by

FIGURE 5–15   A wet glade in Monongahela National Forest, West Virginia. The typical bog-type ground cover consists of sphagnum moss, cinnamon fern, viburnum, and various grasses. Red spruce invade the edge of the glade.

sea water are so stringent and require such physiological specialization of their inhabitants that few intertidal species extend below the lowest tide level and few subtidal species venture into the intertidal zone.

The *salinity* (salt concentration) of surface waters of the ocean varies with the influences of rivers, evaporation, and precipitation. Ocean water contains an average of 3.5 per cent salt. The Mediterranean Sea is

FIGURE 5–16   The coast of Kent Island, New Brunswick, at the mouth of the Bay of Fundy, at high tide (left) and low tide (right). The daily tidal range is about 25 feet.

nearly cut off from the Atlantic Ocean, except for its narrow mouth at the Straits of Gibraltar. The warm, dry Mediterranean climate evaporates water rapidly from the surface of the sea but produces only a trickle of stream and precipitation input. Because of this local imbalance of precipitation and evaporation, the salinity of the Mediterranean Sea is over 4.0 per cent in areas. Diverting river water for irrigation in the Middle East will further increase salinity close to the mouths of large rivers and could alter the marine environment significantly for biological communities. In contrast to the Mediterranean region, the cold, rainy climate of the Baltic Sea and surrounding Scandinavian countries reduces the salinity of surface waters to 0.5 per cent, even less, in some areas.

Steady east-to-west winds cause upwelling currents along the western coasts of North and South America, and Africa. The guano and anchovy industries of Peru are supported in part by nutrients brought to the surface by upwelling currents. Just as happens off the Pacific coast of Panama, winds push surface water away from the shore and deep water moves up to replace it.

The Microenvironment

The heterogeneity created in a habitat by vegetation, soil structure, detritus, and the activities of animals results in a variety of environmental conditions to which animals and plants become specialized. Ecologists often refer to the particular conditions in one part of a forest or a field — as one would find under a log or deep within clumps of grass — as a *microenvironment*, or *microhabitat*.

Soil acts as an insulating layer which dampens variation in temperature with increasing depth (Figure 5–17). The surface of exposed soil absorbs solar radiation and, in desert habitats, heats up as much as 30°C above air temperature during the day. Some of this heat is conducted to underlying soil layers, but the slow movement of heat through earth delays the temperature cycle. At 20-centimeter depth, maximum daily temperature is not reached until early evening, roughly six hours after the surface temperature begins to drop. Soil type influences magnitude and depth of temperature fluctuations. Sandy soils heat up rapidly because of their low heat capacity (due to their low water content). Wet soils and peats resist temperature change more strongly.

Vegetation adds a biological component to the structural diversity of the environment. A tree creates a variety of microhabitats around its roots, bark, leaves, and flowers. Forests stratify light intensity, temperature, wind velocity, and humidity, creating vertical gradients within the atmosphere under the forest canopy. In tropical rain forests in Surinam, mid-afternoon relative humidity dips to 40 per cent at 30 meters height (just above the canopy), but remains above 60 per cent at 5 meters above ground within the canopy, and above 80 per cent near the forest floor.

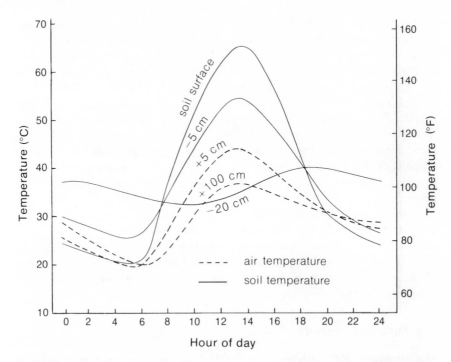

FIGURE 5–17   Diurnal fluctuation in the temperature of a desert soil, and the air above it. Measurements show the depth below the surface of the soil and the height of the air above the surface of the soil.

The relative constancy of the inner forest creates a different environment from the exposed forest canopy.

Small organisms respond to even finer subdivision of the environment. English botanist J. L. Harper and his colleagues have shown that for proper germination, seeds require quite specific combinations of light, temperature, and moisture. These are provided by irregularities in the surface of natural soils, but Harper dramatized the differences by creating an artificially heterogeneous soil environment (Figure 5–18). Three species of plantains sowed in seed beds responded differently to the microenvironments produced by slight depressions, by squares of glass placed on the soil surface, and by vertical walls of glass or wood.

The adaptations of plants and animals to such fine variation in the environment as Harper created in his seed beds emphasize the influence of the environment on the distributions of organisms and upon their form and functioning. Major patterns in environment determined by climate and topography are also reflected in the diversity of vegetation types over the Earth. In the next chapter we shall examine the influence of climate — particularly rainfall and temperature — on the structure of plant communities from the arctic to the tropics.

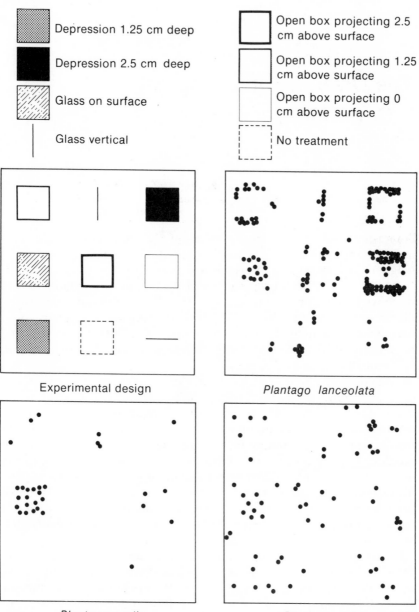

FIGURE 5–18   Germination of seedlings of three species of plantains (genus *Plantago*) with respect to artificially produced variation in the soil surface.

6

The Diversity of Biological Communities

CLASSIFICATION ACCORDING TO STRUCTURE OR CLIMATE

Botanists frequently construct systems of classification for plant communities. Most of these schemes are based on vegetation structure — height of vegetation, leaf or needle structure, deciduousness, and dominant plant form. Such properties of plants are, in turn, adaptations to the physical environment where they live. We should not be surprised therefore to note the close correspondence between vegetation zones and climate. The natural vegetation of the United States, as classified by botanist A. W. Kuchler, conforms closely to variation in temperature and rainfall (Figure 6-1, pages 74-75).

Major vegetation types are clearly discernible: tall forests, shrubland, and prairie are distinct; so are coniferous and broad-leaved forest types. The problem with classifications of vegetation, or with similar schemes for aquatic communities, is that many intermediates occur. In fact, most biological communities intergrade, sometimes almost imperceptibly, as the physical environment changes from one locality to the next. One is tempted to deal with these intermediates by making finer distinctions between communities. This practice can, however, lead to a bewildering variety of names for plant communities. A. W. Kuchler's somewhat conservative scheme lists 116 vegetation types in the United States alone.

Classifications of plant communities are most useful if they relate vegetation types to environment: temperature, rainfall, soil, and topography. In this chapter, we shall examine the influence of environment on vegetation structure and plant classification and survey the diversity of plant communities by way of a photographic essay.

Structural Schemes of Classification

The earliest traditions of vegetation classification were based on attempts to describe the major plant forms of each major association. These classifications embraced complete analysis of the *species* composition of communities (*floristic* analysis) on the one hand, and description of plant *forms*, regardless of the particular species, on the other hand. Floristic analysis proved useful in restricted areas where botanists knew all the species and where minor differences between communities involved the replacement of species by similar species with slightly different ecological requirements. Floristic analysis is completely unworkable on a global scale because biogeographical barriers restrict the distributions of individual species. Functionally similar forests in Europe and the United States have very few species in common. Floristic differences between vegetational counterparts in California and Australia would be even greater. The tropics pose still greater problems for floristic analysis because of their great species diversity. Few botanists can recognize a majority of the hundreds of kinds of trees in a tropical forest and, to make matters more difficult, many species can be distinguished only when they are in flower or fruit.

Difficulties of floristic analysis for world-wide vegetation classification are partly overcome by analysis of form and function of plants rather than their scientific names. Numerous sets of symbols were devised to describe such characteristics as plant size, life form, leaf shape, size, and texture, and per cent of ground coverage. Kuchler worked out a system of letter and number symbols which could be combined into a formula describing the characteristics of any given plant formation. Thus M6iCXE5cD3i6H2pL1c represents an oak–yew woodland and E4hcD2rGH2rL1c(b) a madrone–holly scrub. A similar symbolic method of description, devised by Pierre Dansereau, portrays vegetation formations by use of lollipop and ice-cream cone shaped figures with internal shading and symbols varying according to the nature of the plant (Figure 6–2).

Kuchler's and Dansereau's methods are primarily descriptive. They are too complex to be used as a hierarchical scheme of classification although they have found application where classification of only the predominant features of a plant formation is desired.

In 1903, the Danish botanist Christen Raunkiaer proposed to classify plants according to the position of their buds (regenerating parts), and found that the occurrence of his major categories corresponded closely to climatic conditions. Raunkiaer distinguished five principle life forms (see also Figure 6–3):

phanerophytes (from the Green *phaneros*, visible) carry their buds on the tips of branches, exposed to extremes of climate. Most trees and large shrubs are phanerophytes. As one might expect, this plant form dominates in moist, warm environments where buds require little protection.

chamaephytes (from the Greek *chamai*, on the ground, dwarf) comprise small shrubs and herbs which grow close to the ground (prostrate

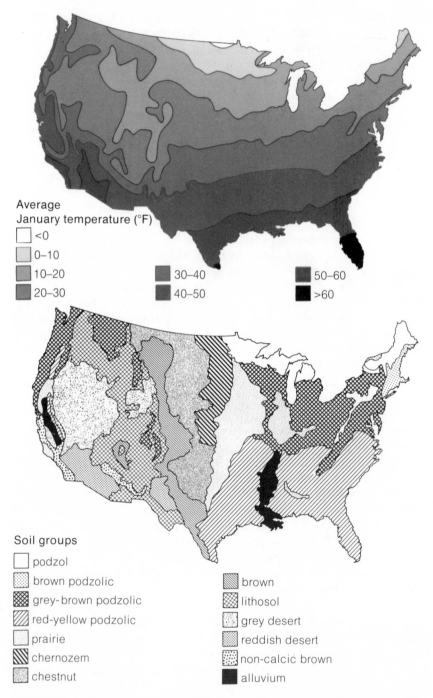

FIGURE 6-1   Temperature, precipitation, soil, and vegetation maps of the United States.

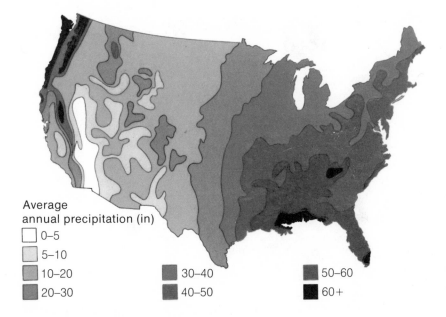

Average
annual precipitation (in)

- ☐ 0–5
- 5–10
- 10–20
- 20–30
- 30–40
- 40–50
- 50–60
- 60+

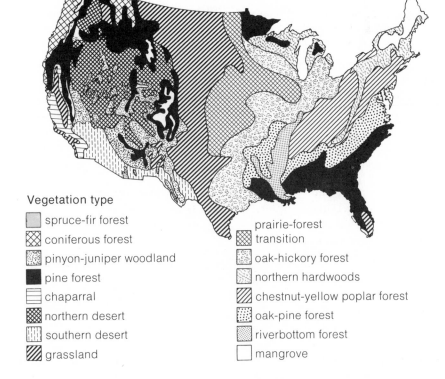

Vegetation type

- spruce-fir forest
- coniferous forest
- pinyon-juniper woodland
- pine forest
- chaparral
- northern desert
- southern desert
- grassland
- prairie-forest transition
- oak-hickory forest
- northern hardwoods
- chestnut-yellow poplar forest
- oak-pine forest
- riverbottom forest
- mangrove

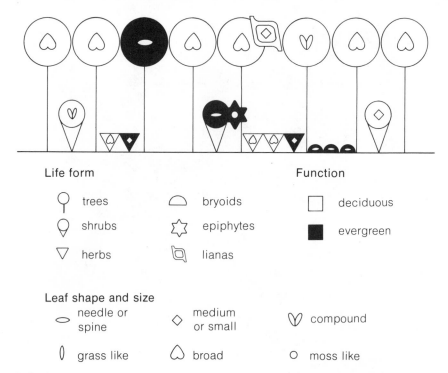

FIGURE 6–2   Symbolic representation of a forest by the Dansereau method of vegetation classification.

life form). Proximity to the soil protects the bud. In regions of heavy snowfall, the buds are protected beneath the snow from extreme air temperatures. Chamaephytes are most frequent in cool, dry climates.

hemicryptophytes (from the Greek *kryptos*, hidden) persist through the extreme environmental conditions of the winter months by dying back to ground level where the regenerating bud is protected by soil and withered leaves. This growth form is characteristic of cold, moist zones.

cryptophytes are further protected from freezing and desiccation by having their buds completely buried beneath the soil. The bulbs of irises and daffodils are representative of cryptophyte plants, and are also found in cold, moist climates.

therophytes (from the Greek *theros*, summer) die during the unfavorable season of the year and do not have persistent buds. Therophytes are regenerated solely by seeds, which easily resist extreme cold and drought. The therophyte form includes most annual plants and occurs most abundantly in deserts and grasslands.

The proportional occurrence of Raunkiaer's life forms in various climatic regions is summarized in Figure 6-4. Life form and climate go

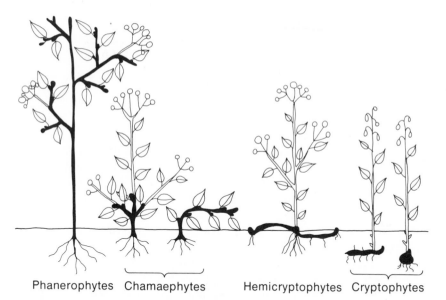

Phanerophytes Chamaephytes   Hemicryptophytes Cryptophytes

FIGURE 6–3 Diagrammatic representation of Raunkiaer's life forms. Un-shaded parts of the plant die back during unfavorable seasons, while the solid black portions persist and give rise to the following year's growth. Proceeding from left to right, the buds are progressively better protected.

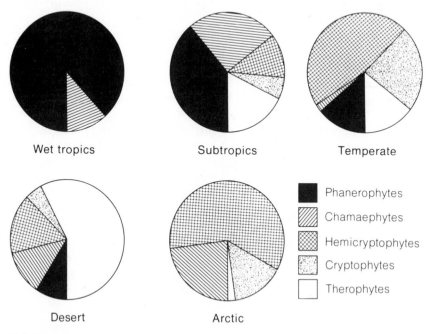

Wet tropics   Subtropics   Temperate

Desert   Arctic

■ Phanerophytes
▨ Chamaephytes
▧ Hemicryptophytes
▨ Cryptophytes
□ Therophytes

FIGURE 6–4 Proportions of plant life forms in various localities according to Raunkiaer's scheme.

closely together. Phanerophytes dominate vegetation forms in warm, moist environments, being replaced by chamaephytes, hemicryptophytes, and cryptophytes in temperate and arctic regions. Deserts have a large proportion of therophytes.

The Holdridge Classification

The botanist L. R. Holdridge has proposed a classification of the world's plant formations based solely on climate (Figure 6-5). Holdridge considers temperature and rainfall to prevail over other environmental factors in determining vegetation form, although soils and exposure may exert strong influences on plants within each climate zone.

Holdridge's scheme incorporates the biological effects of climate on vegetation. As in Thornthwaite's analysis of climate (page 61), temperature and rainfall are seen as interacting to define humidity provinces. The dividing lines between humidity provinces are determined by critical ratios of potential evapotranspiration to precipitation. Potential evapotranspiration is in turn a function of temperature. Therefore the humidity provinces relate temperature and rainfall to the water relations of plants in a way that is meaningful. Holdridge's formula indicates, for example, that the availability of moisture to plants in wet tundra, with an annual rainfall of 25 centimeters (cm) and average temperature near freezing, is similar to that in a wet tropical forest, with 400 cm precipitation and an average temperature of 27°C.

Holdridge relates differences between plant formations to percentage differences between their climates. The temperature or rainfall of each zone is either twice or one-half that of the adjoining zone. Thus 25 cm annual precipitation can make as big a difference in the vegetation of arid regions as 250 cm does between plant formations in the humid tropics. It seems intuitively reasonable that a little rainfall should stimulate desert annuals much more than it would rain forest trees. Experiments under controlled conditions largely support this notion.

Holdridge also constructs his temperature scale with biological considerations in mind. He assumed that biological activity ceases below 0°C. The temperature of any month whose mean was below freezing was set at 0°C for calculating mean annual temperature. Furthermore, because small increases in temperature affect biological systems more at low temperatures than at high temperatures, Holdridge set the temperature boundaries of his life zones at 1.5, 3, 6, 12, and 24°C, each temperature being twice the previous one. A factorial scale of temperature is consistent with increases in rate of evaporation and rate of biological activity in relation to increasing temperature.

Simple climate classifications, such as the Holdridge scheme, are far from ideal. In regions with similar mean rainfall and temperature, differ-

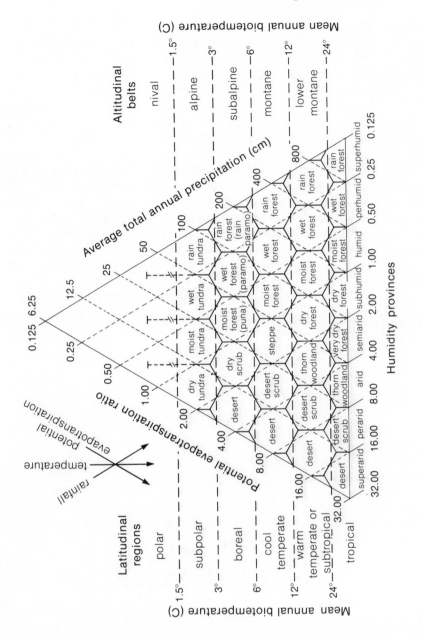

FIGURE 6–5 The Holdridge scheme for the classification of plant formations. *Mean annual biotemperature* is calculated from monthly mean
temperatures after converting means below freezing to 0°C. The *potential
evapotranspiration ratio* is the potential evapotranspiration divided by the
precipitation; the ratio increases from humid to arid regions.

ing seasonal patterns of precipitation and temperature can create differences in vegetation structure. Topography, soil, and fire can also influence the development of vegetation types. Still, it is fair to say that climatic schemes of life zone classification do emphasize the pervasive influence of temperature and moisture on plant formations.

A Survey of Biological Communities

If a random sample of terrestrial localities is placed on a graph according to the mean annual temperature and rainfall of each locality, the points fall within a triangular area whose three corners represent warm moist, warm dry, and cool dry environments (Figure 6–6). Cold regions with high rainfall are conspicuously absent; water does not evaporate rapidly at low temperature and the atmosphere in cold regions has little water vapor. But because of the depressing effect of low temperature on evaporation, a little water goes a long way. In the tropics, 20 inches of rainfall can support little more than a desert scrub-type vegetation, but the same 20 inches permits the development of an impressive coniferous forest in Canada.

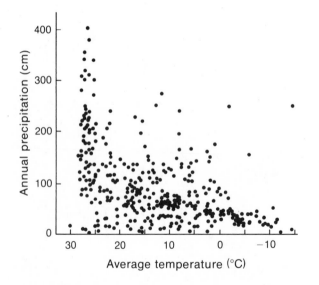

FIGURE 6–6 Average annual temperature and rainfall for a large sample of localities more or less evenly distributed over the land area of the Earth. Most of the points can be enclosed by a triangular region that includes the total range of possible climates on Earth (excluding high mountains).

Plant ecologist, R. H. Whittaker has combined several structural classifications of plant communities into one scheme, which he has transposed onto a graph of temperature and rainfall (Figure 6–7). Within the tropical and subtropical realms, with mean temperatures between 20 and 30°C, vegetation types grade from true rain forest, which is wet throughout the year, to desert. Intermediate climates support seasonal forests, in which some or all trees lose their leaves during the dry season (see Figure 5–7), and low dry forests or scrublands with many thorntrees. As aridity increases, shrubs appear farther apart, exposing large patches of bare ground. This vegetation characteristic is mostly highly developed in true deserts.

The range of plant communities·in temperate areas follows the same pattern as tropical communities, with the same basic vegetation types distinguishable in both. In colder climates, however, the range of pre-

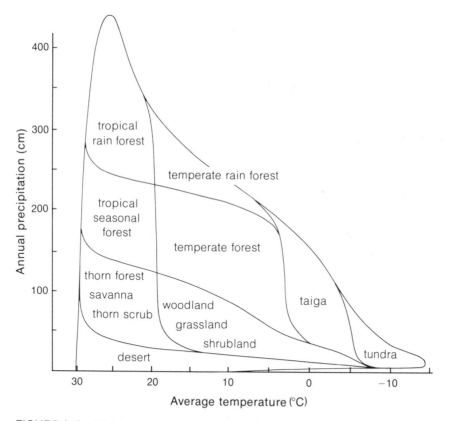

FIGURE 6–7   Whittaker's classification of vegetation types superimposed on the range of terrestrial climates. In climates between forest and desert regions, fire, soil, and climate seasonality determine whether woodland, grassland, or shrubland develops.

cipitation from one locality to another is so narrow that vegetation types are poorly differentiated on the basis of climate. Where mean annual temperatures are below −5°C, Whittaker lumps all plant associations into one type — tundra. The whole scale of moisture gradients represented in the tropics is compressed into a narrow band in the arctic. Which is tundra: rain forest or desert? Water abounds on moist tundra, but because it is frozen most of the year, and permanently frozen a few feet below the soil surface, plants cannot obtain it.

At the drier end of the rainfall spectrum within each temperature range, fire plays a distinct role in shaping the form of plant communities. In the African savannas and midwestern American prairies, frequent fires kill the seedlings of trees and prevent the establishment of tall forest, for which favorable conditions otherwise exist. Burning favors the growth of perennial grasses with extensive root systems that can survive fires. After an area has burned over, the grass roots send up fresh shoots and quickly revegetate the surface.

Like all classifications, exceptions to the scheme outlined in Figure 6-7 appear frequently. Boundaries between vegetation types are at best fuzzy. All plant forms do not respond to climate in the same way. For example, some species of Australian eucalyptus trees can form forests under climate conditions which would support only shrubland or grassland on other continents. Plant communities are affected by factors other than temperature and rainfall. We have already examined the influence of topography, soils, fire, and seasonal variation, but the point deserves emphasis.

A SURVEY OF BIOLOGICAL COMMUNITIES

The photographs on the pages that follow, most of them taken in North and Central America, represent most of the important plant formations. In addition, there are photographs of many important aquatic communities.

Environments Without Life

Life can gain a foothold in regions with almost any combination of temperature and moisture found on Earth, providing the moisture is available and other nutrients are present. But life is excluded from a few extreme environments. Hot lava and noxious gases delay the colonization of the new volcanic island of Surtsey, off the coast of Iceland (above). The extreme cold on the slopes of Mount McKinley, Alaska, (below) freezes life to a standstill. Water occurs only

as ice and is therefore completely unavailable to plants. Water is also a problem on the shifting sand dunes of Death Valley, California (below). The little rain that falls either evaporates or percolates through the coarse sand. Temperatures at White Sands, New Mexico (above), are favorable for life and the region's rainfall supports desert shrubs in the surrounding valley, but the pure gypsum sand (calcium sulfate) does not contain the nutrients needed to support life.

The Humid Tropics

Year-round warm temperatures and plentiful moisture in the humid tropics create conditions for the most luxuriant and diversified communities in the world. Vegetation forms include vines that drape the trees in a lowland forest in Panama (right), and air plants that clothe trees in a mist-enshrouded cloud forest in Guatemala (above). Because soils are impoverished of nutrients except near the surface, root systems of tropical trees tend to be shallow and the trunks of many trees are buttressed for support (left).

Tropical Mountains

Temperature decreases about 6°C for each 1,000-meter increase in elevation. Plant productivity parallels the lower temperatures of montane habitats, creating cold and almost barren deserts in the tropics. The mean annual temperature and rainfall would support a forest or woodland in seasonal temperate climates with warm summers, but the year-round cold of tropical mountains does not permit such luxurious growth. On the paramo of the high Andes in Colombia at about 12,000 feet (above left), the temperature hovers around 5°C throughout the year. One is struck by the paucity of life forms and by the silence, broken only by the relentless wind. At the same elevation in Costa Rica on the Cerro de la Muerte (Mountain of Death, below left), the everpresent fog slips among dwarfed plants whose small thick leaves are clustered tightly around the plant stem for protection from the cold wind (above). The bare patch of rock shows the thinness of the soil layer in the tropical montane habitat.

Subtropical Deserts

Belts of seasonally hot, dry climate girdle the Earth at about 30° north and south of the Equator (see Figures 5-2, 3). These are harsh environments where only a few drought-adapted species of plants and animals thrive. Whereas light and nutrients are critical in the humid tropics, the bare ground exposed in deserts testifies that these resources go wanting where rainfall limits plant growth. Cacti have greatly reduced leaves to decrease water loss. Their thick, succulent stems have taken over the function of photosynthesis. Numerous thorns hinder desert animals from getting at their stored water. Desert shrub habitats of the Sonoran Desert of Arizona and northern Mexico (above left) are among the most diverse vegetation types of arid regions. Giant saguaro cacti and paloverde trees dominate the landscape. The joshua tree, a treelike yucca, occurs primarily in the Mohave Desert of southern California (below left). An extremely dry habitat near the Gulf of California in northern Sonora, Mexico (below) supports only two kinds of large plants. Note the wide spacing between individuals. Desert plants do not tolerate close proximity because their extensive root systems compete for water.

Temperate Woodland and Shrubland

In temperate habitats with better water relations and lower summer temperatures than deserts, succulent cacti are replaced by bushes, shrubs, and small trees. The wide spacing and low growth form of plants in the Great Basin region exemplified in Zion National Park (above) indicate that water is still a critical factor. At higher elevations, in Coconino National Forest of Arizona, an open woodland dominates the landscape (above right). Juniper woodland develops at 6,000 to 7,000 feet elevation in this area, where snow covers the ground for much of the winter and summers are cool. The milder mediterranean climate of the southern California coast, characterized by warm, dry summers and cool, moist winters, supports a characteristic dense shrubland called chaparral (below right). In moist canyons and valleys, oak woodland tends to replace chaparral species, but frequent fires often prevent this natural succession and maintain the fire-adapted chaparral vegetation.

Temperate Forests

Tall forests of broad-leaved, deciduous trees occur throughout the temperate zone where rainfall is plentiful and winters are cold. Oak, beech, maple, hickory, and other hardwoods dominate temperate forests. Seasonal patterns of summer activity and winter dormancy are characteristic. The stand of Indiana hardwoods dominated by white oak has a well-developed understory of sugar maple and smaller shrubs (above). In the Appalachian Mountains of West Virginia, red spruce and hemlock occur with broad-leaved trees to form mixed

forests (below left). In the southeastern United States, sandy soils are too poor for broad-leaved trees. Pines are widely distributed in vast forests that are managed and harvested for paper pulp. In Florida, the palmetto frequently forms a dense understory (above). In the northern United States and Canada, and in mountainous regions of the west, birch and aspen, frequently mixed with spruce and fir, represent the farthest incursion of broad-leaved forests into cold regions (below).

Temperate Grasslands

Grasslands occur under a variety of temperate climates with cold winters and summer drought. True prairie, remnants of which can be found in Kansas (above), Texas (below), and other midwestern states, is characterized by grasses with extensive root systems. Tall grass prairies grow on fertile soil and are

probably maintained by periodic fires which keep trees from becoming established. Farther to the west, lower rainfall supports sparcer vegetation, the shortgrass prairies to the east of the Rocky Mountains (above) and in western interior valleys (below). These grasslands are very delicate and are sensitive to ploughing and overgrazing.

Fresh-Water Habitats

Fresh water covers a small fraction of the Earth's surface, yet fresh-water habitats display remarkable diversity. Variation in water movement, mineral and oxygen content of the water, and size and shape of the stream or lake basin all contribute to this variety. Communities in deep lakes and fast-moving streams consist mostly of phytoplankton and thin layers of diatoms on the surfaces of rocks. Vegetation shows above the surface only where water is shallow and still, as in an artificially flooded marsh in Maine (left), or a cattail marsh in New York (above). Floating water hyacinths choke a deeper channel in Louisiana, buoyed up by gas trapped in their stems (right). The hyacinths are not rooted to the bottom and obtain all their nutrients directly from the water.

Temperate Montane Environments

Montane habitats are much colder and are often drier than the surrounding lowlands. Trees reach their upper limit of elevation at about 10,000 feet in the Cascade Mountains of Oregon (above left), where the extreme topographic heterogeneity creates a mosaic of plant communities in close association. Above timberline, snow persists well into summer in habitats that can support only the low grassy vegetation characteristic of the alpine tundra, as in the Rocky Mountains of Colorado at 12,000-feet elevation (below). Lichens are the first plants to colonize bare rock surfaces in these habitats (right), and start the slow process of soil formation. Wind constitutes a major ecological factor at high elevations. Wind-driven ice strips bark and branches from trees near the timberline in Arapaho National Forest, Colorado, sparing only the protected leeward side of the tree (below left).

Temperate Conifer Forests

Forests of pine, spruce, fir, hemlock, redwood, and others grow under a variety of temperature, soil, moisture, and fire conditions that favor drought-resistant needle-leaf species over less tolerant broad-leaved trees. Poor soils and frequent fires favor pines throughout much of the southeastern United States. Dry summers and cold winters characterize the environments of coniferous forests at high elevations in western mountains. Pinyon pine–juniper–cedar woodland is found near Flagstaff, Arizona, where the climate is too dry to support a closed forest (below left). A moister site in Inyo National Forest, California, is dominated by tall jeffrey pines (above left). Undergrowth is sparse in the dry acid soil. Abundant winter rainfall and cool, foggy summers create ideal conditions for redwoods in the temperate rain forests of northern California (above). These forests are among the tallest plant formations in the world. They bear little resemblance to the humid forests of the tropics because they lack diversity in species and plant forms and are relatively unproductive.

The Arctic Tundra

Permanently frozen soils underlie the arctic tundra habitat. Warm summer temperatures thaw the ground to a depth of a few inches or feet, briefly creating a shallow, often waterlogged layer of soil on which arctic vegetation develops. Repeated freezing and thawing creates characteristic polygonal patterns in the ground surface of some areas (above). At Cold Bay, Alaska, lichens, mosses, and

grasses are found on the hummocky, frost-heaved soil (below left). Kettle lakes formed by the melting of large blocks of ice, left by retreating glaciers, are a prominent feature of the Kuskokwim River Delta, Alaska (above). Montane tundra in the arctic is better drained than lowland habitats and spruce trees occasionally get a foothold in protected valleys, such as these near Mount McKinley, Alaska (below).

The Land Meets the Sea

The topography of coastal areas often determines the character of plant communities at the edge between the land and the sea. Shallow, sloping, sandy shores, like those of Cape Cod, Massachusetts (above), create shifting dune habitats colonized by a few species of plants that stabilize the dune and allow other species to establish a foothold. Salt marshes develop in more protected

bays and in river estuaries, as at Barnstable Harbor on Cape Cod (below left). In the tropics, such protected habitats are usually invaded by mangrove trees, creating forests in standing water at the edge of the ocean, as in Biscayne Bay, Florida (above). At the rockbound coast of Maine, an abrupt meeting of land and sea tolerates little intermingling of the two environments (below).

The Marine Environment

The little explored mantle of water covering most of the surface of the Earth contains a wide variety of habitats and life forms. Open waters create a vast realm for the tiny phytoplankton and zooplankton, the fish that exploit them (above left), and the sea birds that eat the fish (below). Other fish are more like terrestrial grazers and predators, feeding on algae and small animals near the bottom and among corals (above right). Subtidal tropical habitats on the Caribbean coast of Panama are dominated by corals, including the elk horn coral in shallow, rough water (above right) and the important reef-building star coral at a depth of 50 feet (below right). Reef-building corals are restricted to sunlit depths because they rely on symbiotic green algae in their tissues for much of their nutrition.

Primary Production

Organisms need energy to move and grow and to maintain the functions of their bodies. Energy to support these activities enters the ecosystem as light, which plants convert to chemical energy during photosynthesis. The rate at which plants assimilate the energy of sunlight is called *primary productivity*. It is important to realize that primary production underlies the entire trophic structure of the community. The energy made available by photosynthesis drives the machinery of the ecosystem. The flux of energy through populations of herbivores, carnivores, and detritus feeders, and the biological cycling of nutrients through the ecosystem are ultimately tied to the primary productivity of plants. In this chapter, we shall consider how light, temperature, water, and nutrients influence the rate of photosynthesis, and thereby determine the productivity of natural communities.

Photosynthesis

Photosynthesis chemically unites two common inorganic compounds — carbon dioxide and water — to form glucose, a simple sugar. The overall chemical equation for photosynthesis is

$$6\ CO_2 \quad + \quad 6\ H_2O \quad \longrightarrow \quad C_6H_{12}O_6 \quad + \quad 6\ O_2$$

six molecules of carbon dioxide	plus	six molecules of water	yield with an input of energy	one molecule of glucose	plus	six molecules of oxygen

Photosynthesis requires a net energy input in the form of light equivalent to 9.3 kilocalories* per gram of carbon assimilated. Because the light

---

* The kilocalorie (kcal) is a measure of heat energy. It is defined as the energy required to raise the temperature of one kilogram (2.2 pounds or 1.06 quarts) of water one degree

energy is converted to chemical energy, the metabolic breakdown of sugar into its inorganic components makes available 9.3 kilocalories per gram of carbon converted to carbon dioxide; this energy can then be utilized to perform other metabolic functions like muscle contraction and biosynthesis.

Photosynthesis is an uphill process; it requires an input of energy to drive the chemical reaction. The mere presence of the ingredients for photosynthesis is not, however, sufficient to make the reaction occur. Carbon dioxide, water vapor, and sunlight occur together in the Earth's atmosphere, yet we are not deluged by a constant rain of sugar. The probability of the appropriate chemical reaction occurring is very low. In plant cells, pigments (chlorophylls and carotenoids, for example) and enzymes bring molecules and energy together in such a way as to make the chemical reactions of photosynthesis highly probable.

All green plants have identical photosynthetic reactions. Hydrogen atoms obtained from water combine with carbon and oxygen to form a sugar molecule. Some bacteria have evolved alternate biochemical pathways for producing sugars. For example, photosynthetic sulfur bacteria obtain the hydrogen needed to form sugars from hydrogen sulfide (H_2S) rather than from water. The resulting chemical equation is

$$12\ H_2S + 6\ CO_2 \rightarrow C_6H_{12}O_6 + 6\ H_2O + 12\ S$$

Production by sulfur bacteria is, however, a true photosynthetic process because the source of energy is light. Some blue-green algae, especially *Beggiatoa*, have gone one step further and produce sugars in the absence of light. Energy for the process comes from the oxidation of hydrogen sulfide to sulfur, or of sulfur to sulfate (SO_4). Each of these chemical steps releases about 0.5 kilocalories of energy per gram of sulfur oxidized, which is then used to convert carbon dioxide to glucose. Although *Beggiatoa* can produce organic matter in the absence of light, production is restricted except where sulfur is plentiful, and the process is inefficient: about ten grams of sulfur must be oxidized for every gram of carbon converted into sugar.

Production

Plants are made up of more than just glucose. Various biochemical processes use glucose both as a source of energy and as a building block to construct other, more complex, organic compounds. Rearranged and joined together, sugars become fats, oils, and cellulose, the basic struc-

---

Celsius. The kilocalorie, sometimes referred to as the large calorie (Cal), well-known to diet watchers, equals 1,000 small calories (cal). Some familiar benchmark figures may prove useful. Average people utilize 2,000 to 3,000 kcal of energy daily. One kilowatt of power is equivalent to the expenditure of 860 kcal per hour; one horsepower is equivalent to 642 kcal per hour.

tural material of the plant cell wall. Nitrogen, phosphorus, sulfur, and magnesium are combined with carbon, oxygen, and hydrogen to produce an array of proteins, nucleic acids, and pigments. Plants cannot grow unless they have all these basic building materials. In addition to nitrogen and carbon atoms, chlorophyll contains an atom of magnesium, just as hemoglobin contains an atom of iron. Thus, even though all other necessary materials might be present in abundance, a plant lacking magnesium could not produce chlorophyll, and thus could not grow. Plants clearly cannot function in the absence of water, either. The amount of water required by photosynthesis is merely a drop, compared to the bucket needed to balance transpiration. Water makes up the bulk of plant tissues and is the essential medium that makes nutrients available.

The basic equation for production must include mineral inputs as well as the raw materials for photosynthesis. Many biological reactions are involved in production, but we can summarize production by a general word equation:

carbon dioxide + water + minerals
(in the presence of light and within the proper temperature range) →
plant production + oxygen + transpired water

Gross Production and Net Production

The sugars produced by photosynthesis are not all incorporated into plant biomass, that is, all do not serve to increase the size and number of plants. Some of the sugar must be oxidized to release energy for biosynthesis and maintenance. Plants are no different from animals in this respect: they need energy to keep going.

We must distinguish two measures of production. *Gross production* refers to the total energy assimilated by photosynthesis. *Net production* refers to the accumulation of energy in plant biomass, or plant growth and reproduction. The difference between the two measures represents *respiration* by the plant. Gross production (photosynthesis) is thus divided between respiration (maintenance functions) and net production (growth and reproduction). In most studies of plant productivity, particularly in terrestrial habitats, ecologists measure net rather than gross production because the techniques are less difficult and because net production provides a measure of the available resources to heterotrophic consumers in the ecosystem.

Measurement of Primary Production

Net production can be expressed conveniently as grams of carbon assimilated, dry weight, or the energetic equivalent of dry weight. Ecologists use these indices interchangeably. The energy content of an organic compound depends primarily on its carbon and nitrogen content. The proportion of carbon by weight in most plant tissues is close to the proportion of carbon in glucose: 40 per cent.* When plants convert sugars to fats and oils, oxygen is biochemically stripped from the molecules, thereby increasing the proportion of carbon. The fat, tripalmitin ($C_{51}H_{98}O_6$), for example, contains 76 per cent carbon by weight. Fats and oils contain more than twice as much energy per gram as sugars and are therefore widely used by plants and animals for energy storage.

The energy content of a substance is estimated by burning a sample in a device called a bomb calorimeter. The guts of a calorimeter are contained in a small chamber where the sample is burned. Oxygen is forced under high pressure into the chamber to ensure complete combustion. The chamber is surrounded by a water jacket, which absorbs the heat produced. The increase in temperature of a known amount of water in the jacket provides a direct estimate of the heat energy released by combustion.

The photosynthetic combination of carbon dioxide and water requires an energy input of 9.3 kcal for each gram of carbon assimilated. The complete oxidation of a carbon compound to carbon dioxide and water should therefore release exactly 9.3 kcal per gram of carbon oxidized. In practice, the biochemical rearrangements involved in making most complex organic compounds alter energy values slightly. As a result, ecologists rely on measured values for energy content obtained directly from calorimeters. Generally accepted amounts of energy released in oxidation are 4.2 kcal per gram of carbohydrate (sugars, starch, cellulose), 5.7 kcal per gram of protein, and 9.5 kcal per gram of fat metabolized.

The equation for production suggests several possible methods for measuring the primary productivity of natural habitats. Uptake of carbon dioxide and mineral nutrients, production of plant biomass, and release of oxygen are all proportional to productivity. Water flux would not provide a useful measure of photosynthesis because water is too abundant in the plant and the environment, and, depending upon soil moisture, temperature, and humidity, its uptake and transpiration vary independently of the rate of photosynthesis. Uptake of carbon dioxide and production of oxygen and organic matter can be measured more reliably.

Primary production in terrestrial ecosystems is usually estimated by the annual increase in plant biomass (net production). Yearly growth of annual plants is measured by cutting, drying, and weighing the plants at the end of the growing season. The harvest method is commonly used for crop and field plants in temperate regions, where most plants die

---

* The relative weights of atoms of hydrogen, carbon, and oxygen are 1, 12, and 16, respectively: the proportion by weight of carbon in glucose ($C_6H_{12}O_6$) is therefore $72/180 = 0.40$.

back to the ground each year. Because root growth is usually ignored — roots are difficult to remove from most soils — harvesting measures the *net annual aboveground productivity* (NAAP), which is perhaps the most commonly used basis for comparing the productivity of terrestrial communities.

The harvest method has several inherent problems. Herbivores harvest some of the net production. Root growth, as we have just noted, is difficult to measure, though the root systems of annual plants can sometimes be separated from the soil by painstaking washing. But the roots of perennials continue to grow each year, so that *their* biomass represents the accumulation of many years' growth. The difficulty in measurement created by root production in field habitats is compounded by branch and trunk growth in forests. Harvesting leaf fall and clippings of new twigs allows only partial estimate of production. The annual growth of woody parts is often calculated by relating tree girth to total biomass, and then measuring annual increments in the girth of living trees. To arrive at a measurement of total biomass, a series of trees of increasing size is cut down and divided into trunk, branch, and, sometimes, root components, which are then dried in large ovens and weighed (Figure 7-1). The annual increase in girth of living trees can then be converted to an increase in total weight. Growth of leaves, flowers, and fruits, which are renewed each year, is added to the growth of woody parts to complete the estimate of production.

Plant production can be measured by gas exchange in aquatic habitats. The concentration of dissolved oxygen in water is so low that the input of oxygen by photosynthesis adds substantially to the oxygen already present. Under natural conditions, most of the oxygen produced by photosynthesis is either consumed by animals or bacteria, or escapes into the atmosphere. Ecologists control these problems by measuring production within sealed bottles. Samples of water containing phytoplankton are suspended at desired depths beneath the surface of a natural

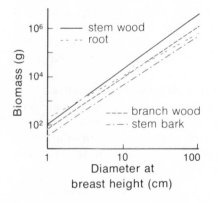

FIGURE 7-1 Relationship between the weight of various plant parts and the diameter of the trunk.

body of water in *light bottles*, which are clear and allow sunlight to enter, and *dark bottles*, which are opaque and exclude light. In the light bottles, photosynthesis and respiration occur together and part of the oxygen released into the water in the bottles is consumed. Photosynthesis does not occur in the dark bottles, but respiration does consume oxygen. The change in oxygen concentration in the light bottle provides a measure of net production; by adding the oxygen removed from the dark bottle, we obtain a value for gross production. The calculations are summarized:

photosynthesis *minus* respiration *equals* net production (light bottle)

plus respiration (dark bottle)

---

photosynthesis *equals* gross production

The estimate for net production obtained from the light bottle includes the respiration of plants, animals, and bacteria. Only the estimate for gross production is strictly valid as a measure of plant productivity.

The light and dark bottle technique is restricted to short-term measurements in small parts of the aquatic ecosystem. The technique cannot be applied easily to benthic algae or to whole systems. Ecologist Howard T. Odum partly solved this problem of measuring the production of entire stream communities. Rather than employ light and dark bottles, he compared the change in the oxygen content of the stream water during day and night periods, correcting for the exchange of oxygen between the stream and the atmosphere. By combining Odum's method with light and dark bottle techniques and conventional harvest methods where large seaweeds grow, ecologists have obtained reasonably accurate measurements of aquatic production.

For measuring photosynthesis in terrestrial ecosystems, carbon dioxide exchange is more useful than oxygen exchange because carbon dioxide is the rarer gas in the atmosphere. Small changes in carbon dioxide concentration are relatively easy to measure (the atmosphere contains only 0.03 per cent CO_2), and leaks in sampling chambers do not produce large errors. Measurement of production by carbon dioxide exchange resembles the light and dark bottle technique. A portion of a habitat, or even of an individual plant, is enclosed in an air-tight chamber and the decrease in carbon dioxide during the day is compared to the increase in carbon dioxide, due to respiration alone, during the night. Gross production can be measured accurately in this way, although attempts to measure production in whole forest canopies have been fraught with technical difficulties, including leakage and problems of air-conditioning large plastic enclosures, which have forced ecologists to fall back on more conventional harvest techniques.

The use of radioactive atoms of carbon, particularly the isotope C^{14}, provides a useful variation on the gas exchange method of measuring productivity. When a known amount of radioactive carbon is added, in the form of carbon dioxide, to an air-tight enclosure, plants assimilate the radioactive carbon atoms in the same proportion in which they occur in the air in the chamber. The rate of carbon fixation is calculated by dividing the amount of radioactive carbon in the plant by the proportion of radioactive carbon dioxide in the chamber at the beginning of the experiment. Thus if a plant assimilates 10 milligrams of C^{14} in an hour, and the proportion of radioactive carbon dioxide in the plant chamber is 0.05 (5 per cent), we calculate that the plant has assimilated carbon at the rate of 200 milligrams per hour (10 ÷ 0.05). Plant respiration eventually releases some of the assimilated carbon back into the air as carbon dioxide, which the plant can reassimilate. Measured over a one- to three-hour period, uptake of radioactive carbon allows a reliable estimate of gross productivity. After one to two days, uptake and release of radioactive carbon approach a steady state, and estimates represent net production more nearly than gross production.

Plants use nutrients other than carbon dioxide and water to synthesize organic compounds. The disappearance of dissolved nitrates and phosphates from aquatic environments can sometimes be used as a relative measure of net production, but only under restricted conditions: growth must occur rapidly and plants must convert inorganic nutrients into biomass much more quickly than they are made available by decomposition of dead plants or by mixing with deep water. When production and decomposition balance each other in a steady state, decomposition releases inorganic nutrients at the same rate that they are assimilated by photosynthesis, and the concentration of dissolved nutrients does not change. Nor are nutrients necessarily accumulated by plants in fixed proportion relative to rates of production. Algae are known to take up more phosphorus when dissolved phosphates are plentiful than when they are scarce. Conversely, plants sometimes leak dissolved minerals into the environment. Many physical and chemical processes, particularly erosion, upwelling, and sedimentation, also influence nutrient concentrations in aquatic systems. Conditions permitting reliable estimation of productivity from the disappearance of inorganic nutrients usually occur only during algal blooms, which follow the quiescent winter period in temperate and arctic lakes and oceans.

A final method for estimating plant production is based on the idea that chlorophyll determines the rate of photosynthesis. Marine algae assimilate a maximum of 3.7 grams of carbon per gram of chlorophyll per hour. The total productivity of a marine area may be estimated if the concentration of chlorophyll at different depths and the decrease in light intensity with depth are known. Although the chlorophyll method lacks the precision of gas exchange methods, it nonetheless provides a simple and rapid index to the productivity of oceans and lakes.

Several methods for measuring the productivity of aquatic ecosystems were compared at Ogac Lake, a landlocked fiord on Baffin Island, Canada. Primary production was measured throughout the growing season by the uptake of radioactive carbon in light bottles suspended beneath the fiord surface, but concentrations of chlorophyll, nitrates, phosphates, and dissolved oxygen were also monitored (Figure 7-2). The daily productivity of the fiord increased rapidly in early summer as ice disappeared from the surface and light began to penetrate the fiord's depths. Chlorophyll concentration paralleled the increase in productivity. Nitrate and phosphate concentrations declined throughout the summer. Dissolved oxygen increased in spring with the burst of plant production, but as the season progressed, increased zooplankton respiration obscured any direct relationship between oxygen and production. A surge in production in late summer was curiously unrelated to any of the factors monitored.

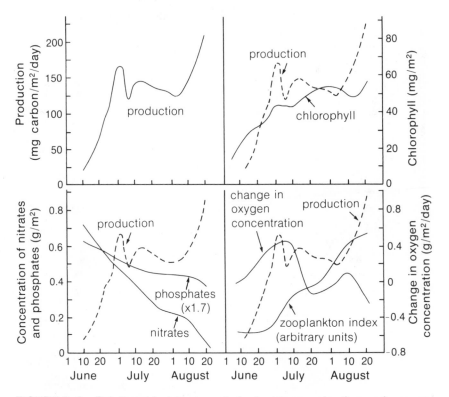

FIGURE 7–2   Relationship between phytoplankton production and concentrations of chlorophyll, nitrates, phosphates, oxygen, and zooplankton in Ogac Lake, Baffin Island.

Light and Photosynthesis

By subjecting the photosynthetic machinery of plants to varied levels of light intensity, plant physiologists have determined the influence of light on productivity.

At relatively low intensities, usually less than one-fourth the intensity of bright sunlight, the rate of photosynthesis is directly proportional to light intensity. Brighter light saturates the photosynthetic pigments, however, and the rate of photosynthesis increases more slowly or levels off. In many algae, very bright light reduces photosynthesis because it deactivates or destroys the photosynthetic apparatus.

Species differ in their response to light intensity. On one hand, photosynthesis reaches a maximum in several groups of marine phytoplankton when light intensity falls between 0.5 and 2 kcal/m²/min (Figure 7-3). On the other hand, although oak and dogwood leaves become light-saturated at intensities similar to those that saturate algae, supersaturation does not depress photosynthetic activity in these species. Loblolly pine is fully light-saturated only on the brightest days.

The sunlight that strikes the surface of a leaf is made up of a spectrum of light of different wavelengths, which we perceive as differ-

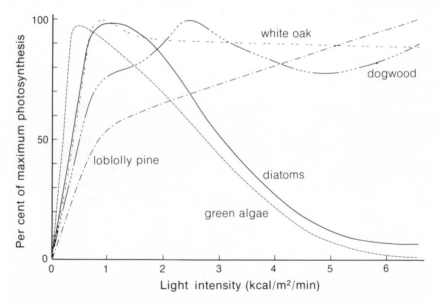

FIGURE 7–3 Relationship between light intensity and photosynthesis in green algae, diatoms, oak, dogwood, and loblolly pine. Photosynthetic rate is expressed as a per cent of the maximum. Light intensities above 7 kcal/m²/min (10,000 foot-candles) are achieved on bright summer days.

ent colors. Not all colors of light are utilized in photosynthesis. Green leaves contain several pigments, particularly *chlorophylls* (green) and *carotenoids* (yellow), that absorb light and harness its energy.

Carotenoids, which give carrots their orange color, absorb light primarily in the blue and green regions of the spectrum (Figure 7-4a) and reflect light with yellow and orange wavelengths. Chlorophyll absorbs light in the red and violet portions of the spectrum and reflects green, the color we perceive in leaves. The absorption spectrum of whole leaves (Figure 7-4b) resembles the combined absorption spectra of photosynthetic pigments, but organic compounds not involved in photosynthesis evidently absorb considerable orange light. Leaves of different species have different absorption spectra. Fig leaves, being thick and heavily pigmented, absorb 85 per cent of green light (550 mμ), the wavelength absorbed *least* efficiently. Tobacco leaves absorb only 50 per cent of green light. Because of the absorptive qualities of leaves, light under the canopy of a forest is relatively rich in the green and infrared, but poor in the red-orange and blue portions of the spectrum (Figure 7-4c).

Although water appears colorless, it absorbs small quantities of light in the red portion of the spectrum and scatters blue light. Thus, near the lower depths of light penetration in the ocean, green light predominates. The photosynthetic pigments of terrestrial and shallow water plants absorb green wavelengths weakly, but some deep-water marine algae contain red pigments (erythrins) that absorb green light strongly (Figure 7-4d). The red appearance of these pigments in direct sunlight gives the red algae their name. Because the red algae can utilize green light, they can occur below the depths to which green algae, having pigment typical of terrestrial plants, are restricted.

Temperature and Photosynthesis

Temperature normally bears a close relationship to light intensity in natural systems, but by controlling these factors in the laboratory, one can assess the separate influences of each of these factors on photosynthesis. Photosynthesis is relatively insensitive to temperature at low light intensities, where light constitutes a limiting factor, but at moderate light intensity photosynthetic rate increases two to five times for each 10°C rise in temperature.

Like most other physiological functions, photosynthesis is greatest within a narrow range of temperature, above which its rate declines rapidly. Because leaves absorb light, their temperature can become great enough during the middle of the day that photosynthesis is effectively prohibited; rate of photosynthesis then reaches a peak in mid-morning

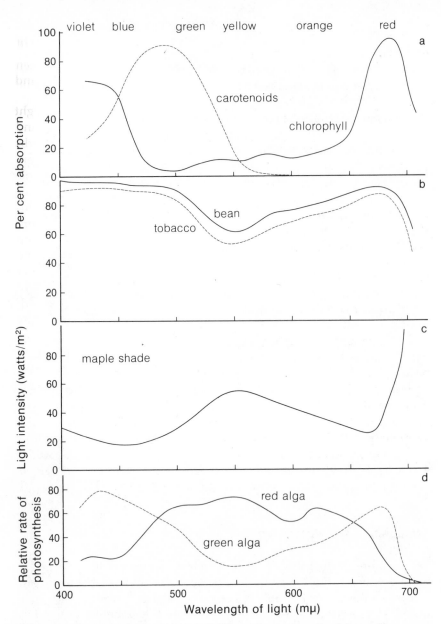

FIGURE 7-4 Absorption of light of different wavelengths (absorption spectrum) within the visible portion of the spectrum. (a) Absorption spectra for two groups of photosynthetic pigments — chlorophylls and carotenoids. (b) Absorption spectrum of whole leaves of the garden bean and tobacco. (c) Spectral distribution of light in the shade of a sugar maple forest. (d) Relative rates of photosynthesis by a shallow water green alga (*Ulva*) and by deep water red alga (*Porphyra*) as a function of the color of light. The red alga is photosynthetically more active in the middle portions of the visible spectrum. Blue and red light are also absorbed by pigments in red algae, but these wavelengths are relatively ineffective in photosynthesis.

and a second peak in mid-afternoon (Figure 7-5). As one would expect, the optimum temperature for photosynthesis varies with the environment, from about 16°C in many temperate species to as high as 38°C in tropical species.

Photosynthetic efficiency is a useful index of rates of primary production of plant formations under natural conditions. The photosynthetic efficiency is the per cent of incident visible radiation that is converted into net primary production during seasons of active photosynthesis. Where water and nutrients do not limit plant production, maximum photosynthetic efficiency varies between 1 and 2 per cent of available light energy.

Water and Transpiration Efficiency

Because photosynthesis requires gas exchange across the surface of the leaf, productivity also parallels the rate of transpiration of water from the leaf surface. As the moisture content of soil decreases, plants have greater difficulty removing water from the soil and leaves must close their stomata to reduce water loss (see page 32). When soil moisture is reduced to the wilting point, leaves are effectively shut off from the surrounding air and photosynthesis slows to a standstill. Rate of photosynthesis is therefore closely tied to the plant's ability to tolerate water loss, to the availability of moisture in the soil, and to the influence of air temperature and solar radiation on rate of evaporation. Humid envi-

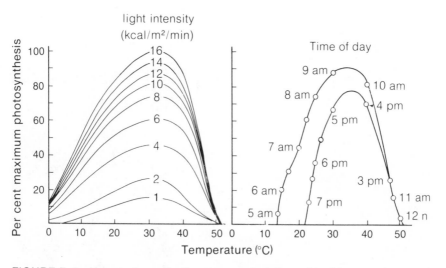

FIGURE 7-5   Net photosynthetic rate as a function of leaf temperature and incident light intensity. The daily course of photosynthetic rate (right) shows a dip in midday because the leaves become excessively hot.

ronments favor high rates of photosynthesis by reducing transpiration from leaves.

Agronomists devised *transpiration efficiency* as an index of the drought resistance of plants. Transpiration efficiency is the ratio between net production and transpiration, expressed as grams of production per 1,000 grams of water transpired. Most plants have transpiration efficiencies of less than 2 grams of production per 1,000 grams of water transpired; some drought-resistant crops have transpiration efficiencies of 4. High efficiencies result from morphological adaptations to reduce evaporation and leaf temperature. Some desert species have hairy leaves; the tiny hairs trap a layer of still air next to the surface of the leaf, thereby reducing evaporation. In cold regions, where water is unavailable to plants when the ground is frozen, most broad-leaved plants drop their leaves in the fall to avoid desiccation, but the evergreen Labrador tea plant has hairy leaves performing the same function as the hairy leaves of many desert plants (Figure 7–6).

Regardless of their adaptations to conserve water, plants cannot escape the physical reality that if they must reduce transpiration during drought, they must also reduce the gas exchange necessary for photosynthesis, and consequently reduce their productivity. Even in some tropical areas, trees drop their leaves during dry seasons to avoid excessive water loss (Figure 5–7).

Nutrient Limitations

Most habitats respond to artificial applications of fertilizers by increased primary production. No matter what the natural fertility of the habitat, nutrient availability interacts with water, temperature, and light to de-

FIGURE 7–6   The dense hair on the underside of the leaves of Labrador tea (right) reduce water loss due to transpiration during the winter when the soil is frozen but the leaves are not yet covered by snow.

termine levels of production. Nutrient limitation is probably most strongly felt in most aquatic habitats, particularly the open ocean, where the scarcity of dissolved minerals reduces production far below terrestrial levels. The abundance of nutrients in lakes and oceans depends on upwelling currents, water depth, proximity to coastlines and rivers, and drainage patterns of nearby land masses. Recent intense fertilization of inland and coastal water by sewage and runoff from fertilized agricultural land has greatly increased aquatic production in some inland and coastal waters. Such relationships will be discussed in greater detail in Chapter 9.

Even oxygen can limit terrestrial plant production under some circumstances. Roots require oxygen in the soil for metabolism and growth. If the soil is nonporous, or if it is completely saturated with stagnant water, available oxygen can be reduced below the point required to support plant growth. In the waterlogged soils of swamps, many plants have structures for obtaining oxygen for their roots directly from the atmosphere. For example, the knees of cypress trees are projections of the roots above the surface of the water. Cypress knees allow free exchange of gases between the root system and the atmosphere (Figure 7–7).

FIGURE 7–7 Cypress knees in a drained swamp in South Carolina. Under normal water levels in the swamp the knees would project above the surface, providing an avenue of gas exchange between the roots and the air.

Production in Terrestrial Ecosystems

The favorable combination of intense sunlight, warm temperature, and abundant rainfall make the humid tropics the most productive terrestrial ecosystem on Earth, square mile for square mile. Low winter temperatures and long nights curtail production in temperate and arctic ecosystems. Lack of water limits plant production in arid regions, where light and temperature are otherwise favorable for plant growth. Within a given latitude belt, where light and temperature do not vary appreciably from one locality to the next, net production is directly related to annual precipitation. In temperate regions, production begins to level off when precipitation exceeds 100 cm (39 in) annually, presumably because the availability of nutrients and light becomes limiting.

Ecologists Robert Whittaker and Gene Likens have recently estimated net primary production for representative terrestrial and aquatic ecosystems (Figure 7–8). The estimates are based on many studies using

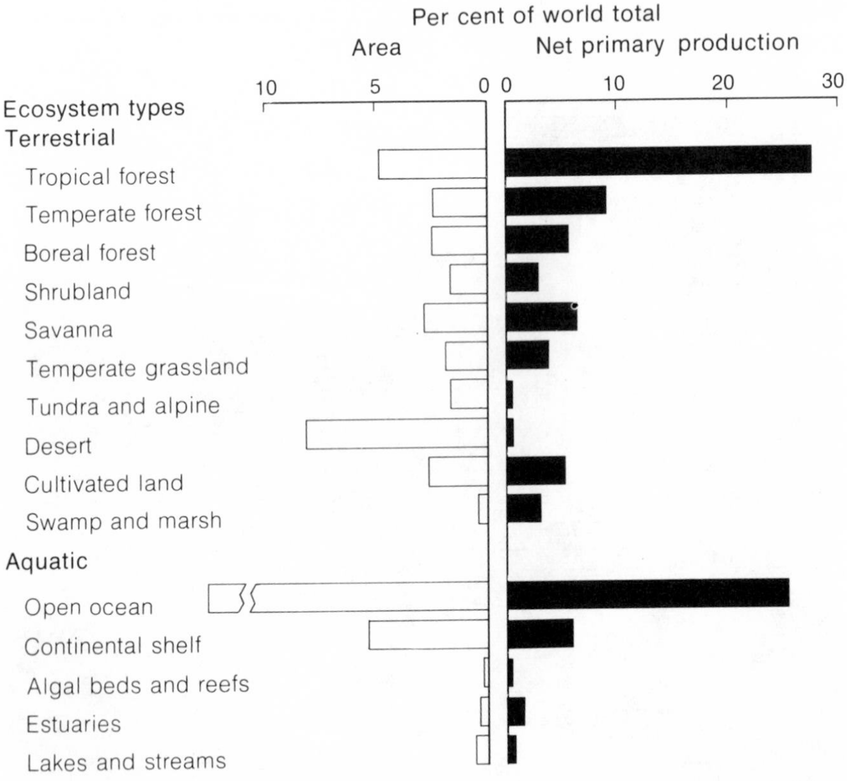

FIGURE 7–8   Surface area and total annual productivity of the major ecosystems. Values are expressed as a percentage of the total for the Earth.

a wide variety of techniques and are probably reasonably close to true values.

Production in terrestrial habitats decreases dramatically from the wet tropics to temperate regions, even more so where the climate is too dry or too cold to support forests. Swamp and marsh ecosystems occupy the interface between terrestrial and aquatic habitats, where plants are as productive as in tropical forests. Maximum rates of production in marshes have been reported to be as high as 4,000 g/m²/yr in temperate regions and 7,000 g/m²/yr in the tropics. Marsh plants are highly productive because their roots are constantly under water and their leaves extend into the sunlight and air, obtaining benefits of both aquatic and terrestrial living. Rapid decomposition by bacteria of detritus that washes into marshes additionally releases abundant nutrients.

The productivity of agricultural land usually falls somewhat below the productivity of natural vegetation in the same area because croplands are plowed each year and laid bare early and late in the growing season when undisturbed habitats continue to be productive. Furthermore, most agricultural plantings consist of a single species, which cannot exploit the resources of the land as efficiently as a mixture of species with differing ecological requirements.

Irrigation and the application of fertilizers can increase agricultural yields two- to threefold over world averages. Sugar cane, a common tropical crop, has a world average production of about 1,700 g/m²/yr. Intensively cultivated sugar cane in the Hawaiian Islands has double the average world yield and a maximum productivity of about 7,000 g/m²/yr. Poor crop management, conversely, can lead to soil deterioration and reduced production.

Net primary production of temperate zone cereal crops (what, corn, oats, and rice), hay, and potatoes, varies between 250 and 500 g/m²/yr; sugar beets commonly attain twice that productivity. These values are compared to the estimates of Whittaker and Likens for temperate forests (600 to 2,500 g/m²/yr) and temperate grasslands (150 to 1,500 g/m²/yr). They estimate that the productivity of all agricultural land varies between 100 and 4,000 g/m²/yr, depending on the crop, with an average of 650 g/m²/yr.

Production in Aquatic Ecosystems

The open ocean is a virtual desert, where scarcity of mineral nutrients — not water — limits productivity to one-tenth or less the productivity of temperate forests. Upwelling zones (where nutrients are brought up from the depths by vertical currents) and continental shelf areas (where exchange between shallow bottom sediments and surface waters is well-developed) support greater production, averaging 500 and 360 g/m²/yr, respectively. In shallow estuaries, coral reefs, and coastal algae beds, production approaches that of adjacent terrestrial habitats, with averages approaching 2,000 g/m²/yr. Primary production in fresh-water habitats is similar to that of comparable marine habitats.

Availability of nutrients largely determines variation in the production of aquatic ecosystems. Light apparently does not limit production within the euphotic zone. As much as 95 per cent of the incident radiation penetrates the surface of the water and is available to plants. Variation in the depth to which light penetrates, a function of the clarity of the water, influences the depth to which photosynthesis occurs but does not affect total productivity per square meter of ocean surface. In clear water, algae are spread thinly throughout the deep euphotic zone; in turbid water, they are concentrated closer to the surface.

Temperature evidently does not influence the overall productivity of marine habitats. Although the photosynthetic rates of individual plants may be depressed by cold temperatures, marine algae attain great density in cold water, enough that arctic oceans are as productive as warm tropical seas. Marine biologist K. H. Mann has found that in cold temperate waters large seaweeds produce as much biomass per square meter of marine habitat as in the Indian Ocean and Caribbean Sea. In Nova Scotia, the species *Laminaria* alone attains a productivity of about 1,500 g/m²/yr, a respectable value for a temperate forest!

Annual Net Primary Production

The most productive habitats attain photosynthetic efficiences of 1 to 2 per cent, but so much of the Earth's surface lacks optimum conditions for plant growth that only one-tenth of 1 per cent of the light energy striking the Earth's surface is assimilated by plants. The energy value of sunlight reaching the outer atmosphere of the Earth directly under the Sun (the *solar constant*) is about 10^7 (10,000,000) kcal/m²/yr. The angle of incident radiation varies, however, with time of day and season and, over a year's time, all points on the Earth are shrouded by night for an equivalent of six months. If the total energy reaching the outer atmosphere were spread evenly over the surface of the Earth, each square meter would receive one-quarter of the solar constant, or 2.5 × 10^6 kcal/m²/yr. In fact, the Earth's surface does not actually receive that much solar energy during the year and its distribution is not uniform. Perhaps 40 per cent of the total light income of the Earth is absorbed by the atmosphere and re-radiated back into space as heat. Part is also reflected and scattered by particles of dust in the atmosphere and reflected by surfaces of water, rock, and vegetation.

The annual energy income of a particular locality varies with latitude and cloud cover. Temperate localities usually receive between

2×10^5 and 2×10^6 kcal/m²/yr, which represents 2 to 20 per cent of the solar constant. Only half the light energy can be assimilated by plants; the other half lies outside the absorption spectrum of plant pigments.

Whittaker and Likens estimated the total annual primary productivity of the Earth as 162×10^{15} (million billion) grams (about 730×10^{15} kcal), of which terrestrial habitats are responsible for two-thirds. The average productivity of terrestrial areas (720 g/m²/yr or 3,200 kcal/m²/yr) represents assimilation of about 0.3 per cent of the light reaching the surface. The overall photosynthetic efficiency of aquatic habitats is less than one-quarter that of terrestrial communities.

The distribution of production among the major vegetation zones of the Earth is a function of the local productivity of these zones, determined by light, temperature, rainfall, and nutrients, and their total surface area (Figure 7–8). Tropical forests cover only 5 per cent of the Earth's surface, but they account for almost 28 per cent of the total production. Temperature forests and the open ocean, representing 2.4 and 63 per cent of the surface area, are responsible for 9.2 and 25 per cent of the Earth's productivity. Although inshore waters (estuaries, algal beds, and reefs) occupy only 0.4 per cent of the Earth's surface they account for 2.3 per cent of the Earth's productivity, and 6.7 per cent of all aquatic production. Swamps and marshes are similarly productive for their small area. Such productivity emphasizes the importance of preserving these coastal environments and protecting them from pollution and land-fill programs.

In this chapter, we have examined basic patterns of primary production over the surface of the Earth. Productivity is greatest where light, warmth, water, and mineral nutrients are all abundant. Decreasing light and temperature reduce production of habitats distant from the tropical zones. Moisture further restricts production in arid regions. In some aquatic habitats, scarcity of nutrients imposes the greatest limitation to production, particularly in the open ocean, but inshore areas with abundant nutrients can match or exceed nearby terrestrial communities for organic production.

Net productivity measures the overall energy input to the ecosystem. This production is eventually consumed and dissipated by herbivores, by the carnivores that eat them, and by the myriad detritus feeders that scavenge dead debris. The precise pathways that energy follows on its route through the ecosystem and the time required for its journey constitute the subject of the next chapter.

8

Energy Flow in the Community

Photosynthesis and net primary production make energy available to the community. Herbivorous animals eat plants, carnivores eat herbivores, carnivores are in turn eaten by other carnivores, and so on through a series of steps that together form a *food chain*. Each step in the food chain represents a *trophic level.* The feeding relationships of organisms impart a trophic structure to the community through which energy flows.

Energy flux is the only sound currency in the economics of ecosystem function; biomass and numbers are static descriptions of the community frozen in an instant of time. The dynamics of the community are measured in terms of change — rates of energy and nutrient transferral from organism to organism through the structure of the food web. The unique status of energy as an ecological currency has greatly stimulated the study of community energetics. In this chapter, we shall study the pathways of energy through the community and how the physical environment of the community influences energy flow.

The food chain illustrated in Figure 8-1 oversimplifies nature in several ways. First, few carnivores feed on a single trophic level; second, many organisms (or their parts in the case of plants) die of causes other than predation, and their remains are consumed by detritus-feeding organisms; third, most energy assimilated by a trophic level is dissipated as heat because biological processes are energetically inefficient and organisms utilize energy to maintain themselves as well as to grow and reproduce. Energy incorporated into growth and reproduction, the food of the next higher trophic level, represents a small percentage of the total food eaten.

A Word on Words

Ecologists have applied many different terms to the feeding relationships of organisms. Such terms are usually introduced to clear up
confusion and give precise definitions to the activity of organisms. But

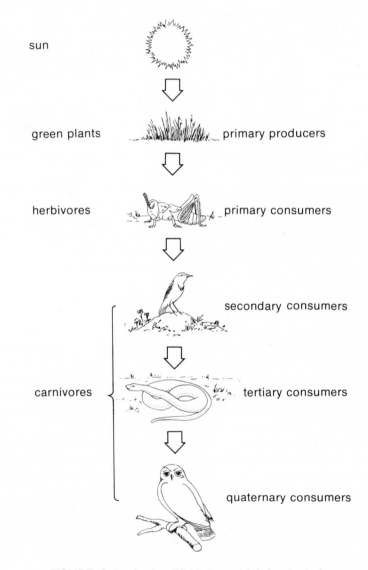

FIGURE 8–1 A simplified terrestrial food chain
showing the sequence of trophic levels.

they often have the opposite effect. Plants and animals have been functioning perfectly well without having names applied to their activities. We should take this cue and concentrate on understanding the feeding relationships within a community, rather than try to categorize them.

We can arrange most of the troublesome terms in pairs with opposite meanings. *Producer* and *consumer* refer to different activities performed by the same organisms, although *primary producer* is usually reserved for green plants and *primary consumer* for herbivores. Production refers to the assimilation of materials into the body of the organism, that is, growth and reproduction. Whether the source of energy for production is sunlight or chemical energy in food, distinguishes primary producers from secondary, tertiary, quaternary, etc., producers. Just as all organisms produce, they also consume. Plants and animals both must metabolize assimilated substances to keep themselves going. Ecologists apparently will not consider the assimilation of sunlight by plants as an act of consumption. Therefore herbivores are designated primary consumers by default.

Autotroph and *heterotroph* distinguish organisms that convert inorganic forms of energy to organic forms from those whose sole source of energy is organic matter. Thus autotrophs are primary producers. Green plants are *photo*autotrophs because they assimilate the energy of sunlight. The blue-green alga *Beggiatoa* is partly a *chemo*autotroph because it can derive energy from the oxidation of hydrogen sulfide to sulfur (see page 111).

Herbivore and *carnivore* refer to eaters of plants and animals, respectively, although carnivore is sometimes reserved for meat eaters. Other terms such as *insectivore* and *piscivore* (fish eater) are used. Herbivores also come in a variety of forms — nectivores, frugivores, grazers, browsers, and so on — depending on what they eat and how they eat it. *Predator* distinguishes organisms that consume whole prey from *parasites,* organisms that live on living prey, or *hosts.* Hence seed eaters are properly called predators because they destroy the tiny plant embryo in the seed, and many grazers and browsers are properly called parasites, because they consume leaves or buds without killing the tree. Mosquitos and vampire bats would fit equally well into the category of browsers.

Biophage and *saprophage* distinguish organisms that eat living prey from those that go the easy route of eating dead prey. Saprophages also may be called *detritivores* as they are in this book. Detritivores, or saprophages, are often referred to as decomposers. While detritivores *do* decompose organic compounds into simpler inorganic compounds, this function is not unique to detritivores. All organisms metabolize organic foods to obtain usable energy and release carbon dioxide and water as end products. All organisms therefore are decomposers.

Food Chains and Food Webs

Trophic levels provide a simple framework for understanding energy flow through the ecosystem. We should be able to determine the trophic level of the energy in a particular chemical bond by the number of times it has changed hands since it was first put together in a green plant. But energy follows complicated paths through the trophic structure of the ecosystem and we would not be able to follow a particular chemical bond very easily.

Many carnivores eat fruit and other plant materials when their normal prey is not readily available. Some carnivores eat other carnivores as well as herbivores. Whether an owl eats herbivorous field mice or insectivorous shrews is of little consequence nutritionally to the owl; both prey have about the same food value and require similar effort to capture. Large owls sometimes eat snakes or small weasels, which themselves feed on mice and shrews. Owls are not known to feed on plant material, but foxes, which normally prey on small mammals, often eat fruit. It is therefore not unusual for predators to feed on three or four trophic levels. Plants are not always the *victims* of animals; some partly carnivorous species are known — Venus's fly traps, pitcher plants, and sundews. Most herbivores, however, particularly those that eat leafy vegetation, are so specialized in their food habits that they are incapable of carnivory. A cow's teeth can grind tough plant material to shreds but could not get a grip on a rabbit's flesh, much less tear it apart. (As if a cow could even catch a rabbit.)

The great variety of feeding relationships within the community links species into a complex *food web*. Each species feeds on many kinds of prey (Figure 8-2). Within the context of this complexity, trophic levels become abstract concepts, useful for understanding general patterns of structure and energy flow through the ecosystem, but usually quite useless as categories for assignment of individual species.

Pyramids of Productivity, Biomass, and Numbers

The amount of energy metabolized by each trophic level decreases as energy is transferred from level to level along the food chain. Green plants, the primary producers, constitute the most productive trophic level. Herbivores are less productive, and carnivores still less. The productivity of each trophic level is limited by the productivity of the trophic level immediately below it. Because plants and animals expend energy for maintenance, less and less energy is made available, through growth and reproduction, to each higher trophic level. Of the light energy assimilated by a plant, 30 to 70 per cent is used by the plant itself

FIGURE 8–2   Some of the feeding relationships in a simple food web.

for maintenance functions and the energetic costs of biosynthesis. Herbivores and carnivores are more active than plants and expend even more of their assimilated energy. As a result, the productivity of each trophic level is usually no more than 5 to 20 per cent of that of the level below it. The percentage transfer of energy from one trophic level to the next is called the *ecological efficiency,* or *food chain efficiency,* of the community.

If we represent each trophic level in the community by a block whose size corresponds to the productivity of the trophic level, and then stack the blocks on top of each other with the primary producers at the bottom, we obtain a characteristic pyramid-shaped structure (Figure 8-3). The structure of the pyramid will vary from community to community, depending on the ecological efficiencies of the trophic levels. In the particular case pictured in Figure 8-3, these efficiencies are 20, 15, and 10 per cent. Herbivore production is therefore 20 per cent of plant production; first-level carnivore production is 15 per cent of herbivore production and only 3 per cent of plant production (15 per cent of 20 per cent equals 3 per cent); second-level carnivore production is only 0.3 per cent of plant production. These values are probably unrealistically high compared to most natural communities, but they illustrate the universal decrease in the availability of energy at progressively higher trophic levels.

As energy availability decreases, the biomass and numbers of individuals on each trophic level usually decrease as well, although no law of energetics prevents a reverse trend. The biomass structure of the

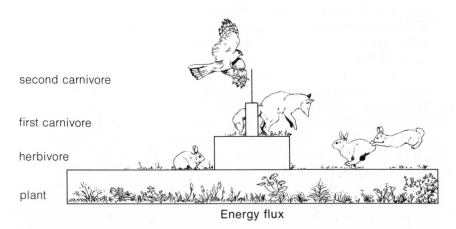

Energy flux

FIGURE 8-3 An ecological pyramid representing the net productivity of each trophic level in the ecosystem. This particular structure represents ecological efficiencies of 20, 15, and 10 per cent between trophic levels, but these values vary widely between communities.

community resembles the pyramid of productivity in most terrestrial communities. If one were to collect all the organisms in a grassland, the plants would far outweigh the grasshoppers and ungulates that eat the plants. The herbivores in turn would outweigh the birds and large cats at the first carnivore level, and these too would outweigh their predators, if there were such. An individual lion may be heavy, but lions are spread out so thinly that they do not count for much on a gram-per-square-meter basis.

The pyramid of biomass is sometimes turned upside down in aquatic plankton communities. Algae must be more productive than the tiny animals that eat them; the laws of energetics can not be violated. But the phytoplankton are sometimes consumed so rapidly that their numbers are kept small by herbivorous zooplankton. Intensive grazing reduces phytoplankton biomass, but the algae are so productive that they can often support a larger biomass of herbivores under optimum conditions for growth.

The pyramid of numbers is even more shaky than the pyramid of biomass. Disease organisms (parasitic bacteria and protozoa, for example), mosquitos, ants, and others are certainly more numerous than the organisms they feed on, even though all of them together do not weigh as much as their prey or hosts. A single tree may be host to thousands of aphids, caterpillars, and other herbivorous insects.

Detritus Pathways of Energy Flow

Many ecologists attribute a distinctive role to organisms which consume dead plant and animal matter. It is frequently said that these detritus eaters are responsible for breaking down dead organic remains, which would otherwise accumulate, and for releasing their nutrients so they can be used again by plants. Detritus-consuming organisms *do* have this function in the ecosystem, but this view of a special role in the community is misleading for two reasons. First, as we have seen, all organisms "decompose" organic matter. Terrestrial mammals consume 20 to 200 grams of food for every gram of body weight produced. The remainder is metabolized and used as a source of energy. Required minerals and other nutrients are retained and incorporated into body tissue, but most injested food is returned to the environment in an inorganic form: carbon dioxide and water are exhaled; water and various mineral salts are excreted in sweat and urine.

Second, detritivores have no special purpose different from any other organism in the overall function of the ecosystem. The individual detritivore obtains energy and nutrients in its food just like herbivores and carnivores. And just like all other organisms, detritivores leave behind undigestible remains, breakdown products of metabolism, and excess minerals that they cannot use. Detritivores eat the garbage of the ecosystem for the same reason herbivores and carnivores eat fresh food: to make a living. It's all a matter of taste.

Detritivores include such diverse species as carrion eaters — crabs, vultures, and the like — whose freshly dead food differs little from the live prey eaten by carnivores — and bacteria and fungi, which are biochemically specialized to consume certain organic materials and waste products that are particularly difficult for most organisms to digest.

From the standpoint of energy use detritus feeders are not readily distinguishable from other kinds of consumers. Detritivores have better pickings in some communities than others. As we shall see below, terrestrial plants produce large quantities of indigestible supportive tissue, most of which is consumed after death by organisms of decay in the soil. More than 90 per cent of the net primary production of a forest is consumed by detritivores, and less than 10 per cent by herbivores. Aquatic plants are more digestible by herbivores, and the detritus pathways are correspondingly less prominent.

The Individual Link in the Food Chain

Once food is eaten, its energy follows a variety of paths through the organism (Figure 8-4). Not all food can be fully digested and assimilated.

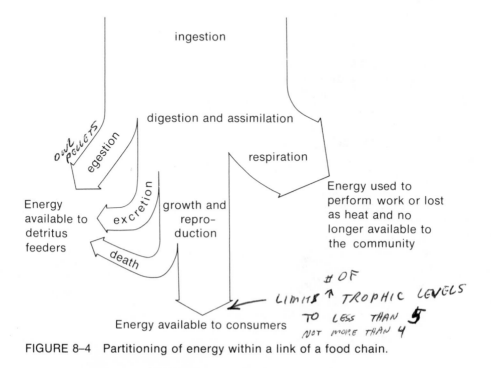

FIGURE 8-4   Partitioning of energy within a link of a food chain.

Hair, feathers, insect exoskeletons, cartilage and bone in animal foods, and cellulose and lignin in plant foods cannot be digested by most animals. These materials are either egested by defecation or regurgitated in pellets of indigested remains. Some egested wastes are relatively unaltered chemically during their passage through an organism, but nearly all are mechanically broken up into fragments by chewing and by contractions of the stomach and intestines and are thereby made more readily usable by detritus feeders.

Organisms use most of the food energy that they assimilate into their bodies to fulfill their metabolic requirements: performance of work, growth, and reproduction. Because biological energy transformations are inefficient, a substantial proportion of metabolized food energy is lost, unused, as heat. Organisms are no different from man-made machines in this respect. Most of the energy in gasoline is lost as heat in a car's engine rather than being transformed into the energy of motion. In natural communities, energy used to perform work or dissipated as heat cannot be consumed by other organisms and is forever lost to the ecosystem.

Proteins create special metabolic problems because they contain nitrogen. Nitrogen in excess of requirements for growth and body maintenance is usually excreted in an organic form — ammonia in most aquatic organisms, urea and uric acid in most terrestrial organisms. Excreted nitrogenous waste products therefore represent a loss of potential chemical energy. But, because these waste products can be metabolized by specialized microorganisms, they enter the detritus pathways of the community. We shall examine the path of nitrogen compounds through the ecosystem in greater detail in the next chapter.

Assimilated energy that is not lost through respiration or excretion is available for the synthesis of new biomass through growth and reproduction. Populations lose some biomass by death, disease, or annual leaf drop, which then enters the detritus pathways of the food chain. The remaining biomass is eventually consumed by herbivores or predators, and its energy thereby enters the next higher trophic level in the community.

Energetic Efficiencies

The movement of energy through the community depends on the efficiency with which organisms consume their food resources and convert them into biomass. This efficiency is referred to as the food chain, or *ecological efficiency*. Ecological efficiencies are determined by both internal, physiological characteristics of organisms and their external, ecological relationships to the environment. To understand the biological basis of ecological efficiency, one must dissect the individual link of the food chain into its component parts (Figure 8–5). Ecological efficiency de-

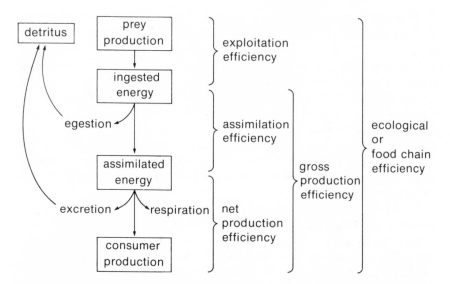

FIGURE 8–5   Diagram of the partitioning of energy in a link of the food chain and the energetic efficiencies associated with each metabolic step through the organism. Detritus produced by egestion and excretion of unusable food remains on the trophic level of the prey and is available for consumption by other organisms.

pends on the efficiencies of three major steps in energy flow: exploitation, assimilation, and net production (Table 8-1). The product of the *assimilation* and *net production efficiencies* is the *gross production efficiency* — the percentage of ingested food converted to consumer biomass. The product of the exploitation and gross production efficiencies is the food chain, or ecological efficiency — the percentage of available food energy in prey converted into consumer biomass.

TABLE 8–1   Definitions of energetic efficiencies.

(1) Exploitation efficiency = $\dfrac{\text{Ingestion of food}}{\text{Prey production}}$

(2) Assimilation efficiency = $\dfrac{\text{Assimilation}}{\text{Ingestion}}$

(3) Net production efficiency = $\dfrac{\text{Production (growth and reproduction)}}{\text{Assimilation}}$

(4) Gross production efficiency = (2) × (3) = $\dfrac{\text{Production}}{\text{Ingestion}}$

(5) Ecological efficiency = (1) × (2) × (3) = $\dfrac{\text{Consumer production}}{\text{Prey production}}$

Categorization — the act of pigeonholing nature into a stilted and often unnatural framework — often produces more ogres than solutions to conceptual problems. Ecological efficiencies are no exception. Egested and excreted energy are the square pegs one tries to fit into round holes in this case. On what trophic level do we place egested and excreted detritus? If put with the prey trophic level, it must be classed as unexploited energy and figured into the exploitation efficiency. If put with the consumer trophic level, egested and excreted matter properly should be figured into gross production efficiency, thereby increasing overall ecological efficiency. Most ecologists either place detritus in a special food category all its own, belonging to no trophic level, or ignore the problem altogether.

Above all, energetic efficiencies, as ecologists define them, correspond to quantities that are readily measured. Practicality is often the architect of concept. The study of energy transformation by plants and animals does, however, provide a useful insight into the basis of ecological efficiency and the trophic structure of the community.

Plants are terrible predators. Perhaps we should say that light is an elusive prey. Light can pass right through a leaf, it can be absorbed by the wrong molecules, or it can be quickly converted to heat and slip away before the plant can harness its energy. As a form of energy, light differs so much from organic molecules that the conversion of energy from one form to the other is inefficient. Once light energy *has* been harnessed, plants can utilize it efficiently for production.

Animals have different problems. With their prey in hand — or in mouth — its energy can be assimilated efficiently by rearranging chemical bonds rather than by converting one form of energy to another. Yet animals expend so much energy in maintaining their delicate — compared to plants — bodies and in pursuing prey that relatively little assimilated energy ends up as production.

Energetic Efficiencies in Plants

The absorption of light by green plants is a form of ingestion and may be used as a basis for calculating the energetic efficiency of the plant trophic level. Probably 20 to 30 per cent of all light striking the Earth's surface in productive habitats is absorbed by green plants. Most of the remainder is either reflected or absorbed by bare earth and water. Because plants incorporate little absorbed energy into biomass, gross production efficiency is usually less than 2 per cent.

Most of the light energy absorbed by plants is converted directly to heat and lost by re-radiation, convection, or transpiration. In a classic study of the energy relations of a corn field, E. N. Transeau determined that only 1.6 per cent of the light energy incident on the field was assimilated by photosynthesis. The remainder was either converted to heat (54 per cent) or absorbed by water, either in the soil or in leaves, in evaporating and transpiring the 15 inches of rainfall received during the growing season (44.4 per cent). Although plants convert 1 to 2 per cent of absorbed light into chemical energy under suitable natural conditions, photosynthesis can achieve a maximum efficiency of 34 per cent under the most favorable laboratory conditions.

Plants use relatively less assimilated energy for maintenance than animals because they do not move or maintain body temperatures, and they "feed" continuously during daylight periods. Estimates of net production efficiency vary between 30 and 85 per cent depending on the habitat (Table 8–2). Rapidly growing vegetation in temperate zones, whether natural fields, crops, or aquatic plants, exhibits uniformly high net production efficiency (75 to 85 per cent). Lower values characterize tropical vegetation. The cause of low net production efficiencies in tropical aquatic communities (40 to 50 per cent) is unclear. Bright light tends to saturate the photosynthetic mechanism and perhaps reduce the rate of photosynthesis without reducing respiration. High temperature also accelerates plant respiration more rapidly than it accelerates photosynthesis. The combination of these factors in tropical waters would increase the rate of respiration relative to photosynthesis and thereby lower net production efficiency.

TABLE 8–2   *Net production* efficiency (net primary production/gross primary production) of several plants and plant communities.

Plant community	Locality	Net production efficiency (%)
Terrestrial		
Perennial grass and herb	Michigan	85
Corn	Ohio	77
Alfalfa		62
Oak and pine forest	New York	45
Tropical grasslands		55
Humid tropical forest		30
Aquatic		
Duckweed	Minnesota	85
Algae	Minnesota	79
Bottom plants	Wisconsin	76
Phytoplankton	Wisconsin	75
Sargasso Sea	Tropical Atlantic Ocean	47
Silver Springs	Florida	42

Wet tropical forests also exhibit low net production efficiencies, probably around 30 per cent. (Measurements of forest productivity is exceedingly difficult and most published figures should be viewed as tentative.) Variation in energetic efficiency of terrestrial habitats roughly parallels the ratio of photosynthetic to supportive tissue in different plants. Leaves comprise 1 to 10 per cent of the aboveground biomass of forest tress compared to 20 to 60 per cent for shrubs and over 80 per cent for most herbs. Nonphotosynthetic living parts of plants respire, though the outer bark and dead cores of trunks and branches do not. Roots certainly contribute substantially to plant respiration.

If the ratio between photosynthetic and supportive tissue determines net production efficiency, young plants that have not yet developed extensive supportive or root tissue should exhibit higher efficiencies than large, mature plants. Comparative data for different vegetation types support this hypothesis. Further evidence comes from studies of seasonal change in the net production efficiency of a field of lespedeza (a member of the pea family). The energetic efficiency fell from 75 per cent in April to 15 per cent in August, paralleling a similar decline in growth rate and increase in the proportion of root and stem biomass.

Energetic Efficiencies in Animals

Assimilation efficiency varies with quality of the diet of animals. In particular, animal food is digested more easily than plant food. Assimilation efficiencies of predatory species vary between 60 and 90 per cent of the food consumed, with insectivores occupying the lower end and meat and fish eaters the upper end of the range. The tough chitinous exoskeleton of insects, which constitutes a large portion of the weight of many species, resists digestion and therefore reduces the assimilation efficiencies of insect-eating species.

The proportion of cellulose, lignin, and other indigestible materials influences the nutritional value of plant foods. Trunks and branches of trees consist mostly of cellulose and lignin, which lack nitrogen and many essential minerals. Species of wood-boring beetles that live off the woody parts of plants either grow very slowly or restrict their feeding to the layer of nonwoody, living cells immediately under the bark.

Leaves contain between 2 and 4 per cent protein and thus are more suitable than wood as a food for herbivores, though plants have devised many defensive mechanisms to protect leaves, including a variety of toxins. The leaves of oaks and other trees have tannins that prevent herbivores from digesting their proteins. Seeds are the most desirable plant food because they are provisioned with the nutrients to get a plant started in life — nutrients that are equally well-suited to sustaining herbivores. Pine nuts, for example, contain about 50 per cent oil, 30 per cent protein, and 5 per cent sugars.

Assimilation efficiencies of herbivores parallel the nutritional quality of their food: up to 80 per cent for seed diets, up to 60 per cent for
young foliage, 30 to 40 per cent for most mature foliage, and 10 to 20 per
cent or less for wood, depending on its state of decay.

Net production efficiency of animals is inversely related to activity.
Maintenance activities, and heat production by warm-blooded animals,
require energy that otherwise could be utilized for growth and reproduction. Active, terrestrial, warm-blooded animals exhibit net low production efficiencies: birds less than 1 per cent because they maintain
uniformly high activity; small mammals with high reproductive rates
(rabbits and mice, for example) up to 6 per cent. Man maintains the net
production efficiency of beef cattle at as much as 11 per cent by
slaughtering them soon after, or even before, growth is completed.
More sedentary, cold-blooded animals, particularly aquatic species,
channel as much as 75 per cent of their assimilated energy into growth
and reproduction, which approaches the maximum biochemical efficiency of growth.

The efficiency of biomass production within a trophic level (gross
production efficiency) is the product of assimilation efficiency and the
net growth efficiency (Figure 8–6). Gross production efficiencies of few
warm-blooded, terrestrial animals exceed 5 per cent, and those of some
birds and large mammals fall below 1 per cent. The gross production
efficiencies of insects lie within the range of 5 to 15 per cent, and some

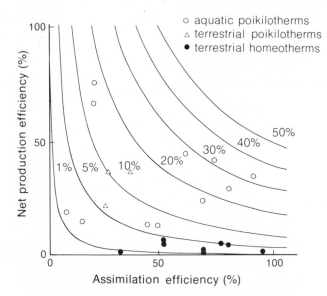

FIGURE 8–6 Relationships between assimilation
efficiency and net production efficiency for a variety of animals. Gross production efficiencies are
indicated by the curved lines on the graph.

aquatic animals exhibit efficiencies in excess of 30 per cent. Net production efficiency tends to be inversely related to assimilation efficiency, especially in aquatic animals, but ecologists do not fully understand the basis for this relationship.

Exploitation Efficiency

The efficiency of energy conversion with a trophic level (gross growth efficiency) does not fully describe the flow of energy between trophic levels. One must also include the efficiency with which consumers exploit their food resources. Unless organic material is steadily accumulating in an ecosystem, exploitation efficiencies on each trophic level, including detritivores, account for all the net production of the next lower trophic level. Peat bogs are an exception to this rule; a large fraction of plant production sinks to the bottom of the bog where acid, anaerobic conditions prevent its decay. Nonetheless, most ecosystems exhibit steady-state conditions in which all production is eventually consumed or transported out of the system by wind or water currents.

Energy flow between trophic levels represents the sum of the feeding activities of many species. Individual populations usually consume only a small fraction of available food resources. Carnivores, seed eaters, and aquatic herbivores are most efficient; the commonest species consume 10 to 100 per cent of the food available to them. Terrestrial herbivores usually consume only 1 to 10 per cent of the leafy vegetation.

Different prey are caught with different success. A study in Alberta, Canada, revealed that red-tailed hawks captured 20 to 60 per cent of local ground squirrel populations during the summer months, but only 1 per cent of snowshoe hares and 1 to 3 per cent of ruffed grouse. Other predators — foxes, weasels, owls, snakes — also hunt the same prey and ensure the transfer of their biomass to the next trophic level. Plants and prey that escape herbivores and predators eventually die and their chemical energy enters the next trophic level by way of the detritus pathway.

One predator's failure is another's success, but the relative feeding efficiency of consumers influences two important aspects of community energetics: the proportion of energy that travels through detritus pathways in the ecosystem, and the time energy remains in the system before it is dissipated as heat.

Detritus Pathways in the Ecosystem

Detritus feeders consume remains of dead plants and animals, undigested or partially digested fecal matter, and excreted nitrogenous waste products of protein metabolism — any nonliving organic material that can be metabolized to provide energy. Detritus feeding is inversely related to the digestibility of fresh food materials. Detritivores are not so prominent in planktonic aquatic communities as in terrestrial communities, where they consume as much as 90 to 95 per cent of net primary production.

Large carrion eaters and scavengers — vultures and crows on land, and crabs in the sea — draw one's attention to detritivores as members of natural communities, but most dead organic matter is consumed by the unnoticed worms, mites, bacteria, and fungi that teem under the litter of the forest floor and in the mucky sediments at the bottom of streams, lakes, and the sea.

Of all the detritus-based communities, the organisms that consume the litter of leaves and branches on the forest floor are probably best known. Herbivores consume less than 10 per cent of the production of a broad-leaved forest — except during outbreaks of defoliating insects, such as gypsy moths. The remainder of the production drops to the forest floor each year as old leaves and branches, or accumulates in roots, trunks, and branches where it escapes consumers until the tree finally dies.

The breakdown of leaf litter occurs in three ways: (1) leaching of soluble minerals and small organic compounds from leaves by water, (2) consumption of leaf material by large detritus feeding organisms (millipedes, earthworms, woodlice, and other invertebrates), and (3) eventual breakdown of organic compounds to inorganic nutrients by specialized bacteria and fungi.

Between 10 and 30 per cent of the substances in newly fallen leaves dissolve in cold water; leaching rapidly removes most of these from the litter. As soil microorganisms decompose the litter further, they produce many small organic and inorganic molecules, which are also exposed to leaching if they are not first assimilated by detritivores.

The role of large organisms in the breakdown of leaf litter has been demonstrated by enclosing samples of litter in mesh bags with openings large enough to let in microorganisms and small arthropods such as mites and springtails, but small enough to keep out large arthropods and earthworms (Figure 8-7). Large detritus feeders assimilate no more than 30 to 45 per cent of the energy available in leaf litter, and even less from wood. They nonetheless speed the decay of litter by microorganisms because they macerate the leaves in their digestive tracts, breaking the litter into fine particles and exposing new surfaces for microbial feeding.

Leaves from different species of trees decompose at different rates, depending on their composition. In eastern Tennessee, weight loss of leaves during the first year after leaf fall varies from 64 per cent for mulberry, to 39 per cent for oak, 32 per cent for sugar maple, and 21 per cent for beech. The needles of pines and other conifers also decompose slowly. Differences between species depend to a large extent upon the

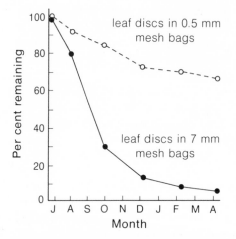

FIGURE 8-7 Percentage consumption of leaf area by detritus feeders. Leaves were enclosed in mesh bags with either large (7 mm) or small (0.5 mm) openings. The small openings admitted bacteria, fungi, and small arthropods, but excluded most large detritus feeders such as earthworms and millipedes.

lignin content of the leaves. Lignin is the substance that gives wood many of its structural qualities, and it is even more difficult to digest than cellulose. Conifer needles typically contain 20 to 30 per cent lignin, broad leaves 15 to 20 per cent.

The toughness of some types of plant litter, particularly wood, points up the unique role of the fungi as detritivores. The familiar mushrooms and shelf fungi are merely fruiting structures produced by the mass of the fungal organism deep within the litter or wood (Figure 8-8). Most fungi consist of a vast network, or mycelium, of hyphae, thread-like elements, which can penetrate the woody cells of plant litter that bacteria cannot reach. Fungi are also distinguished by secreting enzymes and acids into the substrate itself, digesting organic matter even at a distance. The fungal hypha is like a biochemical blowtorch, cutting its way deep into wood and opening the way for bacteria and other microorganisms. Fungi are prominent in woody litter that is not attacked readily by larger detritus feeders. Bacteria occur more abundantly where detritus has been mechanically broken up by earthworms and large arthropods.

The Time Scale of Energy Flow

Characteristics of plants that inhibit digestion by animals slow the passage of assimilated energy through the ecosystem. For a given level of productivity, the *transit time* of energy in the community and the storage of chemical energy in living biomass and detritus are directly related: the longer the transit time, the greater the accumulation of living and dead biomass. Energy storage in the ecosystem has important implications for the stability of the biological community and its ability to withstand perturbation.

FIGURE 8–8   Shelf fungi speed the decomposition of a fallen log. The brackets are fruiting structures produced by the fungal hyphae, together called the mycelium, that grow throughout the interior of the log, slowly digesting its structure.

The average transit time of energy in living organisms in the community is equal to the energy stored in the system as biomass divided by the rate of energy flow through the system, or

$$\text{transit time (yrs)} = \frac{\text{biomass (g/m}^2)}{\text{net productivity (g/m}^2\cdot\text{yr})}$$

(The transit time defined by this equation is sometimes referred to as the *biomass accumulation ratio*.) According to Whittaker and Likens, wet tropical forests produce an average of 2,000 grams of dry matter per square meter per year and have an average living biomass of 45,000 g/m². Inserting these values into the equation for average transit time we obtain 22.5 years (45,000/2,000). Average transit times presented for representative ecosystems in Table 8–3 vary from more than 20 years in forested terrestrial environments to less than 20 days in aquatic planktonic communities. These figures underestimate the total transit time of energy in the ecosystem, however, because they do not include the accumulation of dead organic matter in the litter.

An estimate of the transit or residence time of energy in accumulated litter can be obtained by an equation analogous to the biomass accumulation ratio

$$\text{transit time (yrs)} = \frac{\text{litter accumulation (g/m}^2)}{\text{rate of litter fall (g/m}^2 \cdot \text{yr})}$$

The average transit time for energy varies from a minimum of 3 months in wet tropical forests to 1 to 2 years in dry and montane tropical forests, 4 to 16 years in pine forests of the southeastern United States, and more than 100 years in montane coniferous forests. Warm temperature and abundance of moisture in lowland tropical regions create optimum conditions for rapid decomposition. Because most of the energy assimilated by forest communities is dissipated by detritus feeders (most of a tree being unavailable to herbivores for the several reasons we have seen), the average transit time of energy in the litter must be added to the transit time in living vegetation to obtain a complete estimate of the persistence of assimilated energy in the ecosystem.

More direct estimates of the rate of energy flow can be obtained by using radioactive tracers. Energy itself cannot be followed directly, but organic compounds containing energy can be labeled with a radioactive element, and their movement followed. In radioactive tracer studies, plants (or water if an aquatic system is being studied) are labeled with a radioactive isotope, usually of phosphorus (P^{32}), applied in a phosphate solution. Consumer species are collected at intervals after the initial labeling and are examined with a radiation counter.

Eugene Odum and his coworkers at the University of Georgia have followed the movement of radioactive phosphorus through components of an old-field community. They labeled the dominant plant, telegraph weed (*Heterotheca*), with drops of radioactive phosphate solution placed

TABLE 8–3   Average transit time of energy in living plant biomass (biomass/net primary production) for representative ecosystems.

System	Net primary production (g/m²/yr)	Biomass (g/m²)	Transit time (yrs)
Tropical rain forest	2,000	45,000	22.5
Temperate deciduous forest	1,200	30,000	25.0
Boreal forest	800	20,000	25.0
Temperate grassland	500	1,500	3.0
Desert scrub	70	700	10.0
Swamp and marsh	2,500	15,000	6.0
Lake and stream	500	20	0.04 (15 days)
Algal beds and reefs	2,000	2,000	1.0
Open ocean	125	3	0.024 (9 days)

directly on the leaves, and then collected insects, snails, and spiders at intervals of several days for about five weeks (Figure 8–9). Certain herbivores, notably crickets and ants, began to accumulate radioactive phosphorus within a few days of its initial application, and attained peak amounts within two weeks. Other herbivorous insects and snails accumulated peak amounts of the tracer at two to three weeks. Ground-living, detritus-feeding insects (carabid beetles, tenebrionid beetles, and gryllid crickets) and predatory spiders did not accumulate peak amounts of tracer until three weeks after the start of the experiment. Thus the phosphorus label appeared in herbivores first and in detritivores and predators later, as the investigators undoubtedly had hoped.

The movement of radioactive phosphorus through components of the old-field community gives some indication of the time required by labelled substances to reach various trophic levels. Most of the energy assimilated into a trophic level is dissipated by respiration before it reaches the next level. Respired energy therefore has a shorter residence time in the ecosystem than the food energy that eventually reaches higher trophic levels. Energy appears to move between trophic levels via

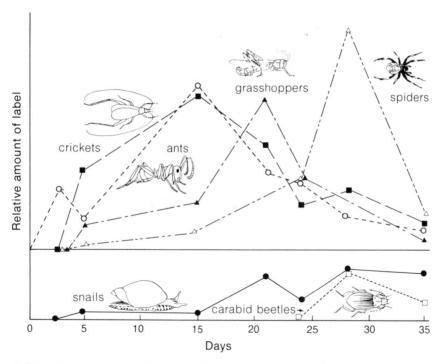

FIGURE 8–9 Accumulation of radioactive phosphorus by animals on different trophic levels after initial labeling of plants. The experiment was performed in an old-field ecosystem.

the herbivorous insect pathway in an average of a few weeks. Most plant production in terrestrial communities is not, however, consumed by herbivores, rather it is stored and consumed by detritus feeders over a prolonged period. Based on the biomass accumulation ratio, a minimum estimate of average transit in temperate grasslands is on the order of three years (Table 8–3).

A radioactive tracer experiment performed on a small trout stream in Michigan gave results comparable to those of Odum's old-field study. Radioactive phosphate was added to the stream at one point and its accumulation in plants and animals downstream from the release site was monitored for two months. The median time for each population to accumulate its maximum concentration of radioactive phosphorus varied from a few days for aquatic plants to one to two weeks for filter feeders and other herbivores, three to four weeks for omnivores, four to five weeks for detritus feeders, and four weeks to more than two months for most predators (Figure 8–10). The results suggest that most of the energy assimilated in aquatic ecosystems is dissipated within a few weeks, although a small portion may linger for months in the predator food chain, and perhaps for years in organic sediments on the bottoms of streams and lakes.

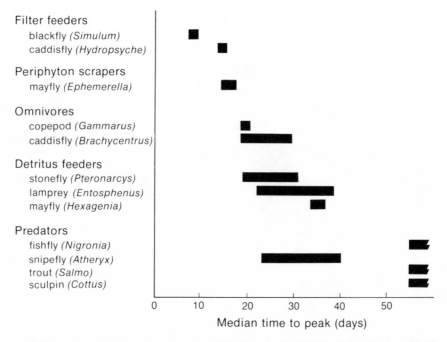

FIGURE 8–10 Median time required for the accumulation of peak levels of radioactive phosphorus in various components of a stream ecosystem after addition of the tracer.

Community Energetics

The measurement of energy flow through an entire community is a complex and virtually impossible task. Yet the total flux of energy and the efficiency of its movement determine the basic trophic structure of the community: number of trophic levels, relative importance of detritus and predatory feeding, steady-state values for biomass and accumulated detritus, and turnover rates of organic matter in the community. Each of these properties, in turn, influences the inherent stability of the community. Detritus pathways stabilize communities because detritivores do not affect the abundance of their "prey" directly. Accumulation of biomass and detritus, coupled with low turnover rates, stabilizes communities by ironing out short-term fluctuations in the physical environment that affect energy flux. Ecologists have not yet agreed on the influence of predation and competition at higher trophic levels on community stability. Some argue that the increased complexity of a diverse and trophically stratified community adds stability to the system. Others contend that biological interactions build lag times into the response of the system to the physical environment, thereby promoting instability. Ecosystem stability will be considered more fully toward the end of this book. In any event, the trophic structure of the community constitutes an important component of the regulation of ecosystem function.

The flow of energy through a community is not unlike the flow of money through a checking account. Income must equal or exceed expenditure over a long period to keep the checking account open and the community viable. The balance undergoes normal periodic fluctuations corresponding to pay day and bills due, just as the community undergoes regular daily and seasonal fluctuations in energy income and expenditure. An unexpected disaster shakes the bank account just as a hurricane or cold snap shakes the community. Under such circumstances, a large balance and hockable assets are as helpful as stored biomass and detritus in meeting disaster.

The red and black ink of community energy balance can be summarized by the following sources of income and expenditure:

Income	Expenditure
Assimilated light energy	Respiration
Transport in	Transport out

All the energy assimilated or transported into the community is either dissipated by respiration or is carried out of the system by wind, water currents, or gravity.

It is difficult to avoid drawing boundaries when we discuss community structure and function, particularly when we wish to distinguish aquatic and terrestrial communities. The lake community clearly extends to the edge of the lake, and no further. Smaller distinctions of communities within the aquatic or terrestrial realms are drawn more arbitrarily. We may find it difficult to point to the exact place where the desert shrub community stops and the grassland community takes up. We can be sure however, of the boundaries around a study area, and ecologists often define the community pragmatically as a place with a characteristic vegetation type.

Regardless of the kind of boundary we imagine around a community, energy frequently finds its way in and out of communities across these boundaries. Movement, or transport, of energy between communities accounts for a varying proportion of the community's income and expenditures. Terrestrial communities receive most of their energy income from light, although they lose energy when plant and animal detritus falls, blows, or washes into rivers and lakes. Transport of material between different terrestrial communities is probably negligible.

Some isolated communities rely primarily on detritus produced elsewhere and transported into the community. For example, in Root Spring, near Concord, Massachusetts, herbivores consume 2,300 kcal/m²/yr, but only 655 kcal are produced by aquatic plants; the balance is transported into the spring by leaf fall from nearby trees. Life in caves and abyssal depths of the oceans, to which no light penetrates, subsists entirely on energy transported into the system by wind, water currents, and sedimentation.

The relative importance of predatory- and detritus-based food chains varies greatly among communities. Predators are most important in plankton communities; detritus feeders consume the bulk of production in terrestrial communities. As we have seen earlier, the proportion of net production that enters the herbivore–predator segment of the food web depends on the relative allocation of plant tissue between structural and supportive functions on one hand and growth and photosynthetic functions on the other. Herbivores consume 1.5 to 2.5 per cent of the net production of temperate deciduous forests, 12 per cent in old-field habitats, and 60 to 99 per cent in plankton communities (Table 8–4). The low level of herbivory in old fields may not truly represent natural habitats because most large North American herbivores became extinct during the last 15,000 years. Large grazing mammals today consume betwen 28 and 60 per cent of the net production of African grasslands; beef cattle consume 30 to 45 per cent of the net production of managed range lands in the United States.

In 1942, Raymond Lindeman made one of the earliest attempts to describe the energy flow in an entire community. He chose Cedar Bog

Lake, Minnesota, a relatively small, self-contained system, for his study. Lindeman estimated energy flow from the harvestable net production of each trophic level and from laboratory determinations of respiration and assimilation. The animals and plants collected at the end of the growing season constituted the net production of the trophic levels to which each species was assigned:

Trophic level	*Harvestable production* (kcal/m²/yr)
Primary producers (green plants)	704
Primary consumers (herbivores)	70
Secondary consumers (carnivores)	13

Lindeman estimated the energy dissipated by respiration from ratios of respiratory metabolism to production measured in the laboratory: 0.33 for aquatic plants, 0.63 for herbivores, and 1.4 for the more

TABLE 8–4 Exploitation of net primary production as living vegetation by primary consumers.

Community	*Characteristics of primary producers*	*Exploitation by herbivores (%)**
Mature deciduous forest	Trees; large amount of non-photosynthetic structure; low turnover rate	1.5 to 2.5
Thirty-year-old Michigan field	Perennial forbs and grasses; medium turnover rate	1.1
Desert shrub	Annual and perennial herbs and shrubs; low turnover rate	5.5**
Georgia salt marsh	Herbaceous perennial plants; medium turnover rate	8
Seven-year-old South Carolina fields	Herbaceous annual plants; medium turnover rate	12
African grasslands	Perennial grasses; high turn-over rate	28 to 60**
Managed rangeland	Perennial grasses, high turn-over rate	30 to 45**
Ocean waters	Phytoplankton; very high turn-over rate	60 to 99

* Aboveground production only for terrestrial systems.
** Grazing by mammals only.

TABLE 8–5   An energy flow model for Cedar Lake Bog, Minnesota

Energy (kcal/m²/yr)	Trophic level		
	Primary producers	Primary consumers	Secondary consumers
Harvestable production*	704	70	13
Respiration	234	44	18
Removal by consumers			
assimilated	148	31	0
unassimilated	28	3	0
Gross production (totals)	1,114	148	31

* Does not include net production removed by consumers. Actual net production, includ-ing removal by consumers, was 879 kcal/m²/yr for primary producers, 104 kcal/m²/yr for primary consumers, and 13 kcal/m²/yr for secondary consumers.

active carnivores in the lake. The gross production of carnivores was calculated as the sum of their harvestable production (13 kcal/m²/yr) and respiration (13 × 1.4 = 18 kcal/m²/yr), or 31 kcal/m²/yr (Table 8–5). Lindeman determined that predation on secondary consumers was neg-ligible. The gross production of primary consumers was similarly calcu-lated as the sum of their harvestable production (70 kcal/m²/yr), respira-tion (70 × 0.63 = 44 kcal/m²/yr), and the consumption of primary consumers by secondary consumers (34 kcal/m²/yr). (Lindeman as-sumed that because secondary consumers have assimilation efficiencies of 90 per cent they must consume 3 kcal/m²/yr over and above their gross production of 31 kcal/m²/yr.) Therefore, the gross production of primary consumers was 148 kcal/m²/yr which corresponded in turn to the re-moval of net primary production by herbivores. Assuming the assimila-tion of herbivores feeding on plant material to be 84 per cent, Lindeman calculated that herbivores consumed, but did not assimilate (0.16/0.84) × 148 = 28 kcal/m²/yr, making the gross primary production 1,114 kcal/m²/yr.

Studies of community energetics have come a long way since Lindeman's pioneering venture. A more recent study by Howard T. Odum on another small aquatic ecosystem at Silver Springs, Florida, employed more refined techniques. Gross production of aquatic plants was estimated by gas exchange rather than by the harvest method. Odum also accounted for the inflow of energy in the form of detritus from tributary streams and the surrounding land. The community energetics of Cedar Bog Lake and Silver Springs are compared in Table 8-6. The more southern location and warmer temperatures of Silver Springs probably accounts for its greater primary production and the lower net production efficiencies of its inhabitants. Herbivores consumed little of the net primary production of Cedar Bog Lake; most was deposited as organic detritus in lake sediments. Consequently, the exploitation efficiency of primary consumers was lower in Cedar Bog Lake than in Silver Springs. In spite of high respiratory energy losses in Silver Springs and quantities of production transported out of the system at Cedar Bog Lake, exploitation efficiencies varied in both locations between 15 and 40 per cent and the overall ecological efficiency of energy transfer between trophic levels varied between 5 and 17 per cent.

Ecological efficiencies are usually lower in terrestrial habitats and a useful rule of thumb states that the top carnivores in terrestrial communities can feed no higher than the third trophic level on the average, whereas aquatic carnivores may feed as high as the fourth or fifth level. This is not to say that there can be no more than three links in a terrestrial food chain; some energy may travel through a dozen links before it is dissipated by respiration. These high trophic levels probably do not, however, contain enough energy to fully support a predator population.

We can crudely estimate the average length of food chains in a community from the net primary production, average ecological efficiency, and average energy flux of predator populations. Because the energy reaching a given trophic level is the product of the net primary production and the intervening ecological efficiencies, the appropriate

TABLE 8–6   A comparison of energy flow models for Cedar Lake Bog, Minnesota and Silver Springs, Florida.

	Cedar Lake Bog	*Silver Springs*
Incoming solar radiation (kcal/m²/yr)	1,188,720	1,700,000
Gross primary production (kcal/m²/yr)	1,113	20,810
Photosynthetic efficiency (%)	0.10	1.20
Net production efficiency (%)		
Producers	79.0	42.4
Primary consumers	70.3	43.9
Secondary consumers	41.9	18.6
Exploitation efficiency (%)*		
Primary consumers	16.8	38.1
Secondary consumers	29.8	27.3
Ecological efficiency (%)		
Primary consumers	11.8	16.7
Secondary consumers	12.5	4.9

* Based on assimilated energy rather than ingested energy. Assimilation efficiencies were probably above 80 per cent, so values are not much below actual exploitation efficiencies.

equation for calculating the average trophic level that a community can support is

$$\text{trophic level} = 1 + \frac{\log (\text{predator ingestion} \div \text{net primary production})^*}{\log (\text{average ecological efficiency})}$$

Using this equation and some rough estimates for the values needed on the right-hand side of the equation, we can calculate average number of trophic levels as about seven for marine plankton communities, five for inshore aquatic communities, four for grasslands, and three for wet tropical forests (Table 8–7). These estimates should be taken with a grain of salt, to be sure, but they do indicate how measurements of energetics for individual species and trophic levels can be used to determine the overall trophic structure of the community.

We have seen how the quality of food and allocation of energy to various functions by organisms create patterns of energy flow through communities. These patterns differ most between aquatic and terrestrial

TABLE 8–7   Community energetics and the average number of trophic levels in various communities. Values for production, predator energy flux, and ecological efficiencies are rough estimates based on many studies.

Community	Net primary production (kcal/m²/yr)	Predator ingestion (kcal/m²/yr)	Ecological efficiency (%)	Number of trophic levels
Open ocean	500	0.1	25	7.1
Coastal marine	8,000	10.0	20	5.1
Temperate grassland	2,000	1.0	10	4.3
Tropical forest	8,000	10.0	5	3.2

* We note that the energy $E(n)$ available to a predator on the nth trophic level may be calculated by the equation

$$E(n) = NPP \cdot Eff^{n-1}$$

where NPP = net primary production. Eff is the geometric mean ecological efficiency of trophic levels 1 to n, and $n - 1$ is the number of links in the food chain between trophic levels 1 and n. We now rearrange the equation above to the following form

$$Eff^{n-1} = \frac{E(n)}{NPP}$$

take the logarithm of both sides of the equation

$$(n - 1) \log (Eff) = \log (E[n]/NPP)$$

and rearrange to obtain

$$n = 1 + \frac{\log (E[n]/NPP)}{\log (Eff)}$$

environments because of basic differences in the adaptations of organisms to each of these realms. In aquatic ecosystems, energy flows rapidly and is transferred efficiently between trophic levels, thereby permitting long food chains. In terrestrial ecosystems, some energy is dissipated rapidly, making energy transfer between trophic levels relatively inefficient, and the remainder is stored for long periods as supportive tissue in plants and as organic detritus in the soil.

Each parcel of energy assimilated by plants travels through the ecosystem only once. Energy is dissipated as heat, a form which plants cannot harness for primary production. As we shall see, however, in the next chapter, mineral nutrients are continually recycled through the ecosystem. These cycles differ greatly among the elements, but they share two fundamental properties: first, the cycles are tied to, and are driven by, energy flux through the ecosystem; second, nutrients alternate between inorganic and organic forms through the complimentary processes of assimilation and decomposition.

9

Nutrient Cycling

Nutrients, unlike energy, are retained within the ecosystem where they are continually recycled between living organisms and the physical environment. Because plants and animals can use only those nutrients that occur at or near the surface of the Earth, persistence of life requires that the materials assimilated by organisms eventually become available to other organisms. Each chemical element follows a unique route in its cycle through the ecosystem, as we shall see below, but all cycles are driven by energy and their elements alternate between organic and inorganic forms.

Nutrient cycles of communities sometimes become unbalanced and nutrients accumulate in, or are removed from, the system. For example, during periods of coal and peat formation, dead organic materials accumulate in the sediments of lakes, marshes, and shallow seas where anaerobic conditions prevent their decomposition by microorganisms. Under intensive cultivation or after removal of natural vegetation, erosion can wash away nutrient-laden layers of soil that take years to develop. Most ecosystems exist in a steady state, in which outflow of nutrients from the system is balanced by inflow from other systems, from the atmosphere, and from the rock beneath the system. Furthermore, gains and losses are usually small compared to the rate at which nutrients are cycled within the system.

A Compartment Model of Nutrient Cycling

Exchange of nutrients between living organisms and inorganic pools is about evenly balanced in most communities. Carbon and oxygen are recycled by the complementary processes of photosynthesis and respiration. Nitrogen, phosphorus, and sulfur follow more complex paths through the ecosystem, aided along the way by microorganisms with specialized metabolic capabilities.

We may describe the ecosystem as being divided into *compartments* through which material passes and within which material may remain for varying periods (Figure 9–1). Most of the mineral cycling in the ecosystem involves three *active* compartments: living organisms, dead organic detritus, and available inorganic minerals. Two additional compartments — indirectly available inorganic minerals and organic sediments — are peripherally involved in nutrient cycles but exchange between these compartments and the rest of the ecosystem occurs slowly compared to exchange among active compartments.

The processes responsible for the movement of mineral nutrients within the ecosystem are indicated in Figure 9–1. Assimilation and

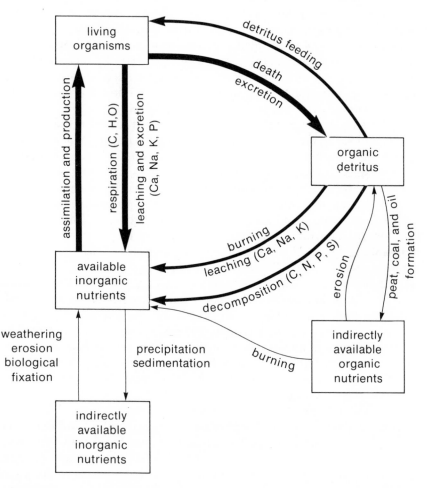

FIGURE 9–1   A compartment model of the ecosystem with some of the most important routes of mineral exchange indicated.

production cause minerals to move from the inorganic to the organic compartment: primary production by plants is the most important component of this step in the cycling of carbon, oxygen, nitrogen, phosphorus, and sulfur, but animals also assimilate many essential minerals, such as sodium, potassium, and calcium, directly from the water they drink.

Respiration returns some carbon and oxygen directly to the pool of available inorganic nutrients, perhaps after being cycled within the living biomass compartment several times in predator food chains. Calcium, sodium, and other mineral ions are excreted or leached out of leaves by rainfall or the water surrounding aquatic organisms and are also recycled rapidly. Most of the carbon and nitrogen assimilated into living biomass is transferred by death and excretion into the detritus compartment. Some nutrients in detritus may be returned to the biomass compartment by detritus feeders, but all eventually are returned to the pool of available inorganic minerals by leaching and decomposition. Exchange between the actively cycled pools of minerals and the vast reservoirs of indirectly available nutrients locked up in the atmosphere, limestone, coal, and the rocks forming the crust of the Earth, occurs slowly, primarily by geological processes.

Nutrient elements usually occur in different forms in the air, soil, water, and living organisms (often referred to as the atmosphere, lithosphere, hydrosphere, and biosphere). For example, oxygen occurs as oxygen molecules (O_2) and as carbon dioxide (CO_2) both in gaseous form in the atmosphere and in dissolved form in water, but it also combines with hydrogen to form water (H_2O). Oxygen appears in the form of oxides (iron oxide, Fe_2O_3) and salts (calcium carbonate, $CaCO_3$) in the lithosphere. The rate at which an element is transferred among its inorganic forms, and its availability in inorganic forms to living organisms, vary greatly. The largest pool of oxygen — more than 90 per cent of the oxygen near the surface of the Earth — occurs as calcium carbonate in sedimentary rocks, particularly limestone. Except for minute quantities released by volcanic activity, the oxygen in limestone and other sedimentary rock is virtually unavailable to the biosphere. In contrast to oxygen, nitrogen is most abundant in its gaseous form (N_2) in the atmosphere, but plants assimilate nitrogen primarily from nitrates (NO_3^-) in the soil or in water. In spite of its abundance, atmospheric nitrogen plays a minor role in nutrient cycling.

The assimilation and decomposition processes that cycle nutrients through the biosphere are closely linked to the acquisition and release of energy by organisms. The paths of nutrients therefore parallel the flow of energy through the community. (As we saw in the last chapter, radioactive elements incorporated into organic compounds can be used to follow the path of energy through the community.) The carbon cycle is most closely linked to the transformation of energy in the community because organic carbon compounds contain most of the energy assimilated by photosynthesis. Most energy-releasing processes, of which respiration is the most important, release carbon as carbon dioxide. When organisms metabolize organic compounds containing nitrogen, phosphorus, and sulfur, these elements are often retained in the body for the synthesis of structural proteins, enzymes, and other organic molecules that make up structural and functional components of living tissue. Consequently, nitrogen, phosphorus, and sulfur pass through each trophic level somewhat more slowly than the average transit time of energy.

Movement of oxygen and hydrogen in the ecosystem is overwhelmingly influenced by the water cycle. Organisms lose water rapidly by evaporation and excretion; body water may be replaced hundreds or even thousands of times during an organism's lifetime. When discussing the oxygen cycle, ecologists usually distinguish pathways involving chemical assimilation of oxygen into organic compounds and those involving movement of water.

The water, or hydrological, cycle does, however, demonstrate the basic features of all nutrient cycles — that they are approximately balanced on a global scale, and that they are driven by energy — and so we shall let it provide a model.

The Water Cycle

Although water is chemically involved in photosynthesis, most of the water flux through the ecosystem occurs through evaporation, transpiration, and precipitation. Evaporation and transpiration correspond to photosynthesis in that light energy is absorbed and utilized to perform the work of evaporating water and lifting it into the atmosphere. The condensation of water vapor in the air, which eventually causes rainfall, releases the potential energy in water vapor as heat, much as respiration by plants and animals releases energy. The water cycle is outlined in Figure 9–2.

Because more than 90 per cent of the Earth's water is locked up in rocks in the core of the Earth and in sedimentary deposits near the Earth's surface, this water enters the hydrological cycle in the ecosystem very slowly through volcanic outpourings of steam. Hence the great reservoirs of water in the Earth's interior contribute little to the movement of water near the Earth's surface.

Precipitation over the land surface of the Earth exceeds evaporation and transpiration from terrestrial habitats. The oceans exhibit a corresponding deficit of rainfall compared to evaporation. Much of the water vapor that winds carry from the oceans to the land condenses over mountainous regions and where rapid heating and cooling of the land

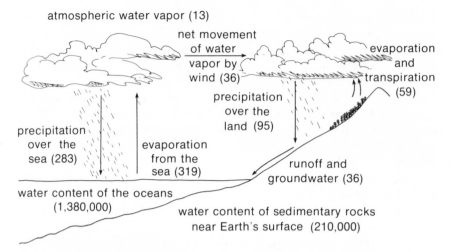

FIGURE 9–2 The water cycle, with its major components expressed on a global scale. All pools and transfer values (shown in parentheses) are expressed as billion billion (10^{18}) grams, and billion billion grams per year.

create vertical air currents. The net flow of atmospheric water vapor from ocean to land areas is balanced by runoff from the land into ocean basins.

We can estimate the role of plant transpiration in the hydrological cycle from the organic productivity of terrestrial habitats and the transpiration efficiency of plant production (see page 121). The primary production of terrestrial habitats is about 1.1×10^{17} grams (g) of dry material per year and approximately 500 g of water are transpired for each gram of production. Terrestrial vegetation, therefore, transpires 55×10^{18} g of water annually, nearly the total evapotranspiration from the land. Although this figure may overestimate transpiration somewhat, plants clearly play the major role. The influence of vegetation on water movement is best shown by removing it. At Hubbard Brook, New Hampshire, experimental cutting of all trees from small watersheds increased the flow of streams draining the clear-cut areas by more than 200 per cent. This excess would normally have been transpired, as water vapor from leaves, directly into the atmosphere.

We may calculate the amount of energy required to drive the hydrological cycle by multiplying together the energy required to evaporate 1 g of water (0.536 kcal) and the total annual evaporation of water from the Earth's surface (378×10^{18} g). The product, about 2×10^{20} kcal, represents about one-fifth of the total energy income in light striking the surface of the Earth.

Evaporation, not precipitation, determines the flux of water through the hydrological cycle. Ignoring the relatively minor inputs of energy to create wind currents and to heat water vapor in the atmosphere, the absorption of light energy by liquid water represents the major point at which an energy source is geared to the water cycle. Furthermore, the ability of the atmosphere to hold water vapor is limited. An increase in the rate of evaporation of water into the atmosphere eventually results in an equal increase in precipitation.

The water present as vapor in the air at any one time corresponds to an average of 2.5 centimeters (1 inch) spread evenly over the surface of the Earth. An average of 65 cm (26 in) of rain falls each year, representing 25 times the water present in the atmosphere at any one time. The steady-state content of water vapor in the atmosphere, referred to as the atmospheric pool, is therefore recycled 25 times each year. Conversely, water has an average transit time (see page 144) of about two weeks. The water content of soils, rivers, lakes, and oceans is a hundred thousand times greater than that of the atmosphere. Rates of flux through both pools are the same, however, because evaporation equals precipitation. The average transit time of water in its liquid form at the Earth's surface (about 3,650 years) is therefore 100,000 times longer than in the atmosphere.

The Oxygen Cycle

Next to nitrogen, oxygen is the most abundant element in the atmosphere, accounting for 21 per cent of its weight. But because oxygen is abundant and ubiquitous in the terrestrial environment, ecologists do not pay as much attention to the oxygen cycle as they do to the cycles of scarcer nutrients — carbon, nitrogen, phosphorus, and others. The oxygen cycle is relatively simple, but it exhibits the basic characteristics of nutrient cycling in the ecosystem.

The atmosphere contains about 1.1×10^{21} g of oxygen. Much more is bound up in water molecules, mineral oxides, and salts in the Earth's rocky crust, but this tremendous pool of oxygen is not directly available to the ecosystem. Terrestrial plants probably assimilate close to 10^{17} g of carbon in gross production. Photosynthesis releases two atoms of oxygen for each atom of carbon fixed. Because oxygen weighs 16/12 as much as carbon, atom for atom, green plants release about 2.7×10^{17} g of oxygen each year. This amount corresponds to about 1/2,500 of the oxygen in the atmosphere, and therefore the cycling (transit) time of oxygen in the atmosphere is about 2,500 years, if exchange of oxygen between the atmosphere and surface water is ignored.

The oxygen cycle is actually somewhat more complicated than the complementary equations for photosynthesis and respiration suggest. Water enters into the complex biochemistry of photosynthesis and although it is released in equal amounts by respiration, water molecules do not survive these processes intact. The oxygen molecule (O_2) produced by photosynthesis derives one atom from carbon dioxide and one from water; the oxygen molecule consumed in respiration supplies one atom to carbon dioxide and the other to water (Figure 9–3).

The Carbon Cycle

The biological cycling of carbon in the ecosystem is more direct than that of oxygen. The cycle involves only organic compounds and carbon dioxide (Figure 9–3). Photosynthesis and respiration fully complement

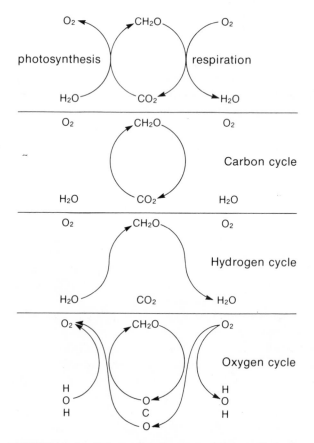

FIGURE 9–3 Schematic diagram of the cycling of hydrogen, oxygen, and carbon by photosynthesis and respiration. The arrows follow the paths of individual atoms. Water and carbon dioxide are represented by structural formulas (HOH and OCO, respectively) in the oxygen cycle to clarify the paths of individual oxygen atoms.

each other. Photosynthesis assimilates carbon entirely into carbohydrate; respiration converts all the carbon in organic compounds to carbon dioxide. Large inorganic pools of carbon — atmospheric carbon dioxide, dissolved carbon dioxide, carbonic acid, and carbonate sediments — enter the carbon cycle to different degrees (Figure 9–4). The carbon in igneous rocks, calcium carbonate (limestone) sediments, coal, and oil is exchanged with other, more active pools so slowly that these pools have little influence on the short-term functioning of the ecosystem.

Plants assimilate about 105×10^{15} g of carbon each year, of which about 32×10^{15} g are returned to the carbon dioxide pool by plant respiration. The remainder, 73×10^{15} g, supports the respiration and production of animals, bacteria, and fungi in herbivore- and detritus-based food chains. Anaerobic respiration (without oxygen) produces a small quantity of methane (CH_4) that is converted to carbon dioxide by a photochemical reaction in the atmosphere. Plants and animals annually cycle between 0.25 and 0.30 per cent of the carbon present in carbon dioxide and carbonic acid in the atmosphere and oceans, hence the total active inorganic pool is recycled every 300 to 400 years (1/0.003 to 1/0.0025). Because the atmosphere and oceans exchange carbon dioxide

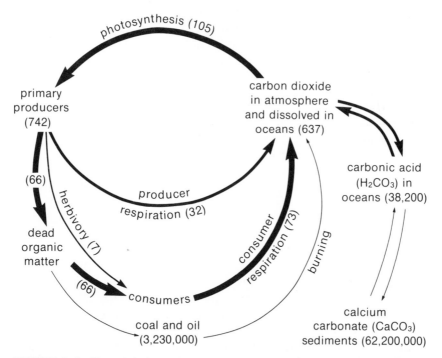

FIGURE 9–4   The global carbon cycle, with some estimated pools and annual transfer rates. Values are in million billion (10^{15}) grams.

slowly, they may be considered as separate pools over short periods. Terrestrial ecosystems annually cycle about 12 per cent of the carbon dioxide in the atmosphere; the transit time of atmospheric carbon is therefore about eight years (1/0.12).

The combustion of coal and oil adds carbon dioxide to the atmosphere. Although man's present use of fossil fuels amounts to less than two per cent of the carbon cycled through the ecosystem each year, combustion of fuels adds carbon dioxide to the atmosphere over and above that produced by photosynthesis. The carbon dioxide content of the atmosphere has in fact risen measurably during this century and can be expected to rise even more rapidly in the future. Scientists cannot agree on the implications of increased atmospheric carbon dioxide for air temperature or for plant production, yet man is unquestionably shifting the steady-state balance of the ecosystem.

The Nitrogen Cycle

The path of nitrogen through the ecosystem differs from that of carbon in several important respects. First, the immense pool of nitrogen (N_2) in the atmosphere (3.85×10^{21} g) cannot be assimilated by most organisms. Second, nitrogen is not directly involved in the release of chemical energy by respiration; its role is linked to protein molecules and nucleic acids, which provide structure and regulate biological function. Third, the biological breakdown of nitrogenous organic compounds to inorganic forms requires many steps, some of which can be performed only by specialized bacteria (Figure 9–5). Fourth, most of the biochemical transformations involved in the decomposition of nitrogeneous compounds occur in the soil, where the solubility of inorganic nitrogen compounds influences the availability of nitrogen to plants.

Living tissues contain slightly more than three per cent of the nitrogen in active pools in the ecosystem. The rest is distributed between detritus and nitrates in the soils and ocean, with smaller amounts in the intermediate stages of protein decomposition — ammonia and nitrites (Table 9–1). Plants assimilate 86×10^{14} g of nitrogen annually, less than one per cent of the active pool; the overall cycling time of nitrogen therefore exceeds 100 years.

The nitrogen cycle involves the stepwise breakdown of organic nitrogen compounds by many kinds of organisms until nitrogen is finally converted to nitrate. Of the forms of nitrogen in the soil that are available to plants, ammonia (NH_3) or the ammonium ion (NH_4^+) would seem to be the most desirable because their conversion to organic compounds requires the least chemical work. Ammonia is, however, unsuitable as a source of nitrogen in the soil because it is toxic to plant tissues in high concentrations and it is not persistent in the soil. Ammonia dissolves

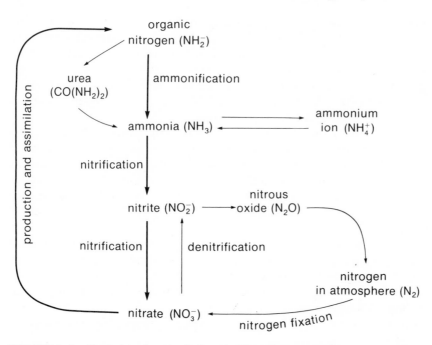

FIGURE 9–5 Basic biochemical steps in the nitrogen cycle.

easily in water and is quickly leached out of soil. Under acid soil conditions, ammonia is converted to ammonium ion. Although the positively
charged ion can adhere to the surface of the clay–humus micelle (see
page 41), it is easily displaced by hydrogen ions in acid soils and thus it,
too, is readily leached out by water. (Some types of clay minerals
actually adsorb ammonium ions into their crystal framework so tightly
that they are inaccessible both to leaching and to plants.) The soil
ammonia that escapes leaching is readily attacked by specialized bac-

TABLE 9–1 Percentage distribution of nitrogen among active pools, and annual transfer rates. The pools taken together contain about 10^{18} grams of nitrogen.

Pool	Nitrogen (%)	Transfer rate (% per year)
Organic forms		
Plants	11	25
Animals	11	
Detritus	6,100	1.4
Inorganic forms in soils and oceans		
Ammonia (NH_3)	286	30
Nitrites (NO_2^-)	138	63
Nitrates (NO_3^-)	4,180	2.1

teria, which obtain energy when they oxidize the nitrogen in ammonia to nitrites (NO_2^-) and nitrates (NO_3^-). Negatively charged nitrite and nitrate ions do not bind to clay particles at all, hence they are susceptible to leaching. Once nitrates are produced in the soil they are quickly assimilated by plant roots. Storage of nitrogen in terrestrial ecosystems occurs primarily in organic detritus. Most of the nitrogen in aquatic ecosystems occurs as dissolved nitrates.

The biochemical reactions of nitrogen compounds are highly varied because nitrogen can combine with other elements in several different ways. The most important processes in the nitrogen cycle are the breakdown of organic nitrogen compounds by *ammonification* and *nitrification*, the reduction of nitrate and nitrite to nitrogen (N_2) and its release into the atmosphere by *denitrification*, and the biological assimilation of atmospheric nitrogen by *nitrogen fixation*.

Denitrification removes nitrogen from active pools in the soil and surface waters and releases it into the atmosphere; nitrogen fixation brings atmospheric nitrogen back into active circulation through the ecosystem. Although these processes are minor compared to the overall cycling of nitrogen in the ecosystem, nitrogen fixation often assumes local importance where soils lack adequate nitrogen for normal plant growth.

In its organic form, nitrogen occurs in amine groups (NH_2), or some variation, combined with other organic molecules. Animals rid their bodies of excess nitrogen by detaching the amines from organic compounds and excreting them relatively unchanged primarily as ammonia, NH_3, or urea, $CO(NH_2)_2$. Soil microorganisms readily convert urea to ammonia by hydrolysis

$$CO(NH_2)_2 + H_2O \rightarrow 2NH_3 + CO_2$$

This reaction does not, however, release energy to perform biological work.

Some specialized but ubiquitous bacteria can release the chemical energy contained in the amine group by a series of nitrifying steps that require oxygen. *Nitrosomonas* transforms the ammonia ion to nitrite; *Nitrobacter* completes the nitrification process by oxidizing nitrite to nitrate.

Nitrification represents a critical step in the nitrogen cycle, ultimately determining the rate at which nitrates become available to green plants and thereby influencing the productivity of the habitat. Any soil condition that inhibits bacterial activity — high acidity, poor soil aeration, low temperature, and lack of moisture — also inhibits nitrification. Slow nutrient release in the soil of cold and dry regions may depress plant productivity over and above the direct effect of these conditions on photosynthesis. Furthermore, if the organic detritus in the soil has a low nitrogen content compared to that of carbon, bacteria assimilate all the nitrogen into their cell structure rather than use some as a substrate for metabolism. Nitrogen thus becomes tied up in bacteria biomass rather than made available to plants. The ratio of carbon to nitrogen (C/N ratio) in detritus exerts an important influence on the rate of bacterial decomposition (Table 9-2). At one extreme, mulberry leaves (C/N ratio = 25) support an abundant microflora of bacteria and fungi and decompose rapidly. At the other extreme, loblolly pine (C/N ratio = 43) inhibits the activity of microorganisms and decomposes slowly. White oak occupies an intermediate position.　　　└ *BUT HAS MYCORHYZIA*

Denitrification transforms nitrate to nitrogen in a series of steps

$$NO_3^- \rightarrow NO_2^- \rightarrow N_2O \rightarrow N_2$$

each of which releases oxygen. (The bacterium *Pseudomonas* uses denitrification to provide oxygen for respiration when the soil lacks free oxygen.) Nitrous oxide (N_2O) and nitrogen molecules (N_2) escape into the atmosphere and leave the active nitrogen pools. Denitrification also occurs purely by chemical means; independently of microorganisms. For example, the reaction of nitric acid (HNO_3) and urea occurs in acid soils

$$2\ HNO_3 + CO(NH_2)_2 \rightarrow CO_2 + 3\ H_2O + 2\ N_2$$

Nitrogen fixation is energetically expensive and requires considerable chemical work. For the price of the chemical energy in a glucose molecule ($C_6H_{12}O_6$), some blue-green algae and the bacterium *Azotobacter* can assimilate eight nitrogen atoms (ignoring the inefficiency of biochemical transformations). Nitrogen fixation requires specialized biochemical machinery that is apparently unavailable to higher plants. Nonetheless, many species, like alfalfa, peas, and their relatives (legumes), plus a few species scattered in other plant groups, have entered into symbiotic relationships with nitrogen-fixing bacteria. Peas

TABLE 9–2 Average weight loss, carbon/nitrogen (C/N) ratio, and microflora in four species of decaying leaves in forest litter at Oak Ridge, Tennessee; November 1960 to November 1961.

Species	Weight loss (%)	C/N ratio	Bacteria colonies (million/g dry weight)	Fungi colonies (thousands/g dry weight)	Bacteria / Fungi
Red mulberry	90	25	698	2,650	264
Redbud	70	26	286	1,870	148
White oak	55	34	32	1,880	17
Loblolly pine	40	43	15	360	42

develop clusters of nodules throughout their root system that become infected by *Azotobacter* (Figure 9-6). The relationship benefits both parties: the plant furnishes glucose to the bacteria and the bacteria assimilate nitrogen from the soil atmosphere for plant uptake.

In some nitrogen deficient habitats, nitrogen fixation is a critical factor in plant production. The nitrogen-fixing capabilities of some plants have become widely exploited in agriculture to restore soil fertil-

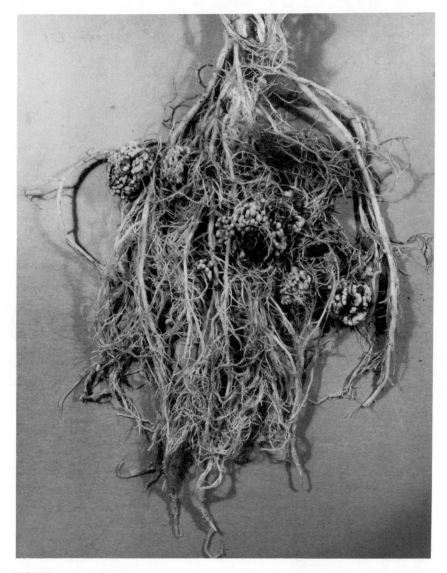

FIGURE 9–6   The root system of an Austrian winter pea plant, showing the clusters of nodules that harbor symbiotic nitrogen-fixing bacteria.

ity after farmland has been planted with soil depleting crops such as corn. Nitrogen-accumulating plants (usually peas or alfalfa) are planted in rotation with corn and then ploughed under the soil, increasing its nitrogen and humus content and water retention.

Assimilation of nitrogen from the atmosphere probably constitutes a more important agent promoting soil fertility than ecologists realize. Nitrogen-fixing microbes are widespread in natural habitats, even on the leaves of trees. If we stop to consider for a moment that parental rocks underlying most soils are completely devoid of nitrogen, we realize that most of the nitrogen in active pools in the ecosystem must have originated through nitrogen fixation.

The Phosphorus Cycle

Oxygen, carbon, and nitrogen cycles demonstrate the basic features of mineral cycles in the ecosystem, but other elements — particularly phosphorus, potassium, calcium, sodium, sulfur, magnesium, and iron — play important roles in ecosystem function. Still other elements, such as cobalt, aluminum, and manganese, may influence ecosystem dynamics in ways still undiscovered.

Ecologists have studied the role of phosphorus in the ecosystem most intensively because organisms require phosphorus at a high level (about one-tenth that of nitrogen) as a major constituent of nucleic acids, cell membranes, energy-transfer systems, bones, and teeth. Phosphorus is important for a number of other reasons. It is thought to limit plant productivity in many aquatic habitats, and the influx of phosphorus to rivers and lakes in the form of sewage (particularly phosphate detergents) and runoff from fertilized agricultural lands stimulates the production of aquatic habitats to undesirably high levels. Also, ecologists can measure concentrations of phosphorus easily and use one of its isotopes as a radioactive tracer in the ecosystem.

The phosphorus cycle has fewer steps than the nitrogen cycle: plants assimilate phosphorus as phosphate ion ($PO_4^{\equiv}$) directly from the soil or water; animals eliminate excess organic phosphorus in their diets by excreting phosphorus salts in urine; phosphatizing bacteria convert the organic phosphorus in detritus to phosphate in the same way. Phosphorus does not enter the atmosphere in any form other than dust. The phosphorus cycle therefore involves only the soil and water of the ecosystem.

In spite of the relative simplicity of the phosphorus cycle, many environmental factors influence the availability of phosphorus to plants. In the presence of abundant dissolved oxygen, phosphorus readily forms insoluble compounds that precipitate and remove phosphorus from the pool of available nutrients. If such conditions persist, deposits of phosphate accumulate and eventually form phosphate rock, which returns to active pools in the ecosystem very slowly by erosion — or by artificial fertilization of crops and disposal of phosphate detergents in sewage.

Acidity also affects the availability of phosphorus to plants. Phosphate compounds of sodium and calcium are relatively insoluble in water. Under alkaline conditions phosphate ions ($PO_4^{\equiv}$) readily combine with sodium or calcium ions to form insoluble compounds. Under acid conditions, phosphate is converted to highly soluble phosphoric acid. At intermediate levels of acidity, phosphate ions form compounds with intermediate solubility, as shown below:

Increasing acidity $\longrightarrow$

Ionic form	$PO_4^{\equiv}$	$\rightarrow$	$HPO_4^{=}$	$\rightarrow$	$H_2PO_4^-$	$\rightarrow$	H_3PO_4
	$\downarrow$		$\downarrow$		$\downarrow$		
Salt	Na_3PO_4		$CaHPO_4$		NaH_2PO_4		None
Solubility	(slightly soluble)		(insoluble)		(soluble)		(very soluble)

Although acidity increases the solubility of phosphate in the laboratory, high acidity reduces the availability of phosphorus in the ecosystem, due to its reaction with other minerals. In acid environments, aluminum, iron, and manganese become soluble and reactive, forming chemical complexes that bind phosphorus and thereby remove it from the active pool of nutrients. The acid conditions in bogs and in soils of cold, wet regions remove phosphorus in this way, and thereby reduce the fertility of these habitats. Phosphorus is most readily available in a narrow range of acidity, just on the acid side of neutrality. In agricultural practice, soils that are too acid are neutralized by adding lime ($CaCO_3$) to increase phosphorus availability. Liming the soil brings about the following reaction which reduces hydrogen ion concentration:

$$CaCO_3 + 2\,H^+ \rightarrow Ca^{++} + CO_2 + H_2O$$

Conversely, adding hydrogen ion to the soil in the presence of carbon dioxide and water (carbonic acid) leaches calcium and increases soil acidity:

$$Ca^{++} + H_2CO_3 \rightarrow CaCO_3 + 2\,H^+$$

Nutrients and Eutrophication

Ecologists classify natural bodies of water between two extremes on the basis of their nutrient content and organic productivity. On the one hand, *oligotrophic* (from the Greek, meaning little-nourished) habitats have low nutrient content and harbor relatively little plant and animal life. The water in oligotrophic lakes and rivers is clear and unproductive. On the other hand, *eutrophic* (from the Greek, meaning well-nourished) lakes and rivers are rich in nutrients and support an abundant flora and fauna. Eutrophic habitats are excellent fisheries because of their high productivity. In fact, lakes and ponds are often artificially fertilized to increase fish production: man harvests between 1 and 7 pounds of fish per acre per year (0.2 to 1.6 kcal/m²/yr) from the oligotrophic Great Lakes; small eutrophic lakes in the United States yield up to 160 lbs/acre/yr (36 kcal/m²/yr); in Germany and the Philippines, where ponds are artificially fertilized, yields of 1,000 lbs/acre/yr (200 kcal/m²/yr) are not uncommon. Primary production of the phytoplankton increases with eutrophy in a similar progression: 7 to 25 g carbon/m²/yr in oligo-trophic lakes, 75 to 250 g C/m²/yr in naturally eutrophic lakes, and 350 to 700 g C/m²/yr in lakes polluted by sewage and agricultural runoff.

Eutrophication does not, by itself, constitute a major problem for the aquatic ecosystem. High rates of production and rapid nutrient cycling are to be expected where the basic mineral resources for production are abundant. In naturally eutrophic lake and stream ecosystems, most of the energy and minerals come from within the system — *autochthonous* inputs — in the form of primary production. In culturally eutrophic lakes and streams, external sources of mineral nutrients and organic matter — *allochthonous* inputs — contribute to the productivity of the system. (In the course of attaching special names to almost everything, ecologists have frequently turned to Greek or Latin roots for terms — because they either ran out of modern language terms or desired to give a name more impact by making it more exotic. *Chthonos* is Greek for "of the earth," *auto* means "the same," and *allo*, "other" or "different," referring to internal and external inputs.)

Whereas naturally eutrophic systems are usually well-balanced, the addition of artificial nutrients can upset the natural workings of the community and create devastating imbalances in the ecosystem. Algal blooms are among the most noticeable of these effects. The combination of high nutrient loads and favorable conditions of light, temperature and carbon dioxide stimulate rapid algal growth. Algal blooms are a natural response of algae to their environment. But when the environment changes and no longer can support dense algal populations, the algae that accumulate during the bloom die and begin to decay. The ensuing rapid decomposition of organic detritus by bacteria robs the water of its oxygen, sometimes so thoroughly depleting the water of oxygen that fish and other aquatic animals suffocate. The nutrient imbalance in culturally eutrophic systems stems from the addition of nutrients at seasons when nutrients are less available in naturally eutrophic waters, primarily during the summer peak of plant production. During less productive seasons, phosphorous is readily absorbed by benthic bacteria and sediments at the bottoms of lakes, and its concentration in lake water is thus quickly reduced.

Early studies of eutrophication suggested that algal blooms occurred only when the concentration of phosphorus was greater than 0.01 milligrams (mg) per liter (l) of water. These findings strongly implicated phosphorus as the prime factor limiting the productivity of many aquatic communities. Although ecologists have since argued this point, recent experiments on small Canadian lakes demonstrate that adding carbon (as sucrose) and nitrogen (as nitrate) does not stimulate algal blooms without simultaneously adding phosphate (Figure 9-7).

In spite of the disturbing effects of outside enrichment, culturally eutrophied lakes can recover their original condition if inputs are shut off. Apparently the sediments at the bottom of most lakes have a high affinity for phosphates and quickly remove them from active pools in the surface waters of the lake. Diversion of sewage from Lake Washington, in Seattle, quickly reversed the eutrophication process, which was turning the lake into a stinking organic soup. Similar recovery could be expected from other bodies of water if inputs from sewage and agricultural runoff were reduced.

In their normal development, lakes usually proceed from oligotrophic to eutrophic stages. New lake basins are deep and devoid of nutrients. As sediments wash into a lake, nutrients are added to the water. As the lake begins to fill in and become shallow, exchange between the bottom and surface water accelerates, returning the nutrients to the active pool near the surface and enriching the water. In extreme old age, lakes fill in completely and are succeeded by the local terrestrial vegetation.

Cation Exchange in Temperate Forests

Calcium, potassium, sodium, and magnesium are not chemically incorporated into organic compounds, although they are abundant as dissolved positive ions — called *cations* — in cellular and extracellular fluids. The cycles of cations in the ecosystem are only loosely associated with the assimilation and release of energy, but they are tremendously important to cell function.

Cation cycles have been studied most intensively in temperate forests, within which the ions exhibit remarkable mobility. The forest is an *open system*, one which freely exchanges minerals with other parts of the environment (Figure 9-8). Leached cations are washed out of the system through runoff, both in streams and groundwater. Minerals enter the system in precipitation, in wind-born dust and organic debris, and by weathering of the parent rock over which the forest lies.

FIGURE 9-7 Experimental lake demonstrating the crucial role of phosphorus in eutrophication. The near basin, fertilized with carbon (in sucrose) and nitrogen (in nitrates), exhibited no change in organic production. The far basin, separated from the first by a plastic curtain, received phosphate in addition to carbon and nitrogen and was covered by a heavy bloom of blue-green algae within two months.

Detailed cation budgets have been obtained for small watersheds by measuring the inputs in rainwater collected at various locations in the watershed area (Figure 9-9) and the outputs in water leaving the watershed by way of the stream that drains it (Figure 9-10). Care must be taken to select a watershed with impervious bedrock to eliminate the problem of groundwater movement into and out of the watershed.

Several watersheds in the Hubbard Brook Experimental Forest,

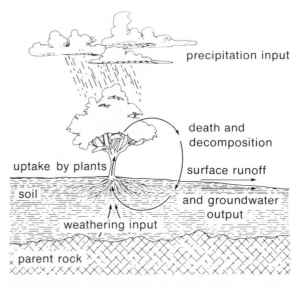

FIGURE 9–8   A generalized cation cycle for a terrestrial ecosystem. If the system is in a steady state, losses due to surface flow and ground water drainage are balanced by gains from precipitation and weathering.

FIGURE 9–9   Rain gauges installed in a ponderosa pine stand in California to intercept precipitation falling through the canopy of the forest and running down the trunks of trees. Analysis of the nutrient content of water collected in sampling programs like this helps determine the overall cation budget of the forest and the specific routes of mineral cycles.

FIGURE 9-10   A stream gauge at the lower end of a watershed at the Coweeta Hydrological Laboratory, North Carolina. The V-shaped notch regulates the flow of water through the weir in such a way that the flow rate is proportional to the water level in the basin.

New Hampshire, have been studied intensively during the past decade to establish patterns of water and nutrient cycles and the effects of disturbances, primarily clear-cutting, on these cycles. The annual distribution of precipitation and runoff for the Hubbard Brook Forest (Figure 9-11) shows a fairly uniform distribution of rainfall during the year, typical of moist temperate locations. Precipitation exceeds runoff during the cold winter months due to snow accumulation. This pattern is reversed in spring as melting snow swells the streams. The difference between precipitation and runoff during the summer months, seen in Figure 9-11, is accounted for by the evaporation and transpiration of water from the watershed.

Cation budgets have been calculated for the entire watershed from the concentrations of the minerals in precipitation and stream flow (Figure 9-12). Only potassium exhibited a net gain in the watershed. Net losses of other cations would have to be balanced by weathering of the bedrock for the system to be in a steady state. The cation budgets of watersheds around the world vary tremendously with total precipitation, soil acidity, and the relative abundance of minerals in rainfall and the soil. Where sodium is abundant in the soil or in rainwater, as in

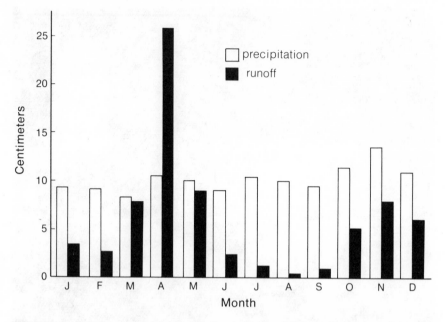

FIGURE 9–11   Average annual distribution of precipitation and runoff for the Hubbard Brook Experimental Forest, New Hampshire, 1955–1969.

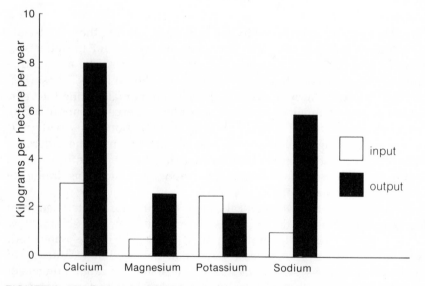

FIGURE 9–12   Cation budgets for the Hubbard Brook Experimental Forest, New Hampshire, during 1963–1964, expressed as kg/ha/yr.

many coastal areas, sodium output in stream flow is great. Similar patterns apply to other cation budgets.

The uptake of cations by plants closely parallels their availability in the soil and hence their general mobility in the ecosystem (Table 9–3). Calcium is most rapidly cycled, magnesium least. In general, uptake by vegetation is 1 to 10 times the annual loss of cations in stream flow; therefore the average transit time (see page 145) of cations in the ecosystem is probably 1 to 10 years. Considering the great mobility of cations in the soil, the long transit times of cations indicate that plants rapidly assimilate free ions in the soil before they are leached out of the system by runoff and groundwater. The role of vegetation in nutrient cycling is dramatized by experiments in which entire watersheds are denuded of trees and shrubs (Figure 9–13). Clear-cutting of small watersheds at the Hubbard Brook Forest increased stream flow several times due to removal of transpiring leaf surface; losses of cations increased 3 to 20 times over comparable undisturbed systems. The nitrogen budget of the cut over watershed sustained the most striking change. Plants assimilate available soil nitrogen so rapidly that the forest usually gains nitrogen at the rate of 1 to 3 kilograms per hectare per year (kg/ha/yr; a hectare = 2.47 acres). In the clear-cut watershed net loss of nitrogen as nitrate (NO_3^-) soared to 54 kg/ha/yr, which is comparable to the annual turnover of nitrogen by vegetation. Precipitation brought only 7 kg/ha/yr of nitrogen into the system, and thus the loss of nitrate represented nitrification of organic nitrogen sources at the normal annual rate by soil microorganisms without simultaneous rapid uptake by plants. These experiments demonstrate the important role of vegetation in maintaining the fertility of the soil and further emphasize that the physical and biological components of the ecosystem cannot exist apart.

In this chapter, we have traced the paths of biologically important elements through the ecosystem. The specific form of the nutrient cycle depends partly on the chemical properties of the element and partly on how it is used by plants and animals. All nutrient cycles are similar in that decomposition balances assimilation in the movement of the min-

TABLE 9–3 Summary of cation budgets for representative temperate forest ecosystems.

	Range of Values (kg/ha/yr)*			
	Precipitation input	*Stream outflow*	*Net loss*	*Uptake by vegetation*
Calcium	2–8	8–26	3–18	25–201
Potassium	1–8	2–13	−1–5	5–99
Magnesium	1–11	3–13	2–4	2–24

* Kilograms per hectare per year. A hectare is equal to 10,000 square meters, or 2.47 acres.

FIGURE 9–13   Clear-cut watershed at the Coweeta Hydrological Laboratory, North Carolina, employed in studies of evapotranspiration and runoff in forest ecosystems.

eral between inorganic and organic pools, and that the cycles are coupled to the assimilation and flow of energy through the community.

Rates of mineral cycling are determined by their usage. If a substance is utilized slowly by organisms at some step in the cycle, the substance tends to accumulate behind the bottleneck, building up a large pool of nutrients. Disruption at any one step of the cycle also affects all others. For this reason, ecologists who are interested in ecosystem stability should pay more attention to nutrient cycling than to energy flow through ecosystems. Disturbances of normal ecosystem function leading to eutrophication or nutrient depletion can probably be traced to disruption of a step in a critical nutrient cycle.

In the preceding chapters, we have examined some large scale measurements of ecosystem function: primary production, energy flow, and nutrient cycling. These properties of structure and function are the sum of the activities of individuals and populations that make up the community. In the remainder of this book, we shall consider structure and function in these smaller components of the ecosystem, beginning with organisms, then populations, and, finally, biological communities.

10

Environment and the Distribution of Organisms

No single type of plant or animal can tolerate all the conditions found on Earth. Each thrives within relatively narrow ranges of temperature, precipitation, soil conditions, and other environmental factors. Moreover, the preferences and tolerances of each species differ so, although the distributions of species broadly overlap one another, no two are found under exactly the same range of conditions. The geographical range of any population corresponds to the geographical distribution of suitable environmental conditions.

This chapter is concerned with the factors that determine the distributions of species' populations. On a local scale, environment plays an overwhelming role in distribution, confining populations to regions and habitats within which the species thrives and perpetuates itself. On a global scale, geographical barriers and historical accidents of distribution assume prominent roles in determining the presence or absence of a particular species. One could easily find localities in Asia, North America, and Europe with closely matched climate and soils, but although the vegetation of each locality would superficially resemble the vegetation of the other two, most of the species would differ. The plants of each region are adapted to tolerate similar environments, but barriers to dispersal, such as oceans, deserts, and mountain ranges, restrict their distributions. That species often can flourish beyond their natural geographical ranges is demonstrated by the success of dandelions, Norway maples, starlings, and honeybees, which were introduced from Europe to the United States, sometimes intentionally and sometimes accidently, by man.

Introduced species may become pleasing ornamentals or destructive pests, and the problems of their propagation or control will receive more attention in later chapters. For the present, we shall examine the ecological factors that determine the local distributions of plants and animals.

The Geographical Distributions of Plants

The range of the sugar maple, a common forest tree in the northeastern United States and southern Canada, is limited by cold winter temperatures to the north, hot summer temperatures to the south (Figure 10–1). Sugar maples cannot tolerate average monthly temperatures over 75 to 80°F or below about 0°F. The western limit of the sugar maple, determined by dryness, coincides closely with the western limit of forest-type vegetation in general. Because temperature and rainfall interact to determine dryness, sugar maples tolerate lower annual rainfall at the northern edge of their range (about 20 inches) than at the southern edge (about 40 inches). To the east, the distribution of the sugar maple is limited abruptly by the Atlantic Ocean.

Within its geographical range, sugar maple is more abundant in northern forests, where it sometimes forms single-species stands, than in the more diverse forests in the south. Sugar maples occur most frequently on moist, podsolized soils that are slightly acid.

The range of the sugar maple overlaps three other tree-sized species of maples: black, red, and silver maple (Figure 10–2). The range of the black maple falls almost entirely within the distributional limits of the sugar maple, indicating similar, but narrower tolerance of temperature and rainfall extremes. In fact, the two maples are so similar that foresters only recently have recognized them as different species, by slight differences in the shape of the fruits and by the presence or absence of hairs on the underside of the leaves. The northern limit of red maple nearly coincides with that of the sugar maple, but red maple extends south to the Gulf Coast, and appears to be less tolerant of drought in the mid-

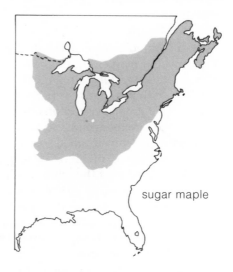

sugar maple

FIGURE 10–1 The range of the sugar maple in eastern North America.

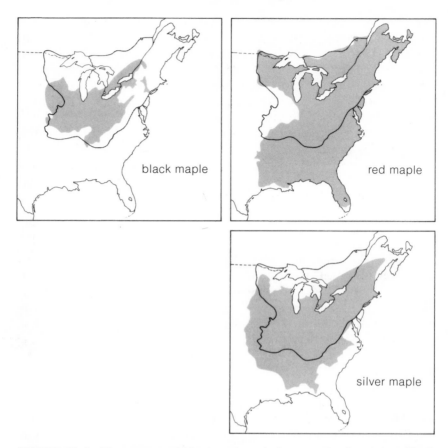

FIGURE 10–2   The ranges of black, red, and silver maples in eastern North America. The range of the sugar maple is outlined on each map to show the area of overlap.

western United States. Silver maple also extends beyond the southern limit of sugar maple, but unlike the red maple, it is found farther to the west, extending along stream valleys where soils are moist.

Where their ranges overlap, maples exhibit distinct preferences for local environmental conditions created by differences in soil and topography. Black maple frequently occurs together with sugar maple, but prefers drier, better-drained soils with higher calcium content (therefore less acidic). Silver maple is widely distributed, but prefers the moist well-drained soils of the Ohio and Mississippi River basins. Red maple is peculiar in prefering either very wet, swampy conditions (it is often called swamp maple), or dry, poorly developed soils. Whether these extremes have some other soil factor in common which the red maple likes, or whether red maple consists of two distinct physiological types, each with different preferences, is not known.

The effects of different factors on the distribution of plants are manifested on different scales of distance. Climate, topography, soil chemistry, and soil texture exert progressively finer influence on geographical distribution. For example, the perennial shrub *Clematis fremontii* variety *riehlii* exhibits a hierarchy of distribution patterns revealed by examining its distribution on different geographical scales (Figure 10–3). Climate restricts this species of *Clematis* to a small area of the midwest-

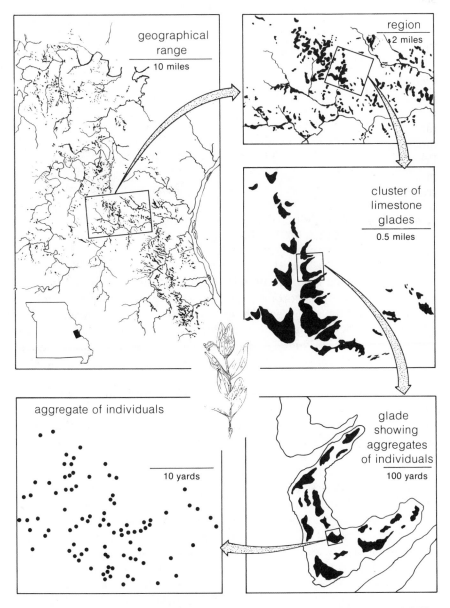

FIGURE 10–3   Hierarchy of distribution of *Clematis fremontii* (variety *riehlii*).

ern United States. Variety *riehlii* is found only in Jefferson County, Missouri. Within its geographical range, *Clematis* is restricted to dry, rocky soils on outcroppings of dolomite, which are distributed with respect to mountain and stream systems. Small variations in relief and soil quality further restrict the distribution of *Clematis* within each dolomite glade to sites with suitable conditions of moisture, nutrients, and soil structure. Local aggregations occurring on each of these sites consists of many, more or less evenly distributed individuals.

The Local Distribution of Plants

Elevation, slope, exposure, and underlying bedrock — factors that greatly influence the plant environment — vary most in mountainous regions. The heterogeneity of such regions breaks the geographical ranges of species into isolated areas with suitable combinations of moisture, soil structure, light, temperature, and available nutrients. Ecologists frequently turn to the varied habitats of mountains to study plant distribution. Along the coast of northern California, mountains create conditions for a variety of plant communities ranging from dry coastal chaparral to tall forests of Douglas fir and redwood. Moisture exerts the strongest influence of all environmental factors on the distribution of forest trees. When localities are ranked on a scale of available moisture from 0 to 100, the distribution of each species among the localities exhibits a distinct optimum (Figure 10–4). The coast redwood dominates the central portion of the moisture gradient and frequently forms pure stands. Cedar and Douglas fir, and two broad-leaved evergreen species with small, thick leaves, manzanita and madrone, are found at the drier end of the moisture gradient. Three deciduous species — alder, big-leaf maple, and black cottonwood — occupy the moister end.

The distribution of species along the moisture gradient may coincide with distribution along an available nutrient gradient. For example, the dry soils in which cedar thrives are also poor in nutrients (Figure 10–5). Because these two ecological factors are so closely related, we cannot determine whether water, nutrients, or a third factor related to both, exerts the greatest influence on the distribution of cedar. In fact, cedars are virtually restricted to serpentine barrens in northern California. There may also be instances of distribution with no apparent relationship between nutrients and moisture. Soil nutrients vary widely over the range of soil moisture conditions that favor the redwood, which occurs in all but the most impoverished soils — the serpentine barrens.

The ecological distributions of some species apparently affect others. Madrone and Douglas fir overlap each other broadly along the moisture gradient, but occur apart on the nutrient gradient. Further-

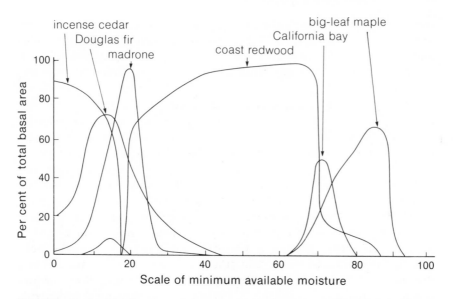

FIGURE 10–4   Distribution of tree species along a gradient of minimum available soil moisture in the northern coast region of California.

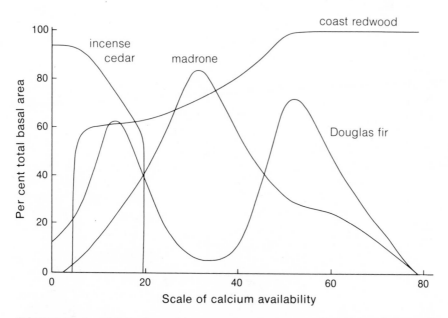

FIGURE 10–5   Distribution of tree species along a gradient of available soil calcium in the northern coast region of California.

more, madrone occupies the central part of the soil nutrient gradient; with Douglas fir forming two peaks of abundance, one on poorer soils and the other on richer soils. The separate peaks of abundance may represent distinct sub-populations of Douglas fir, each with different tolerance ranges.

Environmental variables do not change independently of each other. Increasing soil moisture usually alters the status of available nutrients in the soil. Variation in the amount and source of organic matter in the soil creates parallel gradients of acidity, soil moisture, available nitrogen, and so on. Because these variables interact, one should examine the distribution of plants with respect to all the variables at the same time. We cannot visualize more than three axes on a graph owing to the three-dimensionality of our concepts, and so analyses of plant distributions with regard to more than three dimensions must be left to the computer. We can, however, visualize the interaction between two environmental factors on a simple graph, which will suffice to make the point. In Figure 10–6, distributions of some forest-floor shrubs, seedlings, and herbs in the woodlands of eastern Indiana are related to levels of organic matter and calcium in the soil. These soils contain

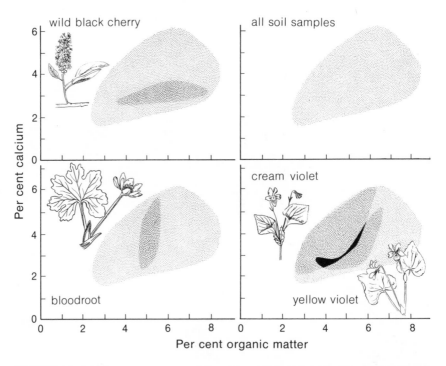

FIGURE 10–6   The occurrence of four forest-floor plants with respect to the calcium and organic matter contents of the soil in woodlands of eastern Indiana.

between two and eight per cent organic matter and between two and six per cent calcium. Furthermore, levels of calcium and organic matter are interrelated: soils that are rich in organic humus tend to be rich in inorganic nutrients. Within the range of soil conditions found in these woods, each species shows different preferences: black cherry is found only within a narrow range of calcium; bloodroot is narrowly restricted by the per cent of organic matter in the soil (Figure 10-6). The distributions of yellow violet and cream violet extend more broadly over levels of organic matter and calcium in the soil, but they do not overlap. Cream violet prefers relatively higher calcium and lower organic matter content than yellow violet. As in the case of madrone and Douglas fir in northern California, the distribution of each violet species is related not only to chemical properties of the environment, but is dependent upon the presence of another species of violet as well.

The Biological Optimum

For each organism there exists some combination of environmental conditions that is optimum for its growth, maintenance, and reproduction. To either side of the optimum, biological activity falls off until the organism ceases altogether to be supported by the environment. We see this pattern over and over again, whether we examine the dependence of photosynthetic rate of leaf temperature, the distribution of plants along moisture and nutrient gradients, the vertical range of seaweeds and marine snails within the intertidal zone, or the size of prey captured by predatory animals. The perpetuation of a population is confined along an environmental gradient more narrowly than the survival of an individual. The successful population requires resources for growth and reproduction in addition to individual maintenance.

Juvenile brook trout survive indefinitely in water containing more than 2 milligrams of oxygen per liter (mg O_2/l). Below about 1.6 mg O_2/l all trout die; survival time decreases with decreasing oxygen concentration from 200 minutes at 1.4 mg O_2/l to 20 minutes at 0.8 mg O_2/l and below. Trout occur at the upper end of the oxygen gradient; their optimum corresponds to the maximum oxygen concentration obtainable. Where trout suffocate, catfish and carp thrive, being adapted to function efficiently in oxygen-poor, stagnant water.

Oysters spend their larval stages in the brackish waters of small bays and estuaries. The oyster larvae grow most rapidly when the concentration of salt in the water remains between 1.5 and 1.8 per cent, about halfway between fresh and salt water. Higher salinity depresses growth slightly. Lower salinity slows growth markedly and causes death: at 1.0 per cent salinity, 90 to 95 per cent of the larvae die within two weeks; at 0.25 per cent salinity, growth ceases and all larvae die within one week.

We may visualize the general relationship of an organism to its environment on a graph on which we relate rate of biological activity (by whatever measure we chose) to a gradient of environmental conditions. This relationship is portrayed in Figure 10-7 as a bell-shaped curve showing a distinct optimum over the middle of the gradient. In nature, these curves can be quite asymmetrical. Oyster development, for example, proceeds best in moderate salinities but is depressed more by low salinity than high salinity (at least up the maximum for sea water). Optimum oxygen concentration for brook trout occurs at one end of the oxygen gradient.

Regardless of the shape of the activity curve, the ecological distribution of a species along the gradient is governed by three levels of tolerance. First, extreme conditions may totally disrupt critical biological functions resulting in rapid death. Such lethal conditions occur beyond points c and c' in Figure 10-7. Along a temperature gradient, for example, temperatures below freezing and above 50°C are lethal for many organisms.

Second, organisms must sustain a certain level of activity to maintain themselves in a steady state for long periods. Within points b and b' on the environmental gradient, the organism can exist indefinitely. Outside these limits, the organism's activity level is too low to be self-maintaining and the organism can venture beyond these limits only briefly.

Third, populations can maintain their size only if reproduction balances death. Reproduction requires resources, hence biological activity, over and above the level needed for self-maintenance. Populations,

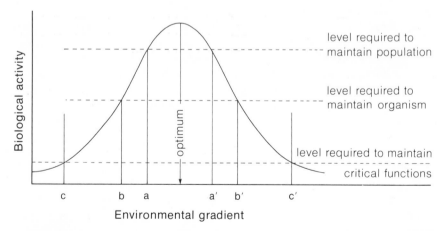

FIGURE 10-7   The general relationship of biological activity to a gradient of environmental conditions. Levels of activity required to maintain critical biological function, the organism, and the population, determine the lethal extremes (*c* and *c'*), the limits of persistence for organisms (*b* and *b'*) and populations (*a* and *a'*).

therefore, persist within a narrower range of conditions than the individual can tolerate. Individuals may live in environments that are inadequate for population maintenance, but their numbers can be maintained only by immigration from populations in more suitable habitats where reproduction exceeds death.

The Suitability of the Environment

Because organisms persist only under characteristic combinations of favorable environmental conditions, one should be able to determine the suitability of a particular place for any species. For example, optimum climates for the Mediterranean fruit fly lie within 16 to 32°C and 65 to 75 per cent relative humidity. Populations thrive under these conditions provided, of course, that food is available. Outside this optimum, fruit flies can maintain populations under conditions between about 10 and 35°C, and 60 to 90 per cent relative humidity; individuals can persist at temperatures as low as 2°C and relative humidity as low as 40 per cent; more extreme conditions are usually lethal (Figure 10–8). The Mediterranean fruit fly is a major agricultural pest, but populations reach outbreak proportions only where conditions are within the biolog-

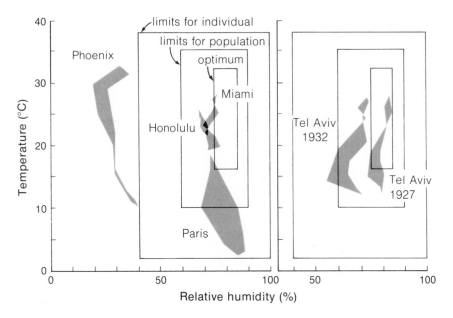

FIGURE 10–8   The seasonal course of air temperature and relative humidity at selected locations in relation to conditions favorable for the Mediterranean fruit fly. The inner rectangle encloses conditions optimum for growth, the middle rectangle encloses conditions suitable for development, and the outer rectangle delimits the extreme tolerance range.

ical optimum most of the year and rarely exceed the limits of tolerance. Thus Tel Aviv, Israel, with annual temperature range of 7 to 31°C and a humidity range of 59 to 73 per cent, is frequently plagued by the fly and requires extensive control measures, although conditions favor outbreaks more in some years than others. The climate of Paris, France is generally too cold for fly populations to reach damaging levels and Phoenix, Arizona is too dry. The climates of Honolulu, Hawaii, and Miami, Florida, where the pest has been accidently introduced, are quite suitable for rapid population growth.

Seasonal changes in climate usually exceed the optimum range of conditions of most species. Consequently, population growth is favored during restricted periods and populations exhibit seasonal cycles of abundance and activity. Temperature governs the seasons in most temperate and arctic regions, restricting the growing season to the warm summer months.

Temperature limits the appearance of adult underwing moths to the period between early July and early October in Connecticut and Massachusetts, but the seasonal activity patterns of different species show that availability of food plants also influences seasonality (Table 10–1). Species feeding on walnut appear later in the summer than species feeding on plants in the oak, rose, and willow families.

In the tropics, rainfall sets the seasonal pace of activity. Arboreal mosquitos are abundant in Panama only during the rainy months of May through December. Most mosquito populations persist through the

TABLE 10–1   Seasonal occurrence and food plant of underwing moths (genus *Catocala*) in Connecticut and southern Massachusetts.

Species	Food plant	July	August	September	October
grynea	rose				50%
ultronia	rose				
antinympha	bayberry				
amica	oak				
concubens	willow				
paleogama	walnut				
residua	walnut				
neogama	walnut				
retecta	walnut				
habilis	walnut				

dry season as desiccation-resistant eggs. Species of arboreal mosquitos whose eggs cannot withstand drying must find permanent sources of water in which to lay their eggs; adults do not live long enough to span the entire dry season.

Seasonal changes in the food value of plants often determine shifts in diet and the occurrence of reproductive cycles in herbivores. In the chaparral regions of California, winter rainfall and warming temperatures in the early spring stimulate plant growth, greatly increasing the abundance and protein content of food plants for deer (Figure 10-9). The dry summer months cause range quality to decline. Deer require 13 per cent protein in their diet for optimum growth and reproduction; 7 per cent protein barely provides a maintenance diet. Chamise is a common chaparral shrub used extensively by deer, but it does not provide a maintenance diet during most of the year and cannot adequately support a deer population without supplemental foods. Mature chaparral vegetation, including all species of plants used by deer, provides an adequate maintenance level of protein during most of the year and supports moderate reproduction and growth in early spring. Wildlife biologists have found, however, that the chaparral habitat cannot support large populations of deer unless old vegetation is cut or burned, and young protein-rich shrubs and herbs allowed to grow up in their place. The Indians of the eastern forests understood this principle and regularly burned the undergrowth of forests to encourage the growth of food plants for deer. Conservationists continue this practice in many areas to improve range land.

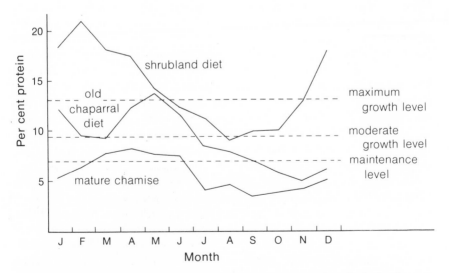

FIGURE 10–9 The protein content of plants used by deer in Lake County, California.

Adaptation and Environment

The adaptations of an organism — its form, physiology, and behavior patterns — cannot easily be separated from the environment in which it lives. Organism and environment go hand in hand. Insect larvae from stagnant aquatic environments in ditches and sloughs can survive longer without oxygen than related species from well-aerated streams and rivers; species of marine snails that occur high in the intertidal zone, where they are frequently exposed to air, can withstand a greater degree of desiccation than species from lower levels.

In general, the forms of organisms are closely linked to their ways of life and to the environmental conditions in which they live. Compare the leaves of deciduous forest trees with those of desert species. The former are typically broad and thin, providing a large surface area for light absorption and for water loss. Desert trees have small, finely divided leaves — or sometimes none at all (Figure 10–10). Leaves heat up in the desert sun. Structures lose heat by convection most rapidly at their edges; the more edges, the cooler the leaf and the lower the water loss. Small size means a great portion of each leaf is given over to its edge. One may find many exceptions to the relationship between leaf size and moisture availability, but the pattern holds for most species. In the mountains that rise from the desert floor of Arizona, several species of oaks replace each other along a gradient of increasing moisture with increasing elevation (Figure 10–11). At lower elevations, the small leaves of long-leaved oak and Emory oak are adapted to resist desiccation. Leaf size increases with elevation to a maximum in the Gambel oak, whose leaves are as large as those of oaks in the deciduous forests of the eastern United States. Like its eastern counterparts, the Gambel oak leaf is designed to take advantage of abundant moisture during the summer months. The leaves do not resist desiccation and they are shed during the winter when soil water freezes and becomes unavailable to plants.

The water relations of coastal sage and chaparral plants in southern California demonstrate divergent courses of adaptation. Chaparral habitats occur at higher elevation than the coastal sage habitats and thus are cooler and moister. Both vegetation types are exposed to prolonged summer drought, but water deficiency is greater in the sage habitat. Plants of the coastal sage habitat are typically shallow-rooted with small, delicate deciduous leaves (Table 10–2). Chaparral species have deep roots, often extending through tiny cracks and fissures far into the bedrock. Their leaves are typically thick and their waxy outer covering (cuticle) reduces water loss. Leaf morphology influences photosynthetic rate in conjunction with its influence on transpiration. The thin leaves of coastal sage species lose water rapidly, but also carry on photosynthesis rapidly when water is available to replace transpiration losses. When leaves are clipped from plants and placed in a chamber where transpira-

FIGURE 10–10 Leaves of some desert plants from Arizona. Mesquite leaves (top) are subdivided into numerous small leaflets, which facilitates the dissipation of heat when the leaves are exposed to sunlight. The paloverde carries this adaptation even farther (center); its leaves are tiny and the thick stems, which contain chlorophyll, are responsible for much of the plant's photosynthesis (hence the name paloverde, which is Spanish for green stick). Cacti rely entirely on their stems for photosynthesis; their leaves are modified into thorns for protection. Unlike most desert plants, limberbush (bottom) has broad succulent leaves, but limberbush plants leaf out only for a few weeks during the summer rainy season in the Sonoran Desert. Photographs are about one-half size.

tion and photosynthesis can be measured, both functions decline as the leaves dry out and their stomata close to prevent further water loss. Coastal sage species, such as the black sage, have high photosynthetic and transpiration rates at the beginning of the transpiration experiment, but shut down quickly due to rapid water loss (Figure 10-12). Chaparral species such as the toyon (rose family) have maximum photosynthetic rates which are only one-fourth to one-third those of coastal species, but they resist desiccation and continue to be active under drying conditions for longer periods.

When chaparral and coastal sage species grow together near the adjoining edges of each other's range, they exploit different parts of the

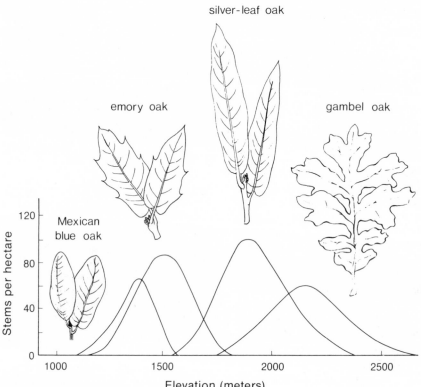

FIGURE 10–11   Altitudinal distribution of four species of oak in the Santa Catalina Mountains of southern Arizona. The characteristic leaf size of each species increases with elevation in response to increased availability of moisture.

environment: deep perennial sources of water and shallow, ephemeral sources of water. In spite of these differences and the corresponding adaptations of leaf morphology and drought response, the annual productivity of both types of species is about the same where there are intermediate levels of water availability. In drier habitats, the prolonged seasonal absence of deep water tips the balance in favor of the adaptations found in coastal sage. Increasing availability of deep water at higher elevations favors chaparral vegetation.

Ecotypes

Although in general organisms are best adapted to function within a narrow range of environment conditions, we often find sub-populations of a given organism with different ecological tolerances corresponding to different ecological conditions within the geographical range of the species. Adaptations that serve the species well in one part of its distribution may not be equally well-suited elsewhere. In particular, we should recall such species as the Douglas fir (Figure 10-5), which is found in two distinct regions along a gradient of soil nutrient abundance. If individuals in the two sub-populations of fir have unique physiological adaptations, they have not yet been discovered. Botanists have, however, long recognized that many species have distinctive forms corresponding to the habitat in which they are found. In the hawkweed, *Hieracium umbellatum*, for example, woodland plants generally have an erect habit, those from sandy fields are prostrate, and those from sand dunes are intermediate in form. Leaves of woodland plants are broadest, those of dune plants are narrowest, and those of sandy fields are intermediate. Plants from sandy fields are covered with fine hairs, a trait the others lack.

TABLE 10–2   Characteristics of chaparral (left) and coastal sage (right) type vegetation in southern California.

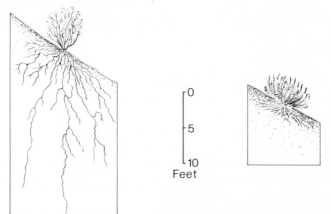

| Characteristics | Vegetation Type | |
	Chaparral	Coastal sage
Roots	Deep*	Shallow**
Leaves	Evergreen	Summer-deciduous
Average leaf duration (months)	12	6
Average leaf size (cm²)	12.6	4.5
Leaf weight (g dry wt/dm²)***	1.8	1.0
Maximum transpiration (g H_2O/dm²/hr)	0.34	0.94
Maximum photosynthetic rate (mg C/dm²/hr)	3.9	8.3
Relative annual CO_2 fixation	49.8	46.8

* Species figured is chamise.
** Species figured is black sage.
*** dm = decimeter; 1 dm² = 100 cm².

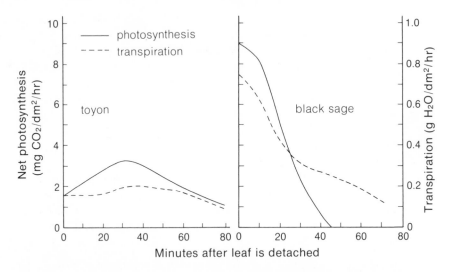

FIGURE 10–12   Time-course curves for photosynthesis and transpiration under standard drying conditions for a chaparral species (left) and a coastal sage species (right). Note that transpiration continues well after photosynthesis has been shut off; hence leaf dormancy is an ineffective long-term solution to drought.

About fifty years ago the Swedish botanist Göte Turesson collected seeds of hawkweed plants from a variety of habitats and cultivated them in his garden. He found that, even when grown under identical conditions, differences in form of plants obtained from different habitats persisted generation after generation. Turesson called these differing forms *ecotypes*, a name that persists. He further suggested that ecotypes represent genetic strains of a population with unique adaptations to the special conditions under which they are found.

Jens Clausen, and his co-workers David Keck and William Hiesey, examined ecotypic differentiation in the yarrow, *Achillea millefolium*, in California. *Achillea*, a member of the sunflower family, grows in a wide variety of habitats ranging from sea level to more than 10,000 feet elevation. Clausen collected seed from plants at various points along the altitude gradient and planted them at Stanford, California, near sea-level. Although the plants were grown under identical conditions for several generations, individuals from montane populations retained their distinctively small size and low seed production (Figure 10-13), thereby demonstrating ecotypic differentiation within the population. Such regional and habitat differences in adaptations undoubtedly broaden the ecological tolerance ranges of many species.

In this chapter we have seen, first, that organisms are restricted within narrow ranges of suitable environmental conditions, second, that

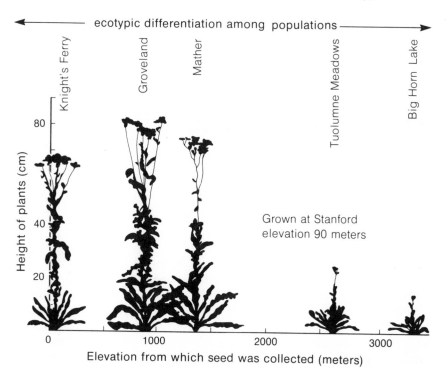

FIGURE 10–13 Ecotypic differentiation in populations of the yarrow, *Achillea millefolium*, demonstrated by raising plants derived from different elevations under identical conditions in the same garden.

the adaptations of plants and animals closely complement the conditions in which they are found, and third, that sub-populations sometimes exhibit distinctive adaptations to local environmental conditions: in other words, the close parallel between organism and environment. Adaptations are not, however, fixed. Plants and animals respond to changes in their environments by altering their form, physiological function, and behavior. In the next chapter, we shall see how such changes keep organisms within optimum ranges of environmental conditions or improve their relationship to a changed environment.

11

Homeostatic Responses
of Organisms

Change pervades an organism's surroundings — the annual cycle of the seasons, daily periods of light and dark, and frequent unpredictable turns of climate. The survival of the organism depends on its ability to cope with change in the environment. We have seen in the last chapter that peak biological activity occurs over a narrow range of conditions. By adjusting their structure and function in response to change, organisms can maintain peak activity over a greater, or a different, range of conditions.

Organisms are able not only to change their adaptations over a long term, but also to respond to environmental changes over a short term. Chameleons, for example, are well-known for their ability to change skin color in a matter of seconds to match the color of their background. As we look around us, we are constantly aware of organisms, including ourselves, responding to changes in their environments. Responses include changes in the use of the physiological apparatus and changes in the apparatus itself. When we step from a warm room into the outdoors on a cold day we shiver to generate heat. A few weeks on the beach and our skin darkens to block damaging radiation from the sun. When we shiver, we make use of a muscle response that is always present, and available on short notice. Morphological changes — thickening of the fur at the approaching winter, proliferation of red blood cells at high altitudes, production of pigment granules in the skin in response to sunlight — involve altering the physiological apparatus itself. Such changes require more time.

Responses involving the structure and function of an organism are usually reversible, as they must be to follow the ups and downs of the environment. Plants and animals also exhibit more or less permanent and irreversible developmental responses to the particular environments in which they grow up.

Response and Negative Feedback

If we walk from a dark room into bright sunlight, the pupils of our eyes rapidly contract, which restricts the amount of light entering the eye. A sudden exposure to heat brings on sweating, which increases evaporative heat loss from the skin and helps to maintain body temperature at its normal level. Behavior, too, is a mechanism of response because it serves either to modify the environment or to change the individual's relationship to its environment. Putting on an overcoat serves a regulatory purpose, the maintenance of body temperature. Social behavior may be viewed in the same context. A man confronted by a threatening person may try to appease his antagonist or to escape. If handing over a wallet satisfies the mugger, he, and the threat he poses, will go away. If the victim can run fast enough, he can escape the threat — a behavior not unlike a bird's southward flight to escape the winter cold.

Responses act to maintain the internal integrity of the organism, and sustain its function at optimum levels. To achieve this goal, most responses are controlled in the manner of a thermostat. If a room becomes too hot, a temperature-sensitive switch turns off the heater; if the temperature drops too low, the temperature-sensitive switch turns on the heater. This pattern of response is called *negative feedback*, meaning that if external influences alter a system from its norm, or desired state, internal response mechanisms act to restore that state.

A man driving a car down a straight road embodies a negative feedback system. If a sudden gust of wind forces the car to veer to the right, the driver immediately responds by turning the steering wheel to the left, and the car returns to its normal path. Experience with driving gives us an intuitive appreciation of three basic requirements of negative feedback systems. First, the response must be of proper magnitude. If the steering wheel is turned too far to the left the car will lurch to the other side of the road; too little response causes the car to spill off the road on the right. Second, the response must occur at the proper time. If a timelag separates the gust of wind and the driver's response, the car may go off the road on the right before the driver's correction to the left can do him any good. Third, the response must be made at the proper rate. If the driver responds immediately to the gust of wind, but turns the car too quickly, he will lose control again.

Rate of Response

To wait for an evolutionary change in the size of the eye's pupil, in response to a change in light intensity, makes no sense whatsoever. By the same token, organisms need not maintain the physiological machin-

ery for rapid response if the environment changes slowly and predicta-
bly. For organisms in the temperate northern hemisphere, freezing
temperatures are not a problem in July but they are a predictable part of
the environment in January. The rate of change in the environment
largely determines the appropriate response. Behavioral responses are
most rapid. One can seek shade more quickly than the skin can darken
in response to bright sunlight. Shivering and sweating are appropriate
responses to sudden changes in temperature; growth and shedding of
fur adjust the heat balance of mammals more slowly to seasonal trends
in temperature.

A persistent environmental change can lead to a series of responses.
For example, when we move to a high altitude, we immediately respond
to the thin air by increasing our rate of breathing and heart beat and
reducing activity. Within a week or so, new red blood cells are produced
which increase the oxygen carrying capacity of the blood and enable us
to return to near normal levels of activity.

Developmental responses are suitable when individuals in a population settle in different habitats or when successive generations or
progeny are exposed to different environmental conditions. For plants
whose seeds germinate under varied soil conditions, developmental
flexibility makes good sense. With many generations of offspring each
year, flies might be expected to exhibit developmental responses to
seasonal change.

The leaf morphology of the arrowleaf plant varies according to
whether individuals grow on land, partly submerged in water, or completely underwater (Figure 11-1). Underwater leaves are long and flexible, resembling the ribbonlike shape of many seaweeds. Because they
lack a waxy waterproof cuticle, the leaves can absorb nutrients directly
from the water, as aquatic plants usually do. Out of water these leaves
collapse because they do not contain the rigid cell walls needed for
support on land. In contrast, aerial leaves are broad and rigid. They also
have a thick cuticle, which reduces water loss. Terrestrial arrowleaf
plants have enlarged root systems compared to individuals growing in
water because terrestrial plants must obtain nutrients from the soil.
Differences in growth form are caused by developmental responses of
the arrowleaf to the particular conditions in each habitat.

Insects sometimes respond to seasonal trends in the environment
by modifying their development. Species of African grasshoppers and
locusts grow up to be different colors depending on the season of the
year. In the wet season, vegetation is lush and the habitat appears
green. During the early part of the dry season, dying vegetation turns
brown and the red-brown earth often shows through. At the peak of the
seasonal drought, fires frequently blacken vast expanses. During the
course of the year, the grasshopper's habitat changes from green to

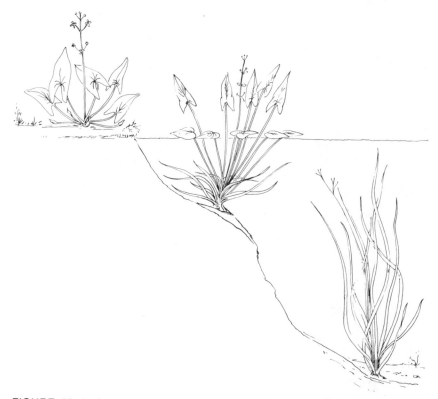

FIGURE 11–1 Variation in the morphology of the arrowleaf plant when it grows on land (left), partially submerged in water (center), and fully submerged (right).

brown to black, and back again to green. The coloration of grasshoppers usually matches that of their background. Laboratory experiments have shown that the color response of the grasshopper is controlled by amount and color of light, and by the humidity of the environment. As the grasshopper grows, it deposits pigments in its skin. High light intensity (resulting from sparse vegetation during the dry season) stimulates the production of brown or black pigments, depending on the color reflected from the ground. Low light intensity, combined with high humidity (wet season conditions) stimulates the production of green pigments. Once pigments are deposited in the developing skin, they cannot be removed. As grasshoppers grow, however, they periodically molt their skin, thereby permitting them to repeat the color response several times during their development. Owing to their rapid growth, they may experience only one habitat condition during their lives, and their color will usually match the color of that habitat.

Regulators and Conformers

Response to the environment can follow two divergent courses. Some organisms change their physiology and structure so that optimum conditions for activity more closely correspond to the external environment. Others maintain their internal environments at optimum conditions. We closely regulate our body temperatures around 37°C, though the surrounding air may vary from −50 to +50°C, by putting on or taking off clothing, modifying activity levels, selecting microhabitats with favorable thermal characteristics, and by replacing, or dissipating, heat lost to, or gained from, the environment. Maintaining a constant internal temperature allows biochemical reactions to proceed without inhibition over a wide range of external temperatures. Over the same environmental temperature range, the body temperatures of frogs and grasshoppers conform closely to the environment. Frogs cannot possibly function in extremes of temperature, so their activity is restricted to a narrow part of the temperature range over which warm-blooded animals remain active.

Animals that maintain constant internal environments are usually referred to as *regulators*; those which allow their internal environments to follow external changes are called *conformers* (Figure 11-2). Few organisms are ideal conformers or regulators. Frogs regulate the salt concentration of their blood but conform to external temperature. Even warm-blooded animals are partial temperature conformers: in cold weather, our hands, feet, nose, and ears (in other words, our exposed extremities) become noticeably cool.

Organisms sometimes regulate their internal environments over moderate ranges of external conditions, but conform under extremes. Small aquatic amphipods of the genus *Gammarus* regulate the salt contents of their body fluids when they are placed in water with less concentrated salt than their blood, but not when they are placed in more

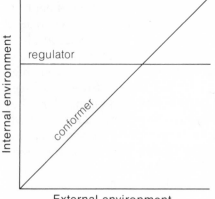

FIGURE 11–2 Relationship between internal and external environments in idealized regulating and conforming organisms. Regulators maintain constant internal environments with homeostatic mechanisms whereas conformers allow their internal environments to follow changes in the external environment. The difference between the curves represents the gradient which regulators maintain between internal and external environments.

concentrated salt water (Figure 11-3). The fresh-water species *G. fas-ciatus* regulates the salt concentration of its blood at a lower level than the salt-water species *G. oceanicus*, and thus begins to conform to concentrated salt solutions at a lower level. In their natural habitat, however, neither the fresh-water species nor the salt-water species encounters salt more concentrated than that in their blood.

Whereas most fully aquatic invertebrates cannot regulate the salt concentration of their blood below that of the surrounding water, land crabs and many intertidal invertebrates, which are periodically exposed to air, possess this capability. This seems puzzling until one recalls that land animals must sometimes tolerate the loss of body water. If blood volume decreases, salt concentration would increase dangerously unless the organism could excrete salt. In addition to the excretory function of salt-secreting organs, impermeable outer skins or shells reduce the movement of salt into the body while restricting water loss. Thus biological solutions to problems of water and salt balance imposed by terrestrial environments enable organisms to regulate the salt concentrations of their body fluids to levels below the levels in their environment.

The Cost of Regulation

The benefit derived from increased activity as a result of regulation must be weighed against the cost of maintaining a difference between the internal and external environments. In some cases, the balance favors regulation, in others it weighs against regulation. Giving up a wallet to a robber is, in a very real manner, the cost of maintaining one's well-being. All homeostatic responses similarly impose a price that the organism must pay by the expenditure of energy. The cost of regulation is

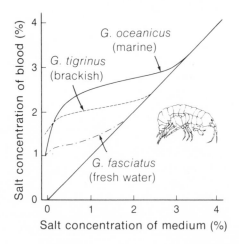

FIGURE 11-3 Salt concentrations in the blood of three gammarid crustaceans from different habitats as a function of the salt concentration of the external environment. The normal salt concentration of sea water is 3.5 per cent.

proportional to the surface area of the organism, across which exchange occurs with the external environment, and, as we shall see in the next paragraph, to the magnitude of the gradient maintained between the body and the environment. Large organisms, because of their relatively low surface to volume ratio, usually regulate their internal environments more closely than small organisms.

The metabolic cost of homeostasis has been measured for temperature regulation by warm-blooded animals (*homeotherms*). Birds and mammals maintain constant body temperatures — generally between 35 and 45°C depending on the species, and well above the temperatures normally encountered in the environment. The gradient between body and air temperature increases as air gets colder, and the rate of heat loss from the skin increases in proportion. An animal that maintains its body temperature at 40°C loses heat twice as fast at an ambient temperature of 20°C (a gradient of 20°C) as it does at an ambient temperature of 30°C (a gradient of 10°C). To maintain a constant body temperature, warm-blooded organisms must replace heat lost across their surface by metabolically generated heat. The rate of metabolism required to maintain body temperature thus increases in direct proportion to the gradient between body and ambient temperature (Figure 11–4).

If metabolism served only to regulate body temperature, metabolic rate would be zero when ambient temperature equals body temperature and heat is neither lost to, nor gained from, the environment. But the organism must maintain other vital functions — heart beat, breathing,

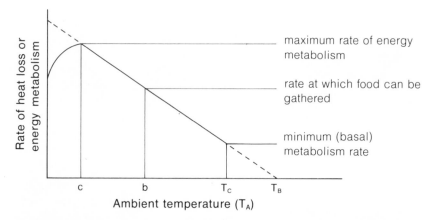

FIGURE 11–4   Relationship between energy metabolism and ambient temperature for a homeothermic bird or mammal whose body temperature is maintained at T_B. T_C is the lower critical temperature, below which metabolism must increase to maintain body temperature. Points c and b, the lower lethal temperature and the lowest temperature at which the organism can maintain itself indefinitely, correspond to points c and b, in Figure 10–7.

muscle tone, kidney function, and so on — at some minimum level that determines the *basal metabolic rate*. Basal metabolism produces enough heat to maintain body temperature when the temperature of the external environment (*ambient temperature*) is above a level known as the *lower critical temperature* (T_c in Figure 11-4); metabolic rate increases at lower temperatures.

Organisms can regulate their body temperature in extreme cold up to limits imposed over short periods by the physiological capacity to generate heat and over long periods by the ecological capacity to gather food needed to sustain a high metabolic rate. Cold temperatures are more likely to cause death by starvation than by freezing. If the temperature of the environment remains between points *b* and *c* in Figure 11-4 for too long, the excess of energy expenditure for temperature regulation over food consumption eventually depletes the organism's energy reserves. If the environment becomes so cold that heat loss exceeds the ability to generate heat (below point *c* in Figure 11-4), the organism can no longer maintain its body temperature, a condition that is fatal to most species.

Arctic birds and mammals avoid death at low temperature by employing energy economy measures such as reducing body-air temperature gradients over all or part of their bodies. The legs and feet of many birds are not insulated by feathers; unfeathered appendages would be major avenues of heat loss in the cold if their temperature was not maintained below that of the feathered body. Birds reduce the temperature of their feet by an ingenious mechanism known as *countercurrent heat exchange*. The arteries carrying warm blood from the body to the feet are cooled by passing close to the veins that carry blood from the feet to the body (Figure 11-5). In this way, heat is transferred to venous blood and transported back to the body rather than lost to the environment across the surface of exposed appendages.

Many marine organisms apply the countercurrent exchange mechanism to blood flowing from the body core to the surface and back. The high thermal conductance of water draws heat from the body faster than aquatic organisms (other than air-breathing mammals and birds) can produce metabolic heat. Some large fish, such as the tuna, maintain temperatures up to 40°C in the center of their muscle masses while they are swimming. Countercurrent heat exchange between vessels carrying blood from the muscles to the skin and back keeps the outer layers of the body from warming up and thereby reduces heat loss.

Because hummingbirds are so small, and their surfaces so large compared to the weight of their bodies, they are faced with tremendous heat conservation problems. In the cold environments of temperate regions and tropical mountains, hummingbirds cannot store enough energy in the form of fat during the day to keep their bodies warm all night. To conserve energy, many hummingbirds become *torpid*. That

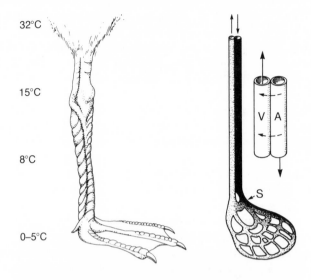

FIGURE 11–5 Skin temperatures of the leg and foot of a gull standing on ice. Countercurrent heat exchange between arterial blood (*A*) and venous blood (*V*) is diagrammed at right. Arrows indicate direction of blood flow and heat transfer (dashed arrows). A shunt at point *S* allows the gull to constrict the blood vessels in its feet, thereby reducing blood flow and heat loss further, without having to increase its blood pressure.

is, they reduce their body temperatures, from 40 to perhaps 20°C, and thereby reduce the body–air temperature gradient across which heat loss occurs. Hummingbirds do not cease to regulate body temperature when they are torpid, they merely change the setting on their thermostats. Mammals that undertake seasonal periods of hibernation reduce their body temperatures to just above the freezing point to reduce heat loss, but still must store large quantities of fat to survive through winter.

Metabolism is not the only source of heat for temperature regulation. Poorly insulated reptiles and small insects cannot afford the metabolic costs of sustained homeothermy, but many so-called cold-blooded (*poikilothermic*) animals make use of "cheap" sources of heat — the Sun and warm rocks — to raise their body temperatures. During the day, lizards can regulate their temperatures precisely by adjusting their orientation to the Sun and their contact with the surface of the ground. Of course, their temperatures drop to near the ambient temperature at night or on cloudy days.

The constancy of body temperature in warm-blooded animals does not mean that they cannot take advantage of microhabitats with favorable temperature conditions. During periods of inactivity, birds and mammals frequently reduce their convective heat loss by resting in places where they are protected from the wind. During the day, desert homeotherms seek refuge from extreme heat in underground burrows or in the shade of vegetation (Figure 11–6).

Temperature Regulation in Hot Desert Climates

Heat stress is one of the most critical factors to an organism's survival. Warm-blooded animals maintain their body temperatures near the upper lethal maximum. In cool environments, animals can quickly dissipate excess heat, generated by activity or absorbed from the Sun, to their surroundings. When, however, air temperature exceeds body temperature, evaporative water loss becomes the primary route of heat dissipation, but desert animals cannot afford to use scarce water to regulate their temperatures. Inactivity, use of cool microclimates, and seasonal migrations provide escape from heat stress for desert animals, but also limit the animal's ability to exploit the desert environment.

FIGURE 11–6 A jackrabbit seeking refuge from the hot sun of southern Arizona in the shade of a mesquite tree. The large ears and long legs of desert jackrabbits effectively radiate heat from the body when the temperature of the surroundings is lower than body temperature.

Temperature regulation and water balance are closely linked. Where fresh water is scarce, organisms have a wide variety of behavioral, morphological, and physiological adaptations for conserving water and using it efficiently to dissipate heat. The daily activity patterns of desert animals are closely tied to problems of temperature regulation. Many small animals, such as kangaroo rats, which eat only dry seeds, appear above ground only at night when the desert is cool. Ground squirrels, in sharp contrast, remain active during the day, but they conserve water by allowing their body temperatures to rise when they are above ground. Before their body temperatures become dangerously high, ground squirrels return to their cool burrows where they dissipate their heat load by convection rather than evaporative heat loss. By alternately appearing above ground and retreating to their burrows, ground squirrels extend their activity into the heat of the day and pay a relatively small price in water loss.

Among vertebrates, birds are perhaps the most successful inhabitants of the desert. They remain active in the heat long after other animals have sought refuge. The success of birds derives from their low excretory water loss and from feeding on insects, from which they obtain some free water. Even some seed-eating birds can persist without water in the desert provided they avoid both full sun and shade temperatures above 35°C. The behavior of the cactus wren, a desert insectivore (Figure 11-7), shows that it too must respect the physiological demands of the hot desert climate. In cool air, wrens lose two to three milliliters (ml) of water each day in the air they exhale. Water loss increases rapidly above 30 to 35°C, to over 20 ml per day at 45°C; active birds might use five times that much water to dissipate their heat load. (The wren's body contains about 25 ml of water.) In the cool temperatures of the early morning, wrens forage throughout most of the environment, actively searching for food in small trees and shrubs, and on the ground. As the day brings warmer temperatures, wrens select cooler parts of their habitat, always managing to avoid feeding where the temperature of the microhabitat exceeds 35°C. When the minimum temperature in the environment rises above 35°C, the wrens become less active. They even feed their young less frequently during hot periods.

Birds lack sweat glands. Dissipation of heat by evaporative cooling occurs primarily from the mouth and respiratory surfaces. At high temperatures, the rapid ventilation needed for heat dissipation draws carbon dioxide from the blood stream passing through the lungs rapidly enough to upset the delicate chemistry of the blood. Panting also requires energy expenditure and thus itself increases the heat load on the organism. Some birds avoid these problems by gular flutter — passing air rapidly in and out of the mouth and throat, without increasing the ventilation of the lungs. Gular flutter localizes evaporative cooling to the oral surfaces. Muscles expand and contract the mouth and throat cavity

FIGURE 11–7 The cactus wren, a conspicuous resident of deserts in the southwestern United States and northern Mexico.

in a regular rhythm like a vibrating rubber band. Because the frequency of gular flutter is adapted to coincide with the natural resonating frequency of muscles lining the mouth cavity, gular flutter requires little energy. It occurs, however, only in such birds as doves and nighthawks, which have large distensible throats for ingesting or storing large quantities of food.

Many desert birds build enclosed nests or place their nests in holes in the stems of large cacti, where the young are protected from the Sun and from extremes of temperature. The cactus wren builds an untidy nest, resembling a bulky ball of grass, with a side entrance. Once a pair of wrens have built their nest, they cannot change its position or orientation. For a month and a half, from the beginning of egg-laying until the young fledge, the nest must provide a suitable environment day and night, in hot and cool weather. Cactus wrens usually nest several times during the period March through September. Early nests are oriented with their entrance facing away from the direction of the cold winds of early spring; during the hot summer months, nests are oriented to face prevailing afternoon breezes, which circulate air through the nest and

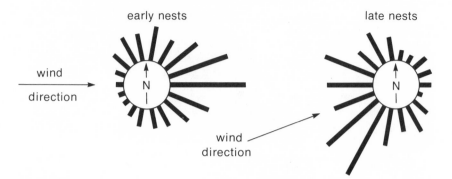

FIGURE 11-8   Orientation of the enclosed nests of the verdin, a small, insectivorous desert bird, during the early (cool) and late (hot) part of the breeding season.

facilitate heat loss (Figure 11-8). Nest orientation is an important component of nesting success in cactus wrens, particularly during the hot summer months. Nests oriented properly for the season are consistently more successful (82 per cent) than nests facing the wrong direction (45 per cent).

Acclimation

If we view the cactus wren's nest as a behavioral extension of the wren's body, we can compare the different orientation of nests built early and late in the breeding season to a major modification, or *acclimation*, of an organism's morphology or physiology in response to long-term environmental change. Many birds have a heavier plumage, with greater insulating properties, during the cold winter months than during hot summer months. These species replace their body feathers only twice each year and each plumage must be suited to the average conditions of the environment between each molt. The willow ptarmigan, a ground-feeding arctic bird, sheds its lightweight brown summer plumage in the fall for a thick white winter plumage, which provides both insulation and camouflage against a background of snow. With increased insulation, ptarmigans require less energy expenditure to maintain their body temperatures during the winter (Figure 11-9). Acclimatory responses in body insulation occur only in species that are active during cold weather. Red squirrels, which hibernate during the winter in Alaska, exhibit no winter increase in the insulation of their fur.

Seasonal change in the body integument of animals and birds is a classic *acclimatory response*: a reversible modification of structure in response to slow and persistent changes in the environment. Adjusting insulation to enhance heat conservation in winter and to facilitate heat

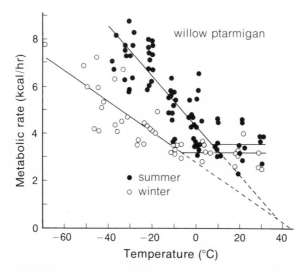

FIGURE 11–9 Metabolic responses of willow ptarmigan acclimatized to summer and winter temperatures. Winter-acclimated birds have thicker plumages providing better insulation than summer birds. Hence their metabolic rates are lower at any given temperature, and their lower critical temperature is also lower.

dissipation in summer, maintains constant body temperature at the least possible cost.

Temperature-conforming animals and plants also acclimate to seasonal changes in their environment. By switching between enzymes and other biochemical systems with different temperature optima, cold-blooded animals adjust their tolerance ranges in response to prevalent environmental conditions. Experiments with lobsters have shown that the upper lethal temperature increases as the temperature to which the lobsters are acclimated increases. Lobsters kept at 5°C die when exposed to 26°C, but tolerate 28°C when acclimated to 15°C, and tolerate 31°C when acclimated to 25°C. Acclimation does not, however, allow an organism to respond infinitely to environmental change. Regardless of their previous temperature experience, lobsters cannot be physiologically acclimated to withstand temperatures above 31°C because their physiological systems are adapted to much colder environments and the capacity of lobsters to acclimate to extreme temperature is limited.

The relationship between swimming speed and water temperature in goldfish (Figure 11–10) shows at once the capabilities and limitations of acclimation. Goldfish swim most rapidly when they are acclimated to 25°C and placed in water between 25 and 30°C, conditions that closely resemble their natural habitat. Lowering the acclimation temperature to

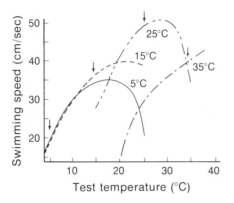

FIGURE 11–10   Swimming speed of the goldfish as a function of temperature. Acclimation temperatures varied from 5 to 35°C and are indicated on the appropriate curve by an arrow.

15°C increases the swimming speed at 15°C but reduces it at 25°C. (Increased tolerance of one extreme often brings reduced tolerance of the other.) Reducing acclimation temperature to 5°C, well beyond the normal lower range of temperatures experienced by goldfish in nature, does not increase swimming speed at 5°C. Different physiological acclimations to low temperature apparently reduce swimming speed at moderate temperature in cold-acclimated fish.

Acclimatory changes in physiology enable a basically warm-water fish, like the goldfish, to better tolerate temperatures near the upper and lower end of the normal temperature range in its environment. When brook trout, a cold-water species, are tested over a range of temperatures, they perform best when acclimated to 15°C, approximately the summer temperature of their environment. The difference between goldfish and trout represents the accumulation of evolutionary modifications to the different temperature ranges of their respective habitats.

Food Storage and Environmental Change

Although homeostatic responses help maintain function in the face of a changing physical environment, environmental changes often plunge organisms from feast into famine. When the environment becomes barely tolerable and small fluctuations in food or water supply can mean disaster, many plants and animals store food and water reserves. Desert cacti absorb water during rainy periods and store it in their succulent stems. Cacti use these reserves during the long, dry intervals between desert rains. Many temperate and arctic animals store fat during periods of mild weather in winter as a reserve of energy for periods when heavy snow covers food sources. Tropical animals sometimes store fat prior to the onset of seasonal dry periods. Instead of accumulating body fat, many winter-active mammals (beavers and squirrels) and birds (acorn woodpeckers and jays) cache food underground or under the bark of trees. During winter months, piñon jays of the western United States normally feed on insect grubs in the soil but they cannot do so when snow covers the ground. In the fall, the jays harvest the vast crops of piñon pine nuts and bury them in the ground. They always place their caches near the base of a tree on the south side of its trunk. After a snowfall, the snow first melts on the south side of tree trunks, thus exposing the stored pine seeds.

All deciduous plants store materials during the summer and early fall to provide energy and nutrients needed for flowering and the early growth of leaves in the spring. Just before trees begin to leaf out, sap rises in the trunk to the tips of the branches, carrying sugars and other nutrients. (This is the time of year when New Englanders tap the sugar maples for their sap.)

Many plants store nutritive materials in their roots to allow recovery after their shoots have been destroyed by fire or by defoliating insects. Every few years in the northeastern United States, tent caterpillars defoliate black cherry trees in the early spring. The cherry trees respond by putting out new sets of leaves, drawing on untapped reserves of nutrients in their roots.

When fires frequently sweep through habitats — as in chaparral habitats of southern California — many plants store food reserves in fire-resistant root crowns, which sprout and send up new shoots shortly after a fire has passed (Figure 11-11). Root sprouting promotes the recovery of vegetation and stabilization of the soil more rapidly than would be possible if the vegetation could grow back only from seed. The seeds of many annual plants are also fire-resistant, and so get an early foothold in burned over areas, growing up and themselves producing seed before shrubby vegetation crowds them out.

Migration and Dormancy

Under conditions of extreme drought or cold, physical conditions may become sufficiently stressful, and food sufficiently difficult to find, that plants and animals can no longer maintain normal activity. Faced with these conditions, some organisms leave the environment, seeking more favorable conditions elsewhere, while others enter a dormant state, sealing themselves off from the rigors of the environment.

Migrations are widespread in nature, particularly among flying animals: birds, bats, and some insects. Many of them perform impressive feats of long-distance navigation each year. Shore birds which breed in the Canadian arctic spend their winters as far south as Patagonia at the southern tip of South America and make yearly round trips up to 25,000 miles. The arctic tern probably holds the record for long-distance

FIGURE 11–11   Root-crown sprouting by chamise following a fire in the chaparral habitat of southern California. The upper photograph was taken on May 4, 1939, six months after the burn. The lower photograph, showing extensive regeneration, was taken on July 16, 1940.

migration with a yearly round trip of 36,000 miles between its North Atlantic breeding grounds and its Antarctic wintering grounds. A few insects, like the monarch butterfly, perform impressive migratory movements each year, but most species of insects overwinter in a dormant state as eggs or pupae.

Each fall, hundreds of species of land birds move out of temperate and arctic North America in anticipation of cold winter weather and dwindling supplies of their invertebrate food. Montane birds similarly make altitudinal migrations of several thousand feet, the mountain quail making its annual trek on foot. Because the Southern Hemisphere winter is less harsh than the Northern Hemisphere winter (see page 51), South American species do not need to escape the southern winter to take advantage of the temperate summer occurring to the north.

Mammals, other than bats, do not have the migratory abilities of birds and some insects, but some mammals do exhibit impressive seasonal movements. The barren-ground caribou of northern Canada migrates from its summer home on the tundra into the spruce forest for the winter because its foods (lichens and mosses) are covered by snow on the tundra, but remain accessible in the protected spruce forest habitat.

Some marine organisms also undertake large-scale migrations to reach spawning grounds, to follow a food supply, or to keep within suitable temperature ranges for development. The migration of salmon from the ocean to their spawning grounds at the headwaters of rivers, and the reverse migration of adult fresh-water eels to their breeding grounds in the Sargasso Sea are striking examples. Lobsters undertake less conspicuous annual migrations up to several hundred kilometers off Long Island and Massachusetts, from deep waters in summer to shallow in winter, always staying within zones of cool temperature. Predatory gastropods and some sea urchins are known to make seasonal movements into shallow waters for feeding.

Some populations exhibit irregular or sporadic movements that are tied to food scarcity. The occasional failure of cone crops in coniferous forests of Canada and the mountains of the western United States forces large numbers of birds that rely on seeds to move to lower elevations or latitudes. Birds of prey that normally feed on rodents disperse widely when their prey populations decline sharply. Snowy owls move southward from their arctic hunting grounds into the northern United States and adjacent Canada every four years on the average, corresponding to a periodic scarcity of lemmings. In desert areas, irregular rainfall forces animals such as the budgereegah (an Australian parakeet) into a nomadic existence in a continual search for areas where rain has recently fallen. Even insects are subject to irregular movements. Outbreaks of migratory locusts, from areas of high local density where food has been depleted, can reach immense proportions and cause extensive crop damage over wide areas (Figure 11–12). Whatever the historical association of locust plagues with sin and evil, for the locust, these migrations are simply a necessary search for food.

When the environment becomes so extreme that normal life processes can no longer function, or that the maintenance of normal activity would quickly lead to starvation or desiccation, plants, and animals that are incapable of migration, must enter into a physiologically dormant state. For many small invertebrates and cold-blooded vertebrates, freezing temperatures directly curtail activity and lead to dormancy. Many mammals enter a dormant state because of a lack of food, rather than because of physiological inability to cope with the physical environment.

Lack of water is the key factor in the adoption of the deciduous habit by plants. Many tropical and subtropical trees shed their leaves during seasonal periods of drought (see Figure 5–7). Temperate and arctic broad-leaved trees shed their leaves in the fall to avoid desiccation. Moisture frozen in the soil is unavailable to plants; if these trees kept their leaves through the winter, transpiration of water from the leaves would wilt them as quickly as if the tree were cut down.

FIGURE 11–12   A dense swarm of migratory locusts in Somalia, Africa, in 1962.

Many physiological changes accompany dormancy. The onset of hibernation in mammals is anticipated by the accumulation of a specific type of fat, with a low melting point, that will not harden and cause stiffness at low temperature. Blood chemistry changes to prevent clotting in capillaries at reduced rate of blood flow. Hibernating ground squirrels have heart-beat rates of 7 to 10 per minute compared to 200 to 400 for active individuals. Body temperature falls 30°C to about 6°C as the ground squirrel enters hibernation, and its metabolism drops to 1 to 5 per cent of normal.

Insects enter into a resting state known as *diapause*, in which water is chemically bound or reduced to prevent freezing and metabolism drops to near nought. In summer diapause, drought-resistant insects either allow their bodies to dry out and tolerate desiccation or secrete an impermeable outer covering to prevent drying. Plant seeds and the spores of bacteria and fungi have similar dormancy mechanisms.

Although dormant organisms can tolerate conditions far beyond the limits for active organisms, they still can sense and react to changes in their surroundings. How else would dormant plants and animals be able to break the dormant state when favorable conditions returned? Hibernating mammals respond metabolically to changes in ambient temperature to keep their body temperatures within a narrow range just above freezing. Blood temperature is continuously monitored by temperature-sensitive cells in the brain. When the blood temperature drops too low, the brain stimulates heat production in special fat tissues that form just prior to hibernation.

Homeostasis and Ecosystem Function

The homeostatic mechanisms described in this chapter represent adaptations that enable the organism to maintain its activity at the optimum level in a varying environment. The sum of such mechanisms in the populations of the biological community constitutes a major component of ecosystem homeostasis. The physiological adjustments resulting from acclimations to temperature, for example, help maintain primary production and energy flow through the community at a more uniform level than would be possible in the absence of physiological response mechanisms. Even dormancy, which signals failure to cope with the environment, allows the ecosystem to recover its normal function rapidly when the changing seasons bring suitable conditions. Life need not recolonize the arctic tundra or the deserts each year. Vast populations wait either in dormant states or in more suitable habitats elsewhere, to return the system quickly to its full level of activity.

Substances stored by plants and animals reinforce the stabilizing effects of accumulated detritus and inorganic minerals on perterbations in the system. Even fire, scorching a habitat into apparent lifelessness, leaves fire-resistant seeds and roots that are adapted to preserve themselves and, in so doing, to preserve the system as a whole.

Ecosystem homeostasis is at least partly a function of individual homeostasis. The individual does not, however, solely embody the homeostasis of the community. Populations respond to the environment by changes in number, sex ratio, and age structure — properties that can be appreciated only on the level of the population. Populations are also capable of evolutionary change — modification of the genetic composition of the population through the replacement of individuals with unfavorable traits by individuals with more favorable traits. This evolutionary process, called natural selection, results in new adaptations where changing environments demand them. Ecologists normally regard natural selection as too slow-acting to provide any degree of ecosystem response. On the contrary, all homeostatic mechanisms are in fact themselves adaptations accumulated over geologic eras in response to environmental change. Furthermore, organisms with short generations and rapid population turnover rates are quite capable of rapid adaptive response to environmental change, as the evolution of resistance to drugs and pesticides by bacteria and insects has repeatedly shown. In the next chapter, we shall investigate natural selection as a response mechanism and examine several cases of evolutionary change in response to environmental change.

12

Evolutionary Responses

Plants and animals are remarkably well-suited to their environments. Every detail of their morphology, physiology, and behavior seems capable of meeting the challenge of their surroundings. The close correspondence between organism and environment is no accident. Only those individuals in the population that are well-suited to the environment survive and produce offspring. The inherited traits which they pass on to their progeny are preserved. Unsuccessful individuals do not survive and produce offspring, and their less suitable traits are eliminated from the population. This process, called *natural selection*, allows the population to respond to its environment over long periods and slowly refines the adaptations of organisms to fit the requirements of the environment.

In this chapter we shall examine natural selection as a process and as a mechanism of response to environmental change in natural and disturbed communities.

Natural Selection

Natural selection occurs because individuals that differ genetically tend to leave different numbers of progeny in future generations. In a population, the larger number of progeny of some individuals results in an increase of the frequency of the genetic traits of those individuals. Consequently, natural selection can sustain or improve adaptations to a given environment or can change the genetic composition of the population when an environmental change favors some new trait.

Natural selection can be outlined in general terms and by specific example as follows:

a. The reproductive potential of populations is great, but	a. Rabbits should cover the Earth, but
b. populations tend to remain constant in size, because	b. they don't, because
c. populations suffer high mortality.	c. many are caught by predators.
d. Individuals vary within populations, leading to	d. Some rabbits run faster than others,
e. differential survival of individuals.	e. and escape from predators.
f. Traits of individuals are inherited by their offspring.	f. So do their young.
g. The composition of the population changes by the elimination of unfit individuals.	g. Populations of rabbits, as a whole, tend to run faster than their predecessors.

As it is described above, natural selection occurs because of three properties of organisms and their relationship to the environment: (1) genetic variability, (2) inheritance of traits, and (3) influence of the environment on survival and reproduction.

Genetic variation within populations and the inheritance of traits are facts of genetics obvious to anyone who has noticed both the variability of physique, facial appearance, eye color, and hair among humans and the tendency of related individuals to share many of these traits. The relationship between genetic traits on the one hand and survival and reproduction on the other, is less obvious. By our own technological devices we have surrounded ourselves with an environment largely of our own making, one that seems to have little influence on physique, eye color, blood type, baldness, and other genetic traits. To be sure, some inherited diseases — such as retinoblastoma, achondroplastic dwarfism, aniridia, and sickle-cell anemia — have a profound effect on the individual's survival. These are deleterious, often deadly, traits. They occur rarely and produce a major genetic disruption of body function. But we are reluctant to think *they* are the material of evolution.

The demonstration of natural selection acting to produce evolutionary changes in natural populations did not come until more than half a century after Charles Darwin originally proposed the mechanism. The first evidence that a change in the environment could select a new trait in a population came from early work on biological control of insect pests. Agricultural researchers had found that cyanide gas could be used to control populations of scale insects on citrus crops. By 1914, however, populations in some groves had become tolerant of the gas and fumigation was no longer effective. Laboratory experiments demonstrated not only that tolerance was inherited, but also that some individuals naturally resisted cyanide poisoning in areas where the fumigant had never been used. This resistance was caused by a mutation — an error in the genetic code — which appeared rarely and sporadically in individual scale insects. In the absence of cyanide the mutation for cyanide resistance would be of no value to the organism and could be detrimental if it altered normal body functions. Where it was used, the gas treatment resulted in survival of individuals selected for the resistant trait, while killing all others.

One may view the evolution of resistance to cyanide gas as a homeostatic evolutionary response of the population. But the precise mechanism of cyanide resistance is not known, and we do not know how cyanide gas acts as a selective mechanism. The correspondence between the environment and the selected trait is shown more clearly by a classic study of the evolution of dark forms of moths in polluted forests.

Industrial Melanism

The English have always been avid butterfly and moth collectors, and such enthusiasts look carefully for rare variant forms. Early in the last century, occasional dark (or *melanistic*) specimens of the common peppered moth were collected. Over the succeeding hundred years, the melanistic form, referred to as *carbonaria*, became increasingly common in some industrial areas, until at present it makes up nearly 100 per cent of local populations. The phenomenon aroused considerable interest among geneticists, who showed by cross-mating light and dark forms of the moth that melanism is a simple inherited trait.

In the early 1950's, H. B. D. Kettlewell, an English physician who had been practicing medicine for 15 years and who was also an amateur butterfly and moth collector, changed the pattern of his life to pursue the study of industrial melanism. Several facts about melanism were already known before Kettlewell began his studies. First, the melanistic trait is an inherited characteristic; hence, the widespread occurrence of melanism had been the result of genetic changes in populations. Second, the earliest records of *carbonaria* were from forests near heavily industrialized regions of England. Third, the dark form occurs most frequently in populations near modern industrial centers (Figure 12–1). Fourth, where there is relatively little industrialization, the light form of the moth still prevails. It was also known that dark forms have similarly

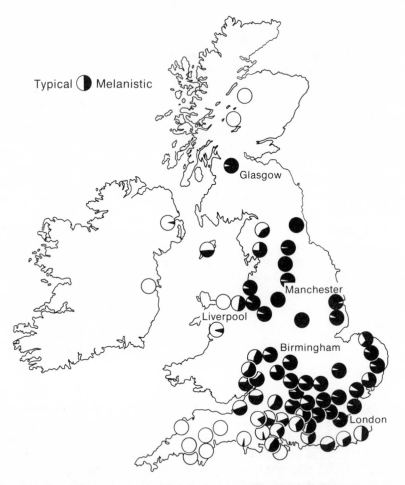

Typical ◐ Melanistic

Glasgow

Manchester

Liverpool

Birmingham

London

FIGURE 12–1   The frequency of melanistic individuals in populations of the peppered moth in various localities in the British Isles. The map summarizes more than 20,000 records, from 83 centers, collected from 1952 to 1956.

appeared in many other moths and other insects. Melanism is not unique to the peppered moth.

Kettlewell knew that the peppered moth inhabits dense woods and rests on tree trunks during the day. He reasoned that where melanistic individuals had become common, the environment must have been altered in some way to give the dark form a greater survival advantage than the light form. Could natural selection have led to the replacement of the "typical" light form by *carbonaria*? To test this hypothesis, Kettlewell had to find some measure of the evolutionary fitness of each form.

To determine whether *carbonaria* had a greater fitness than typical peppered moths in areas where melanism occurred, Kettlewell chose the mark–release–recapture method. Large numbers of individuals of both forms were to be marked with cellulose paint and then released in a suitable forest. The area would be thoroughly trapped for moths, and the number of marked individuals that were recaptured would be noted. Any difference in the survival of the two forms would appear in the relative percentage of the light and dark forms that were recaptured.

Because large numbers of individuals had to be released to guarantee even a small number of recaptures, Kettlewell captured and raised some 3,000 caterpillars to provide adult moths for his experiments. Two patches of woods, one in a polluted area and one in an unpolluted area, were chosen for the experiment. Adult moths of both forms were marked with a dot of cellulose paint and then released. The mark was placed on the underside of the wing so that it would not call the attention of predators to a moth resting on a tree trunk. Moths were recaptured by attracting them to a mercury vapor lamp in the center of the woods and to caged virgin females around the periphery of the woods. (Only male moths could be used in the study because females are attracted neither to lights nor to virgin females.)

In one experiment, Kettlewell released 201 typicals and 601 *carbonaria* in a polluted woods near Birmingham. Here is what happened:

Birmingham (industrial area)	*Typicals*	*Melanics*
Number of moths released	201	601
Number of moths recaptured	34	205
Per cent recaptured	16.0	34.1

These figures indicate that the dark form survived better than the light. Although the results were consistent with Kettlewell's original hypothesis, they could be interpreted otherwise: as differential attraction of the two forms to the traps used or as differential dispersion of the two forms from the point of release.

To test the hypothesis of natural selection unequivocally, Kettlewell ran a control experiment in an unpolluted forest near Dorset with the following results: of 496 marked typicals released, 62 (12.5 per cent) were recaptured; of 473 marked melanics, 30 (6.3 per cent) were recaptured. Thus, in the unpolluted forest, light adults had a higher survival rate than dark adults. The difference in relative survival of the two forms in the different environments confirmed Kettlewell's hypothesis and established that natural selection was responsible for the high frequency of *carbonaria* in industrial areas.

Having demonstrated that natural selection had been responsible for the replacement of typical by melanistic forms of the peppered moth in industrial regions, Kettlewell then sought to determine the specific agent of selection. He reasoned that in industrial areas pollution had darkened the trunks of the trees so much that typicals stood out against them and were easily found by predators. Any aberrant dark forms would be better camouflaged against the darkened tree trunks, and their coloration would confer survival value to them. Eventually, differential survival of dark and light forms would lead to changes in their relative frequencies in a population. To test this hypothesis, Kettlewell had to determine whether tree trunks are, in fact, darker in areas where the melanic form was prevalent and then to demonstrate that camouflage is important to the survival of the moths.

A clean handkerchief rubbed against a tree trunk can satisfy even the most doubting critic that the trunks of trees in polluted areas are darker than those in nonpolluted areas. And as one might have expected, the *carbonaria* stands out against tree trunks in unpolluted woods, whereas typicals are more conspicuous in polluted woods (Figure 12–2).

Kettlewell was certain that the conspicuousness of light-colored forms resting on darkened backgrounds must have greatly increased predation on them by visually oriented animals. The dark forms must have a similar disadvantage in unpolluted woods. To test this, Kettlewell placed equal numbers of the light and dark forms on tree trunk in polluted and unpolluted woods and watched them carefully at some distance from behind a blind. (A blind is a tentlike structure

FIGURE 12–2 Typical and melanistic forms of the peppered moth at rest on a lichen-covered tree trunk in unpolluted countryside (left) and on a soot-covered tree trunk near Birmingham.

intended to conceal an observer from his subjects, more appropriately called a "hide" by the English.) He quickly discovered that several species of birds regularly searched the tree trunks for moths and other insects and that the birds more readily found the moth that contrasted with the bark than the moth that blended with the bark. Kettlewell tabulated the following instances of predation:

	Individuals taken by birds	
	Typicals	*Melanics*
Unpolluted woods	26	164
Polluted woods	43	15

These data are consistent with the results of the mark-release-recapture experiments.

Kettlewell's observations demonstrate the operation of natural selection, which, over a long period, resulted in genetic changes in populations of the peppered moth in polluted areas. Many decades were required for the replacement of one form by the other. The agents of selection were insectivorous birds whose ability to find the moths depended on the coloration of the moth with respect to its background. The evolution of industrial melanism shows clearly how interaction between organism and environment determines the organism's fitness.

The theory of natural selection enables us to predict genetic changes in a population from known changes in the environment. If pollution were to be controlled in industrialized areas, and if this allowed forests to revert to their natural state, we would predict that the frequency of the light form of the peppered moth would begin to increase. In fact, smoke control programs were started in Manchester in 1952. Collections of the peppered moth over the last twenty years in the Manchester area do show a statistically significant increase in the proportion of the light form in the population.

Cultural Disturbance and Evolutionary Change

The rapid evolutionary change in coloration of peppered moths was stimulated by the effects of industrial pollutants on the appearance of tree trunks. Man has a propensity for creating rapid change in the environment. As a result, many plants and animals have undergone remarkable evolutionary adjustments of their adaptations in response to new conditions. A few examples are worth noting briefly.

The house sparrow, a common bird of cities and farms, was introduced to the east coast of the United States from England and Germany in 1852. The species rapidly spread throughout North America from southern Canada to southern Mexico and can be found in regions as varied as desert, prairie, and coniferous forest. In little more than a century, local populations of sparrows have diverged in body size and coloration in response to local variation in the environment over their extensive range. Northern populations are now conspicuously larger than southern populations; birds from hot, arid localities are paler in color than birds from cool, humid localities. Moreover, geographical variation in the house sparrow parallels similar variation in native species such as the song sparrow. Thus, when accidentally introduced to a variety of new habitats, the sparrow populations quickly responded to the new conditions through evolutionary change.

In many mining operations, refuse ore of too low grade to be refined is often dumped at the surface surrounding the mine shaft. If the ore contains toxic concentrations of lead, zinc, or copper, the presence of these metals in the soil that forms on the mine dump grossly affects the local vegetation. If one examines a long-abandoned mine dump, one finds species from the surrounding pastures have colonized the toxic soils and appear to grow perfectly well. Plants of the same species growing in unaffected pastures cannot tolerate high metal concentrations and die if they are transplanted into mine dump soil. Most individuals of these species are clearly not inherently tolerant of high concentrations of heavy metals, but toxic elements must have strongly selected those few individuals with mutations that enabled them to tolerate the mine dump soils.

Plants sampled along a transect that passes from contaminated mine soil to natural pasture reveal an abrupt boundary at which several adaptations change (Figure 12-3). Zinc tolerance is much greater in plants raised from seed collected on the mine dump soils, as we would have expected, but differences in height, degree of self-fertility, and other characteristics are also evident. The evolution of self-fertility is thought to reduce pollination of mine dump plants by plants on natural pastures where zinc tolerance has not evolved. Self-pollination would thereby help to preserve favorable combinations of genes in the progeny of mine dump plants by keeping nontolerant genes out of the population.

Rapid evolution of plant morphology and physiology has appeared in a variety of situations. Plants directly at the base of old, galvanized (zinc-plated) fences have been shown to tolerate high zinc concentrations in the soil, while plants of the same species growing six feet from the fence died when grown in soil containing zinc. Lawn mowers and grazing pressure by cattle have exerted a strong selection for short, rapidly growing varieties of many pasture and lawn plants. Rapid growth improves a plant's prospects for producing seed before becoming fodder.

At times evolutionary responses have directly confounded attempts by man to control populations of injurious pests. The evolution of

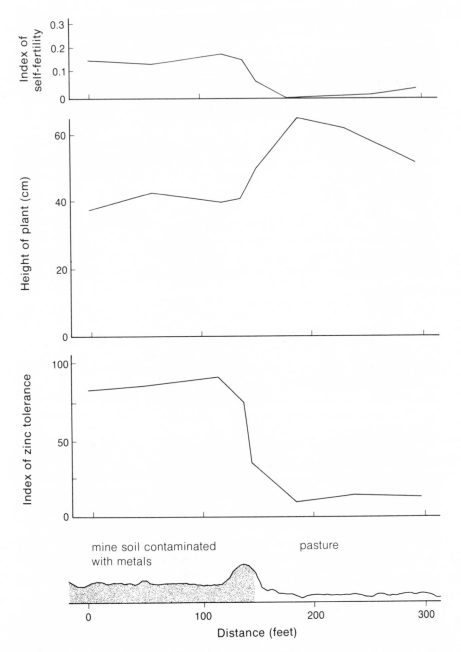

FIGURE 12–3 Location of sampling areas along a transect from a mine dump to natural pasture in North Wales, England. Plant height, index of zinc tolerance, and index of self-fertility in sweet vernal grass change sharply at the boundary between the two soils.

resistance to cyanide by the red scale insect was followed by the appearance of tolerance in several others species of scale. The evolution of resistant strains greatly diminished the value of cyanide gas as an insecticide. Recently developed chemicals, such as DDT, have also stimulated the evolution of resistance in flies and mosquitos where the insecticides have been used repeatedly. Strains of bacteria resistant to antibiotics have similarly reduced the effectiveness of many disease control programs.

For several decades, plant geneticists have been waging an evolutionary battle with wheat rust, a variety of fungus that infects wheat crops and greatly reduces production of grain. Each time a new wheat strain is developed to resist infection by known types of wheat rust, some new variety of rust invariably appears sooner or later with pathogenic effects on the wheat. The genetic characteristics that distinguish new infective strains of rust were undoubtedly present as rare mutations in the rust population. The new strain of wheat created favorable conditions for the proliferation of the infective rust strain, and it flourished.

Myxomatosis and Rabbits in Australia

Evolutionary responses of populations have seriously hampered efforts to control rabbit populations in Australia, but at the same time have demonstrated how natural selection can adjust a population not only to its physical environment, but to the influence of other populations as well.

As Western man immigrated to distant corners of the earth he brought a host of stowaways and invited guests who exceeded their welcome. Australia was not spared. In 1859, twelve pairs of the European rabbit were released on a ranch in Victoria. Within six years, the rabbit population had increased so rapidly that 20,000 were killed in a single hunting drive. By 1900, the several hundred million rabbits distributed throughout most of the continent became a critical problem for the Australians, whose economy has always been based largely on raising sheep. Being efficient grazers, rabbits destroyed range and pasture lands that otherwise would have been utilized for wool production. The Australian government tried poisons, predators, and other possible control programs — all without success. The answer to the rabbit problem seemed to be a myxoma virus, a relative of smallpox, which was discovered in populations of the related South American rabbit. Myxoma produced a small, localized fibroma (a fibrous cancer of the skin). Its effect on South American rabbits was not severe, but a European rabbit infected by the virus died quickly of myxomatosis.

In 1950, the myxoma virus was introduced in one location in Australia. A myxomatosis epidemic broke out in Victoria around Christmas time and then spread rapidly (Figure 12-4). The virus was transmitted primarily by mosquitos, which bite infected skin areas and carry the virus on their snouts. The first epidemic killed 99.8 per cent of the individuals in infected rabbit populations. During the following myxomatosis season (coinciding with the presence of mosquitos), only 90 per cent of the remaining population was killed, and during the third outbreak, only 40 to 60 per cent of rabbits in disease areas succumbed. At present, the myxoma virus has little effect on the rabbits, which have increased in number to the point that they are again a nuisance and an economic problem.

Several factors contributed to the decline in the virulence of the myxoma virus in the rabbits. The few rabbits that survived the disease developed immunity and were unaffected by later outbreaks of the virus. More important, immunity was conferred to the offspring of immune females through the uterus. Over and above the immune response, evolutionary changes occurred which increased the resistance of the rabbit population to the myxoma virus and reduced the virulence of the virus population. Selection was strong on both the rabbit and virus. Genetically determined immunity to a disease reduced mortality, thereby increasing the fitness of the host. At the same time, selection favors virus strains with less virulence, because reduced virulence lengthens the survival time of the rabbits and thus increases the

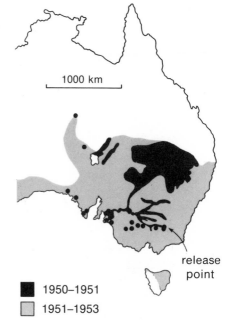

FIGURE 12–4 The spread of myxomatosis in European rabbits in southeastern Australia after its release near Albury (arrow) in 1950. The initial spread of the disease occurred along the Murray-Darling river system where it was transmitted by a mosquito that breeds only in persistent water. The rapid spread of myxomatosis to the north was greatly aided by abnormally wet weather in 1950.

1000 km

release point

■ 1950–1951
□ 1951–1953

mosquito-borne dispersal of the virus. A virus organism that kills its host quickly has a small chance of being carried by mosquitos to other hosts.

Periodic samples of myxoma virus, collected in the field and tested on European rabbits with no previous exposure to the virus, demonstrated that the virulence of natural myxoma strains decreased over the years following its introduction. In 1950 and 1951, all strains of the virus collected in the field produced grade I infections (more than 99 per cent mortality and a mean survival time of less than two weeks). By the 1958–59 season, however, no field strains of the myxoma virus had grade I virulence; and by 1963–64, grade II virulence (99 per cent mortality, 14 to 16 day survival time) was no longer evident. Similarly, resistance of wild rabbits to a particular grade of virus has increased steadily since 1950 (Figure 12–5). When the myxoma virus was first introduced to Australian rabbits, virtually all were mortally susceptible to myxomatosis, but by 1957 somewhat less than half the rabbits tested succumbed to virus with grade III virulence. Studies further indicated that whether a rabbit died from myxomatosis depended largely on the inheritance of genetic immunity from its ancestors.

Reciprocal Evolution Between Populations

Plant and animal populations are caught in an eternal evolutionary struggle to outdo one another. Any new twist in a predator's hunting strategy is countered by an evolutionary response in its prey. Natural insecticides in plants are rendered ineffective when herbivorous insects evolve biochemical mechanisms to detoxify the substance. Each evolutionary advantage is countered by novel adaptations in another population. The natural world is like Alice's looking-glass world where one has to run as fast as possible to stay in the same place. On our

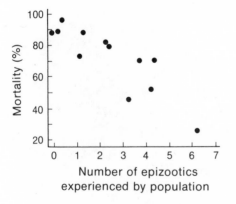

FIGURE 12–5 Decrease in susceptibility of wild rabbits to myxoma virus with grade III virulence (90 per cent mortality in genetically unselected wild rabbits).

limited time scale, organisms seem static and conservative, yet their characteristics are being molded by selective forces in their physical and biological environments.

The protective coloration of many animals provides a striking demonstration of adaptations to the biological world. Each of the species portrayed in Figure 12–6 is uniquely adapted either to blend in with its background and thereby avoid detection by predators, or to announce to predators that it is bad-tasting or toxic and should be left alone.

Plants and herbivorous insects wage less conspicuous chemical battles. For example, the tannins in oak leaves make it nearly impossible for caterpillars to make use of the proteins in the leaves. Tannins prevent the proper digestion of protein in the gut of an insect and thus reduce the food value in leaves. Tannins are usually deposited in vacuoles near the surface of the leaf where they won't interfere with the normal functioning of the plant but are nonetheless consumed by caterpillars. Where caterpillars are stumped by the chemical defenses of oak leaves, leaf-mining beetle larvae have outflanked the defences by burrowing through and consuming the leaf's inner tissues, completely avoiding the tannin-filled vacuoles near the leaf surface.

Most legumes (peas and their relatives) contain substances in their seeds that reduce the digestibility of proteins by inhibiting the functioning of digestive enzymes. Although these inhibitors probably evolved as a general biochemical defense against insects, they are futile against bruchid weevils, which have evolved metabolic pathways that either bypass, or are insensitive to, the inhibitors. Soybeans however, resist attack even by most weevils. If bruchid eggs are laid on soybeans, the larvae die soon after they burrow beneath the seed coat. In experiments, chemicals have been isolated from soybeans which inhibit the development of bruchid weevil larvae. Other legumes are not completely helpless against bruchid attack. In fact, ecologist Daniel Janzen has listed 31 different defensive adaptations that legumes have evolved to reduce the destruction of their seeds by bruchids.

Adaptations of organisms, and therefore the structure and function of the biological community itself, are in a constant state of flux. Attributes of the community change in response to pervasive changes in the physical environment and to shifts in its own biological composition. Natural selection endows the community with a powerful homeostatic mechanism sensitive to changes in the environment over hundreds and thousands of generations. Evolutionary changes enable populations of short-lived insects and microorganisms to respond to the havoc wreaked by man on the Earth's surface. Over longer periods, evolution has shaped the structure and function of present-day ecosystems.

Ecosystems are constantly put under stress by changes in the environment that are too rapid to be countered by evolutionary change or

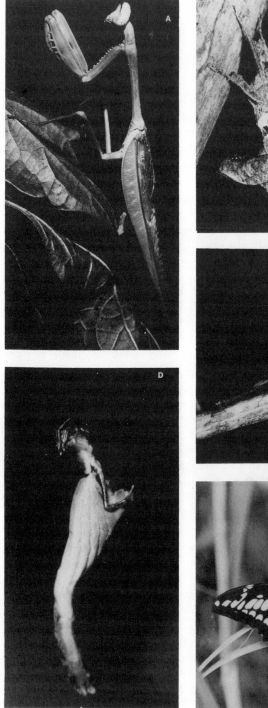

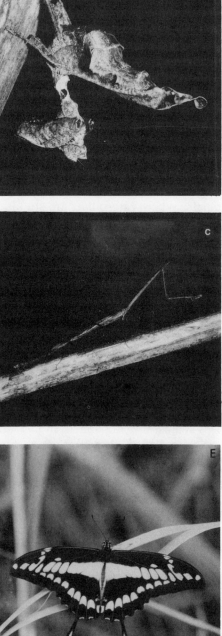

FIGURE 12–6 Some examples of protective coloration and form in forest dwelling animals in Panama. Three species of praying mantids (A, B, and C) avoid detection by blending with different backgrounds. The mantid in A is green and blends in with leafy vegetation; the mantid in B resembles a dead, curled up leaf; the mantid in C has a long, thin body that can easily be confused with a twig. The spider in D has a greatly enlarged, orange-colored abdomen whose appearance resembles flower parts that frequently drop into its web. The swallow-tailed butterfly in E is strikingly colored with a bold, iridescent green and black pattern as a warning to predators that it is unpleasant to taste. The frog in F escapes notice among the leaf litter of the forest floor. The caterpillar in G blends in perfectly with the twig on which it is resting when viewed from above; from the side, it is betrayed by its silhouette. The moth in H folds its wings unevenly over its back to break up its symmetry. The numerous spiny projections covered with toxic hairs on the body of the caterpillar in I tend to dissuade predators from eating it.

too great to be compensated by homeostatic responses of organisms. In the absence of compensating responses, environmental stresses increase mortality, reduce fecundity, or both, and thereby cause populations to decline. The ability of the ecosystem to withstand stress depends, in part, on the ability of populations to recover from declines following the return of normal conditions. In the next chapter, we shall examine the dynamics of populations and discover how populations resist, and recover from, environmental stress.

Population Growth and Regulation

When the resources available to organisms exceed their requirements for maintenance, plants and animals can allocate excess resources to growth and reproduction. When plant production increases in the temperate zone spring in response to increasing light, available water, and temperature, herbivore populations increase either in number through reproduction or in biomass through growth. Predator populations also respond to the increase in their food resources. Long-lived organisms on all trophic levels respond to increased resource availability by channeling energy into spurts of body growth or reproduction. Short-lived organisms, whose populations are reduced during unfavorable seasons, respond to favorable conditions by rapid population growth. Seasonal and year-to-year variation in primary production is thus paralleled by changes in rate of food consumption by herbivores and, in turn, by their predators. Population responses maintain the flux of energy and nutrients through the community at as high a level as possible. In so doing, they reduce periodic accumulations of resources in any one compartment of the ecosystem and increase its overall energetic efficiency.

Although it is the individual that reproduces and the individual that dies, the ecological effect of these events can be appreciated only by examining the entire population. This chapter is about populations — their growth and regulation — but we should bear in mind that population responses to change in the environment are, above all, the accumulation of events in the lives of individuals, of the responses of individuals to environmental change.

Population Structure in Space

Populations are composed of many individuals spread out over the geographical range of the species. Individuals are not, however, evenly distributed. Ecologists recognize three patterns of dispersion in populations: _clumped_ (underdispersed), _random_, and _spaced_ (overdispersed) (Figure 13-1). These types of distribution patterns are caused by different types of individual behavior. On one hand, clumping results from the attraction of individuals to the same place, either an intrinsically suitable environment or a gathering spot for such social functions as mating. On the other hand, an evenly spaced distribution usually signifies antagonistic interactions between individuals. In the absence of mutual attraction or social dispersion, individuals are distributed at random, without regard to the position of other individuals in the population.

Variation in the suitability of the environment causes individuals to congregate where the environment is most favorable. Although the sugar maple occurs widely throughout the eastern United States and Canada (see Figure 10-1), maple trees are most numerous on soils that are slightly acid; on the other hand, they may not be found at all in some hardwood forests within their geographical range. On a smaller distance scale, the salamanders in a maple forest congregate under fallen logs where the humidity of the environment falls within their range of tolerance.

Within suitable parts of the geographical range, behavioral interactions between individuals may space members of the population at regular distances. Amidst the pandemonium of sea-bird breeding colonies, nests are separated just far enough to prevent the adults from pecking each other while sitting on their nests (Figure 13-2). Most land birds, many fish and arthropods, and some reptiles and mammals defend exclusive _territories_ within which they search for food and raise their offspring. All antagonistic behavior, including territorial defense, tends to space individuals evenly over suitable habitats. Plant populations exhibit even spacing of individuals owing to mutual shading and root competition.

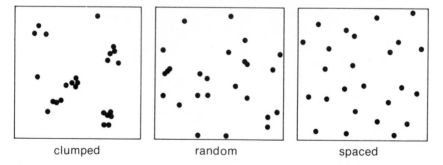

clumped   random   spaced

FIGURE 13-1   Diagramatic representation of the distribution of individuals in clumped, random, and evenly spaced populations.

FIGURE 13–2   A nesting colony of Peruvian boobies on an island near the coast of Peru. The densely packed birds space their nests more or less evenly, the distance being determined by behavioral interactions between individuals. Along the entire length of the Peruvian coast, however, sea bird populations are clumped, during the breeding season, on a few offshore islands.

When we view the spatial distribution of a population at a moment in time, the positions of individuals are fixed as if in a photograph. Of course, most animals continually move. Even plants disperse during reproductive periods either as pollen grains or seeds; phytoplankton float freely in the water. Dispersal behavior is an adaptation of the organism, allowing it to find a suitable habitat to settle in. But dispersal has an important population function as well. Spatial and temporal variation in the suitability of a habitat result in population growth in some areas and population decline in others. Movement of individuals between areas enables a population to respond to local variation in the environment more rapidly than if local reproduction and death were the only avenues of response open to populations.

Age Structure and Sex Ratio

All individuals are not alike. Demographers, people who study the structure and dynamics of populations, distinguish individuals according to sex and age. Because death rates and fecundity rates vary by sex and age, the future growth or decline of a population is influenced by its age and sex structure. For example, the growth potential of a human population with many females between 15 and 35 years of age is greater than a population consisting mostly of old men and preadolescents.

The growth rate of a population — the net result of birth and death rates — also influences the age structure of a population. The size of the human population of Sweden has been relatively constant for many decades. As a result, the proportion of the population in each age class (the *age structure* of the population) parallels the expected percentage survival of newborn babies to each age (the *survivorship curve*) (Figure 13-3). Consider for a moment a population with a constant number of births and an equal number of deaths each year. Because births exactly balance deaths, the size of the hypothetical population is also constant.

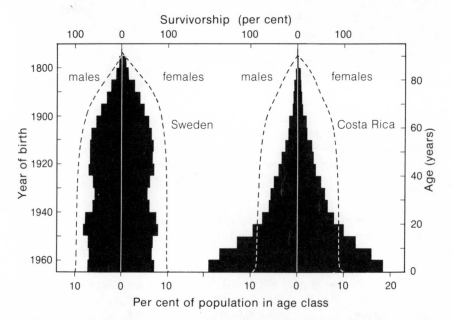

FIGURE 13-3 Population age-structure and survivorship, calculated separately for males and females, in Sweden (1965) and Costa Rica (1963). Because Sweden's population has grown slowly, its age structure resembles the survivorship curve. Declining birth rates during the Depression, and the baby boom which followed World War II, are responsible for irregularities in the age-structure of Sweden's population. Costa Rica's rapid population growth has resulted in a bottom-heavy age structure.

If, in such a population, 95 per cent of newborn babies lived to 45 years of age, the population would contain 95 per cent as many 45-year-olds as newborn. Similarly if 35 per cent of babies lived to be 80, the population would contain 35 per cent as many 80-year-olds as newborn. In Sweden, the correspondence between age structure and survivorship is reasonably close, although the relative dearth of old individuals betrays a slight though persistent excess of births over deaths. In 1965, the population of Sweden was increasing at a rate of 0.6 per cent per year.

In sharp contrast, the population of Costa Rica had a growth rate of about 4.1 per cent in 1963. A high birth rate swelled the young age classes, resulting in a bottom-heavy structure. Although a male baby born in 1963 had an 80 per cent probability of surviving to 45 years of age, the number of 40- to 45-year-old men in the population was only one-fifth the number of 1- to 5-year-old male children. The discrepancy between the size of the age class and survivorship is caused by the smaller number of babies producing the 40 to 45 year age class compared to the 1 to 5 year class. Costa Rican men 40 to 45 years old in 1963 were born between 1918 and 1923, when the population was much smaller.

Any variation in the death rate or number of young born creates irregularities in the age structure of a population. An exceptionally productive year might send a large year-class through the age structure of the population in such a manner that 1-year-olds would predominate the first year, 2-year-olds the following year, and so on. Populations in which the number of young born or their survival during the first year varies greatly from year to year exhibit irregular age structure. Environmental changes that increase adult death rate usually affect all ages equally; although population size may decline, age structure does not change greatly. Subsequent population growth will, however, result in a relative increase in the younger age classes.

Seasonal patterns of reproduction and mortality in short-lived animals create cyclic changes in age structure. Honeybee populations expand rapidly in the spring and assume the bottom-heavy age distribution characteristic of growing populations. As reproduction declines toward the end of the summer, the younger age classes dwindle in number, and finally disappear altogether after reproduction ceases. During winter, surviving honeybees move through progressively older age classes.

The ratio of males to females (the *sex ratio*) in most animal populations is about 1:1. The sex ratio changes with age if births and deaths favor one sex over the other. For example, more males are born into human populations, but females live longer. The sex ratios of some parasitic species and social insects are grossly biased in favor of females. Males have no role in reproduction beyond fertilizing the eggs, and one male can go a long way toward that end. Some fish, lizards, and aquatic invertebrates have abandoned sex altogether: females produce eggs that do not require fertilization to develop. Plants are a demographic night-mare. The flowers of most species have both male and female sexual organs. In species with separate sexes, sex ratios vary from place to place within a population and rarely are evenly balanced. Until recently, population biologists have paid little attention to plants and corre-spondingly little is known about their population structure.

The Life Table

The *life table* is a summary of the vital statistics of a population: the *survivorship* to, and *fecundity* of, each age class. Survivorship to a given age is usually denoted by the symbol l_x, where x is the age specified; fecundity at a given age is denoted by b_x. These statistics are usually computed only for females because the fecundity of males is extremely difficult to determine in populations where mating occurs promiscuously.

The construction of a life table for a population requires that the ages of individuals be known, a straightforward problem for captive or laboratory populations and for (or perhaps including) human popula-tions, but more difficult for natural populations. Age-specific survival rates are normally estimated from the survival of individuals marked at birth or from the deaths of individuals of known age, but both methods have serious limitations. The first requires that individuals do not emi-grate from the area in which they were marked; otherwise they would erroneously be counted as dead. The second method requires that large numbers of individuals be marked to obtain sufficient recoveries of dead individuals. For example, of 180,718 robins banded as adults in the United States between 1946 and 1965, only 2,444 were ever retrapped or found dead in subsequent years; of 7,604 brown pelicans banded as nestlings, only 375 were ever seen again.

When the ages of individuals can be determined by their appear-ance, without knowledge of date of birth, the construction of a survivor-ship table becomes a relatively simple task. Among plants, for example, the age of temperate forest trees can be determined from growth rings in their wood. Among animals, the age of fish can be determined from growth rings in their scales or ear bones; the age of mountain sheep can be estimated from the size of their horns.

Estimated age of death, judged by the size of horns on skulls, was used to calculate survivorship for the Dall mountain sheep (Figure 13-4) in Mount McKinley National Park, Alaska. In all, 608 skeletal remains of sheep were found: 121 of the sheep were judged to have been less than 1-year-old at death; 7 between 1 and 2 years, 8 between 2 and 3 years; and so on, as shown in Table 13-1. We may construct a survivorship table using the following reasoning: all 608 dead sheep must have been alive

FIGURE 13–4   Dall mountain sheep in Mount McKinley National Park, Alaska. The size of the horns increases with age.

TABLE 13–1   Survivorship table for the Dall mountain sheep constructed from the age of death of 608 sheep in Mount McKinley National Park, Alaska.

Age interval (years)	Number dying during age interval	Number surviving at beginning of age interval	Number surviving as a fraction of newborn (survivorship) l_x
0–1	121	608	1.000
1–2	7	487	0.801
2-3	8	480	0.789
3–4	7	472	0.776
4–5	18	465	0.764
5–6	28	447	0.734
6–7	29	419	0.688
7–8	42	390	0.640
8–9	80	348	0.571
9–10	114	268	0.439
10–11	95	154	0.252
11–12	55	59	0.096
12–13	2	4	0.006
13–14	2	2	0.003
14–15	0	0	0.000

at birth; all but 121 that died during the first year must have been alive at the age of 1 year (608 − 121 = 487); all but 128 (121 dying during the first year and 7 dying during the second) must have been alive at the end of the second year (607 − 128 = 480); and so on, until the oldest sheep died during their fourteenth year. Survivorship (right-hand column in Table 13-1) was calculated by converting the number of sheep alive at the beginning of each age interval to a decimal fraction of those alive at birth. Thus, the 390 sheep alive at the beginning of the seventh year represented 64.0 per cent (decimal fraction 0.640) of the original newborn in the sample.

Survivorship provides half the life table. One must also know the age-specific fecundity to fully understand the dynamics of the population. Field biologists have devised techniques for estimating fecundity from embryo counts in mammals, nest checks in birds, ratios of juveniles to adults in many kinds of animals, and direct counts of eggs in amphibians, insects, marine invertebrates, and others. As in the matter of death, the age at which a female produces offspring is all-important. An average litter size of 5 young at age 3 contributes to population growth less than the same size litter born to females at age 2 simply because more females live to age 2 than to age 3. To calculate the total production of young by a population, the average fecundity of each age group must be multiplied by the number of adults in that age group, and the products summed over all ages.

Calculating Population Growth Rate from the Life Table

The growth rate of a population depends on two calculations from the life table: the net reproductive rate and the mean generation time. The *net reproductive rate* is the expected number of female offspring to which a newborn female will give birth in her lifetime. Of course, some individuals die before they attain reproductive age and thus leave no offspring; exceptionally long-lived females give birth to many more young than the population average. But it stands to reason that if females produce more than one female progeny on the average, the population will grow; if females fail to replace themselves on the average, the population will decline.

The rate at which a population grows or declines also depends on the *mean generation time*, which is the average age at which females produce offspring. The earlier young are born, the earlier they, in turn, have offspring and the more rapid population growth will be.

The net reproductive rate, denoted here by the letter R, is calculated by adding the expected production of offspring by a female at each age. In the example presented in Table 13-2, each female could expect to

TABLE 13–2   Life table of a hypothetical population, demonstrating the calculation of the net reproductive rate and mean generation time.

Age (x)	Survivorship (l_x)	Fecundity (b_x)	Expected offspring $(l_x b_x)$	Product of age and expected offspring $(x l_x b_x)$
0	1.00	0.0	0.0	0.0
1	0.50	1.0	0.5	0.5
2	0.40	4.0	1.6	3.2
3	0.20	4.0	0.8	2.4
4	0.10	2.0	0.2	0.8
5	0.00	0.0	0.0	0.0
	Net reproductive rate =		3.1	
	Total weighted age =			6.9

$$\frac{6.9}{3.1} = 2.22$$

$$T$$

$$T = \frac{\Sigma (x \, l_x b_x)}{\Sigma \, l_x b_x}$$

produce 0.5 young, on the average, at age 1 (average fecundity of 1.0 times expected survivorship of 0.5), 1.6 young at age 2 (0.40 times 4.0), and so on, totalling an expected net reproductive rate of 3.1. In other words, at the time of death, females in the population would have left an average of 3.1 female offspring.

The mean generation time (T) is calculated by finding the average age at which a female gives birth to her offspring. In the present example, 0.5 young are produced at age 1, 1.6 young at age 2, 0.8 young at age 3, and 0.2 young at age 4. We calculate the average age by adding the products of each age and the number of offspring produced at the same age, and dividing the sum by the total number of offspring produced. In the example in Table 13–2, the sum of the products is 6.9. Dividing this figure by the net reproductive rate (3.1), we obtain a mean generation time of 2.22 years. Our calculations show that the population will increase by a factor of 3.1 times every 2.22 years. It is difficult to compare population growth rates expressed on a generation basis. For example, is a factor of 3.1 every 2.22 years greater or less than a factor of 1.4 every 1.36 years? To avoid this problem, population growth rates are usually expressed on an annual basis, and are specified by the Greek letter lambda (λ). The population described above would increase by a factor of 1.6 each year.*

---

* The annual growth rate of a population (λ) is calculated by the formula

$$\lambda = R^{1/T}$$

where R is the net reproductive rate and T, the mean generation time. In the example presented in Table 13–2 ($R = 3.1$ and $T = 2.22$), the annual growth rate is

$$\lambda = 3.1^{1/2.22}$$
$$= 3.1^{0.45}$$
$$= 1.6$$

The Growth Potential of Populations

The age-specific survivorship and fecundity of a population are used to calculate its rate of growth. Birth and death rates reflect the interaction of the individual with its environment — its ability to channel resources into reproduction and its ability to avoid predation, starvation, or death by exposure. The mathematics of demography translates the individual's activities into the dynamics of the population. If the environment of the individual is unchanging, the growth rate of the population should not vary. If a population increases by a factor of 1.6 each year, 100 individuals become 160 after 1 year, 256 after 2 years, 410 after 3 years, progressing in a geometric fashion (Figure 13–5).

Populations increase by multiplication (geometric growth) rather than by simple addition (arithmetic growth) because young born into the population themselves grow up to give birth. Population increase thus resembles compound interest on a bank account in which the earned interest (newborn young) is periodically added to the principal (reproducing adults).

The geometric growth rates of populations under conditions that are ideal for growth reveal a basic fact: populations have a tremendous intrinsic capacity for growth. Population increase, if unrestricted by limiting resources or predators, would be sufficient to raise the numbers of most kinds of animals to astronomical figures in a very short time. Charles Darwin appreciated this growth potential more than a century ago, when he wrote in *On the Origin of Species* that "There is no exception to the rule that every organic being naturally increases at so high a rate that, if not destroyed, the earth would soon be covered by the progeny of a single pair."

We can best appreciate the capacity of a population for growth by following its rapid increase when introduced to a new region with a

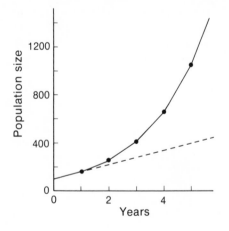

FIGURE 13–5 Growth of a population whose annual rate of increase (λ) is 1.6. The initial population size (time = 0) was arbitrarily set at 100 individuals. The dashed line represents the population size if growth had continued by adding a constant amount (60 individuals) to the population (arithmetic growth) rather than multiplying the population by a constant factor (1.6 times each year) (geometric growth).

suitable environment. The number of colonists is at first so low that crowding and depletion of resources do not hinder population growth. When domestic sheep were introduced to Tasmania, a large island off the coast of Australia, the population increased from less than 200,000 in 1820 to more than 2,000,000 in 1850. The ten-fold increase in 30 years is equivalent to an annual rate of increase of 8 per cent ($\lambda = 1.08$). Two male and six female ring-necked pheasants introduced to Protection Island, Washington, in 1937 increased to 1,325 adults within five years. The 166-fold increase represents a 180 per cent annual rate of increase ($\lambda = 2.80$). Even such an unlikely creature as the elephant seal, whose population had been all but obliterated by hunting during the nineteenth century, increased from 20 individuals in 1890 to 30,000 in 1970 ($\lambda = 1.096$). If we are unimpressed, we should consider that another century of unrestrained growth would find 27,000,000 elephant seals crowding surfers and sunbathers off southern California beaches. The beaches of Florida and New Jersey would quickly succumb. Before the end of the next century, the shorelines of the western hemisphere would give lodging to one trillion elephant seals.

Elephant seal populations do not hold any growth records. Life tables of populations maintained under optimum conditions in the laboratory have shown that potential annual growth rates (λ) may be as great as 24 for the field vole, ten billion (10^{10}) for flour beetles, and 10^{30} for the water flea *Daphnia*. Rapid growth rates are more conveniently expressed in terms of the time required for the population to double in number. Corresponding doubling times are 7.6 years for the elephant seal, 8 months for the pheasant, 80 days for the vole, 10 days for the flour beetle, and less than 3 days for the water flea. Populations of microorganisms (bacteria and viruses) and many unicellular plants and animals can double in a day or a few hours.

Even though natural populations occasionally attain their maximum growth potential, such rates of increase never prevail for long periods, and population growth is, fortunately, brought under control.

The Regulation of Population Size

The number of individuals in a population is limited by the availability of resources. Barnacle populations cannot increase beyond the point that all the available rock surface is covered. The number of pairs of titmice in a forest cannot exceed the number of available nesting sites. Predators cannot become so numerous that they reduce their prey below the level required for their own maintenance.

As populations increase, the physiological effects of crowding and depletion of resources begin to change the life table of decreasing birth rate or increasing death rate, or both, with the result that population growth rate declines (Figure 13-6). If the density of a population is low relative to the abundance of its resources, the birth rate will exceed the death rate, and the population will increase. The growth rate of the population (number of individuals added to or removed from the population) is equal to the birth rate minus the death rate. As the population increases, deaths increase proportionately faster than number of individuals in the population, and the number of births per individual decreases. If population density exceeds the level that the environment can support, the death rate will exceed the birth rate and the population

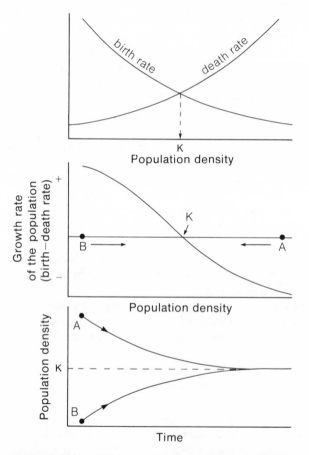

FIGURE 13-6 Diagramatic representation of birth and death rates in a regulated population (top), the resulting growth rate of the population with respect to density (middle), and the approach of population size to equilibrium (*K*) with time (bottom). A and B represent the densities of two populations whose birth and death rates are not in equilibrium.

will decline. For each set of environmental conditions, a level of population density exists at which birth and death rates exactly balance each other and the population neither grows nor declines. This point of population equilibrium is referred to as the *carrying capacity* of the environment (K). Whenever the population falls below or exceeds K, population responses cause population size to return to the carrying capacity. At low population densities (point B in Figure 13-6), population growth rate is positive; at high densities (point A), growth rate is negative. Hence K represents a stable equilibrium.

Laboratory populations maintained at different densities, but under identical conditions, demonstrate the response of birth rate and death rate to density. In a series of experiments on the water flea, *Daphnia pulex*, initial population densities varied between 1 and 32 individuals per milliliter of water. The populations were grown in small beakers with cultures of green algae provided for food. Survivorship and fecundity of females were noted for two months after the beginning of the experiment and the data were used to construct life tables for each population density, shown graphically in Figure 13-7. Fecundity decreased markedly with increasing population density. Survivorship actually increased at densities up to eight individuals per milliliter; high fecundity apparently reduces survivorship, indicating a trade-off between reproductive effort and adult survival. At densities of eight or more individuals per milliliter the body growth of individual water fleas was stunted, suggesting that depletion of food resources ultimately limited birth rates and survivorship in the dense cultures.

Population growth rate (λ), calculated from life table data for the water fleas, decreases with increasing density and falls below 1.0 (equilibrium population size) at a density of about 20 individuals per milliliter (Figure 13-8). Under the conditions of temperature, light, water quality, and food availability provided in the laboratory, water flea populations reach a stable equilibrium of 20 individuals per milliliter, representing the carrying capacity of the environment, regardless of the initial density of the culture.

The Carrying Capacity of the Environment

Populations are regulated by the balance between two opposing forces: the inherent growth potential of the population and the limits to population growth imposed by the environment. The ability of the environment to support a species varies with climate and resource availability, and so then does the equilibrium population density of the species.

Population density is limited by two kinds of resources. On one hand, *nonrenewable resources* such as space or nesting sites can be completely utilized by a population, thereby placing an abrupt upper limit

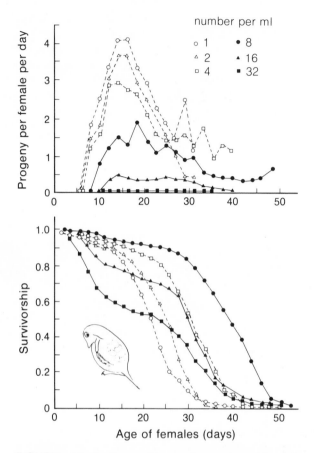

FIGURE 13-7 Fecundity and survivorship of females in laboratory populations of the water flea *Daphnia pulex* maintained at different densities.

on population size. On the other hand, *renewable resources* such as food, water, and light are supplied continually to populations. The resource demands of a large population can reduce renewable resources to levels at which they are so difficult to locate or assimilate that they will not support further population growth, but renewable resources are never completely exhausted. Renewable resources are maintained at an equilibrium level by the balance between exploitation and production. When population size equals the carrying capacity of the environment, the resource requirements of the population just match the renewal rate of its resources. If the population exceeds its carrying capacity, exploitation exceeds renewal, resources are depleted, individuals starve, and the population begins to decline. Conversely, if the rate of resource production increases, the environment can support a larger population. Dense populations are usually found in the most productive habitats, as

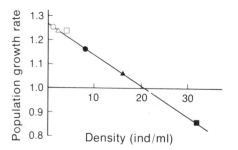

FIGURE 13–8   Values of λ calculated from the life table data portrayed in Figure 13–7. Population growth rate decreases as a function of density. A density of 20 individuals per milliliter represents the apparent carrying capacity of the environment, at which point λ is equal to 1.0 and the population just maintains itself.

shown by the relationship between the population density of birds and primary plant productivity in a variety of habitats (Table 13–3).

Populations of barnacles and mussels on rocky coasts are limited by a nonrenewable resource — space. Individuals become so crowded that larvae have no place to settle (Figure 13–9). Such high densities occur because barnacles and mussels are restricted to flat surfaces at the edge of their food supply, plankton obtained directly from sea water washing over the animals. As a result, these filter feeders have little effect on the abundance of their food, and therefore food does not limit the size of their populations. Other intertidal creatures that feed on algae that grows directly on the rock surface can exhaust their food supplies. Limpets and herbivorous snails are much more sparsely distributed in the intertidal region than are barnacles.

The influence of resource productivity on the carrying capacity of the environment for mule deer populations has been demonstrated by range management experiments. Chaparral habitats in northern California were turned into shrublands by mechanical thinning, controlled burning, and seeding with herbaceous plants. As a result of stimulated plant growth, experimental shrublands had 2.8 times as much edible foliage in woody species and 17 times as much edible foliage in herbaceous species as unaltered chaparral control areas. Deer in shrublands were able to select higher quality foods than in chaparral

TABLE 13–3   Censuses of birds in six habitats listed in increasing order of primary production.

Habitat	Locality	Total breeding pairs per 100 acres	Net primary production (g/m²/yr)
Desert	Mexico	22	70
Prairie	Saskatchewan	92	300
Chaparral	California	190	500
Pine forest	Colorado	290	800
Hardwood forest	West Virginia	320	1,000
Flood-plain forest	Maryland	581	1,500

FIGURE 13–9   A crowded population of mussels.

habitats as well (14 versus 9 per cent protein content). As a result of increased food production, populations of deer in artificially maintained shrublands were more than twice as dense as populations in chaparral; individuals were 5 to 10 per cent heavier in shrubland habitats.

The direct physiological effects of food availability on reproduction were shown in a study of white-tailed deer in New York. Per cent of females pregnant and number of embryos per does (incidence of twins) was directly related to range conditions (Table 13-4). The opening of one Adirondack study area to hunting incidentally demonstrated the relationship between food availability and reproduction. Before hunting was allowed, the overcrowded deer population in the DeBar Mountain

TABLE 13–4   Pregnancy rate and number of embryos per pregnant female in white-tailed deer populations in New York. The localities are listed in decreasing order of range condition.

Region	Per cent of females pregnant	Embryos per pregnant female
Western (best range)	94	1.7
Catskill periphery	92	1.5
Catskill center	87	1.4
Adirondack periphery	86	1.3
Adirondack center (worst range)	79	1.1

region had a pregnancy rate of 57 per cent, with 0.7 embryos per pregnant female. After the population had been reduced by two years of hunting, pregnancy rates rose to nearly 100 per cent with 1.8 embryos per female; nearby areas inaccessible to hunters exhibited no increase in reproductive performance.

Fluctuations in Natural Populations

Natural environments are rarely so constant as the laboratory. Variation in climate and food availability, through their influence on survivorship and fecundity, continually change the direction and rate of growth of natural populations. The degree of variation in the size of a population depends partly on the magnitude of environmental fluctuations and partly on the inherent stability of the population. Populations of large, long-lived, slowly reproducing plants and animals are comparatively insensitive to changing environments because of their inherent homeostatic capacities. For example, after sheep became established on Tasmania, their population varied irregularly between 1,230,000 and 2,250,000 individuals over nearly a century (Figure 13-10). Short-lived organisms with high reproductive capacities are more sensitive to short-term fluctuations in the environment; their populations frequently increase and decrease by factors of hundreds or thousands over days or weeks (Figure 13-11). Because sheep raised for wool production live for five to ten years, the population at any one time consists of young born over a relatively long period, thereby evening out the influence of short-term variations in birth rate on population size. The lifespan of the single-celled algae that constitute the phytoplankton of a lake is measured in days; the individuals in an algae population are replaced rapidly and the population is thus vulnerable to the capriciousness of the environment.

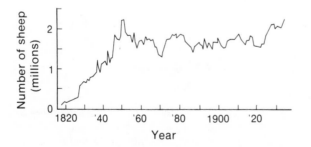

FIGURE 13-10 Number of sheep on the island of Tasmania since their introduction in the early 1800's.

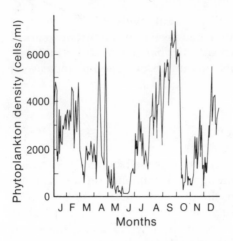

FIGURE 13–11 Variation in the number of phytoplankton in samples of water from Lake Erie during 1962.

In seasonal environments, reproduction occurs only when climate and resources combine to produce favorable conditions. Seasonal changes in temperature, moisture, and nutrients additionally affect mortality, either directly or indirectly. When lifespan is so short that many generations occur within a year, seasonality can greatly influence population size. Near Adelaide, South Australia, populations of the tiny insect pest *Thrips imaginis*, which infests roses and other cultivated plants, undergo regular cycles of increase during seasonally favorable conditions, followed by rapid decline when the environment becomes either too dry or too cold to support population growth (Figure 13-12). Adelaide has a Mediterranean climate: winters are cool and rainy, sum-

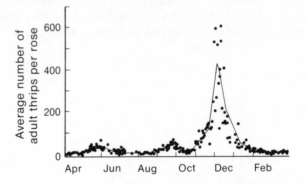

FIGURE 13–12 Number of *Thrips imaginis* per rose from April 1932 through March 1933 near Adelaide, Australia. Dots indicate daily records; the curve is a moving average for successive 15-day intervals.

mers are hot and dry. The winter months are generally too cold to sustain growth of thrips populations; summers are too dry. The spring (October through December in the Southern Hemisphere) brings an ideal combination of moisture, warmth, and plant flowering for population growth.

Thrips subsist mainly on plant pollen, whose abundance varies seasonally with the production of flowers. The climate of Adelaide is mild enough that some flowers are always available and thrips are active all year. During the winter, however, the depressing effect of cool temperature on development rate and fecundity (Table 13–5) brings a marked decline in the number of thrips. Population growth is further checked by high mortality during the immature stages; the period from egg to adult is so long during the winter that most flowers wither and fall off before the thrips reach maturity. The warm weather of spring increases the net reproductive rate of the thrips while shortening the mean generation time. Under these conditions, populations rapidly increase to infestation levels. Increasing mortality caused by ensuing summer drought halts population growth and results in rapid population decline in late November or early December.

Density-Dependence and Density-Independence

Many of the environmental factors that influence the growth rate of thrips populations do so independently of population density. Temperature interacts with development rate to influence mean generation time whether thrips are numerous or few. Hot, dry air robs thrips of their body moisture regardless of their numbers. Factors whose influence on population growth rate is independent of population size are called *density-independent factors*. Although such factors can cause great fluctuation in the numbers of individuals in populations, their action does not lead to a stable equilibrium.

TABLE 13–5 Influence of temperature on development rate, lifespan, and fecundity of *Thrips imaginis*.

	Temperature (°C)	
	8 to 12	23 to 25
Length of adult life (days)	250	46
Total eggs laid per female*	192	252
Daily egg production	1.4	5.6
Development period (days)	44	9

* With pollen (a protein source) in their diet. If adults are raised without pollen, their egg production falls to 20 eggs at 24°C, and life span increases to 77 days.

The influence of *density-dependent factors* on population size changes with the density of individuals (see Figure 13–6). These factors some-how depend on the interaction of individuals. A dense population de-pletes its food supply, utilizes suitable breeding and resting places, attracts predators, and hastens the spread of disease.

Density-dependence is often associated with biotic factors and density-independence with physical factors, although this distinction does not always pertain. For example, the killing effects of a frost (a "density-independent" factor) depend on whether the individual can find a sheltered refuge. If the environment contains a limited number of well-protected living places, the influence of such climatic factors may well be "density-dependent." A severe frost does not affect all individu-als in a population equally; some survive, others die. Climate and other physical factors also indirectly exert a density-dependent influence on population size through their effect on food supply.

The rapid increase of thrips populations in the spring, as well as their decline in the fall, seemingly results from changes in density-independent influences on population size. Thrips rarely infest every flower and they consume only a small portion of the food available to them, even when populations are dense. Population increase is fostered in the spring by favorable conditions of temperature and moisture. Density-independent factors overwhelmingly influence the size of the thrips populations, but they cannot regulate population size over long periods. In fact, the peak annual population of thrips varies little from year to year compared to seasonal variation in population size. It would be quite unlikely that the product of all the daily values of λ for the thrips population would be very close to 1.0 (representing an unvarying long-term population trend) just by chance unless density-dependent factors operate at some season. These presumed factors have not actually been identified for the thrips population, but research on Canadian insect pests, particularly the spruce sawfly, has identified the factors responsible for regulation of some insect populations.

Population Dynamics of the Spruce Sawfly

The spruce sawfly was first introduced to Quebec from Europe in about 1930. By 1938, it had spread throughout most of Quebec, the maritime provinces of Canada, and New England, causing widespread damage, including complete defoliation of spruce forests. Most sawflies are female — fewer than one in a thousand are male — and reproduction is predominately asexual. Females lay eggs singly in slits cut in spruce needles; the larvae feed on the foliage after they hatch (Figure 13–13). After the larvae are fully grown, they drop to the ground and burrow a few inches into moss, where they spin a tough cocoon around them-

FIGURE 13-13   The larch sawfly, a close relative of the European spruce sawfly, feeding on needles of western larch.

selves and metamorphose into adults. The sawfly population goes through several generations each summer, but toward fall, pupae (the developmental stage between larval and adult forms) enter into a diapause state in their cocoons to spend the winter. The Canadian government introduced several parasites in an attempt to control the sawfly, but eventually a virus, accidentally introduced with one of the parasites, proved to be the most effective control agent.

Canadian foresters and entomologists initiated a census program during the height of the sawfly outbreak in the late 1930's. The program was continued for more than 20 years. Larvae were sampled by spreading large canvas sheets on the ground under a tree and vigorously shaking the limbs directly above it. The dislodged larvae were collected and reared individually in vials to determine rates of disease and parasitism in the population. A census of cocoons was taken each year. Wooden trays were filled with moss gathered from an area free of sawflies. The trays were then set out in an infested forest during July, before the first generation larvae began to drop to the forest floor, to serve the larvae in place of naturally occurring moss. The following May, after most sawflies had emerged from their cocoons, the trays were brought back to the laboratory and the cocoons classified as either (a) sound but unemerged, (b) emerged normally, (c) parasitized, (d) preyed upon by wireworms, or (e) preyed upon by small mammals, mostly rodents. Each type of predator and parasite leaves a distinguishing mark on the tough and leathery cocoons.

The size of the sawfly population varied by factors typically as great as 10 and even as great as 100 over a period of a few years (Figure 13-14). Rates of parasitism and disease also varied greatly. Statistical analysis of population trends in relation to biotic and climatic factors identified factors responsible for fluctuations in population size; the analysis also distinguished density-dependent from density-independent factors. When the annual growth rate of the population (λ) was plotted on a graph with respect to population size, growth rate was found to de-

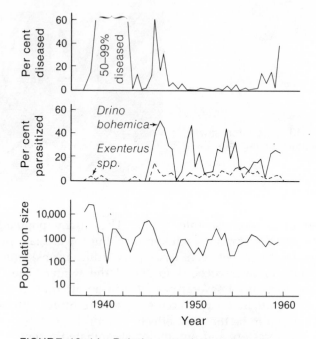

FIGURE 13–14   Relative population size, disease rate, and parasitism rate of European sawfly larvae in a black spruce forest. Note the apparent long-term constancy underlying the short-term fluctuations in population size.

crease significantly with increasing density — an indication of population regulation by density-dependent factors.

Values of λ were at first calculated as the ratio of one year's population of larvae to the previous year's population of larvae. To determine whether parasites and disease organisms contributed to density-dependent population regulation, λ was recalculated as the ratio of one year's larvae to the previous year's emerged adults. By using the previous year's adult stage as the basis of comparison, larval mortality would be eliminated from the relationship between population growth rate and density. If λ still exhibited density-dependence, one would have to look elsewhere for the action of density-dependent factors. If the density dependence of λ disappears by eliminating larval mortality from the analysis, one could pinpoint the regulating factor to the larval stage. Similar analyses were used to test each mortality factor for density-dependent influence.

The sawfly study provided useful insight into the regulation of insect populations. First, although disease and weather were important components of mortality, neither exerted a stabilizing effect on population size. (Disease was, however, largely responsible for bringing serious outbreaks under control.) Second, parasites were found to be the only factor that acted in a density-dependent manner during the larval period. Third, additional unidentified density-dependent factors must have acted at other stages of the life history because larval mortality could not account for all the regulatory influences.

Entomologists have reached similar conclusions from studying other insect populations. For example, whereas climate accounts for 80 per cent of variation in numbers of the diamondback moth, a pest on cabbage, the long-term regulation of population size resides with parasites that strike late in the larval period. In other studies, the critical stage at which mortality occurred, as well as the mortality factor itself, varied from species to species. Only rarely did density-dependent factors cause most of the short-term variation in population size.

Population Cycles

Some populations of long-lived organisms, particularly birds and mammals, fluctuate with great regularity; peaks and troughs in their numbers occur at intervals of anywhere from three to ten years. The most remarkable of the cycles are exhibited by some mammal populations of the New World arctic, where the regularity of population fluctuations is precise enough to predict population size several years in advance. For example, lynx populations of Canada have a cycle of approximately ten years (Figure 13–15), closely following cyclic changes

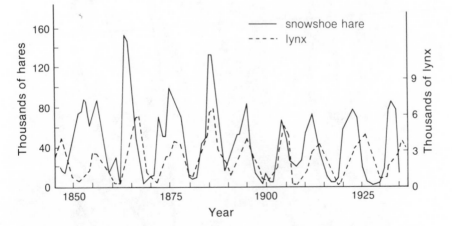

FIGURE 13–15 Population cycles of the lynx and the snowshoe hare in the Hudson Bay region of Canada, as indicated by fur returns from trappers to the Hudson's Bay Company.

in the population of their principle prey, the snowshoe hare (Figure 13–16).

Ecologists are not completely agreed on the cause of population cycles. The populations vary much too regularly for the cycles to be caused solely by variation in the environment. The most reasonable explanation ties the cycles to inherent population properties of predator–prey interactions. In one version of this hypothesis, an increase in the hare population is followed by an increase in the lynx population, which eventually becomes so dense that the hare population can no longer sustain the predation rate and declines. Lynx follow suit as their food supply diminishes. When the lynx become scarce, the hare population can increase and the cycle begins again. This version probably does not adequately explain the lynx and hare population cycles because there are too many bits of contrary evidence. First, the reproductive potential of the hare is so much greater than that of the lynx that the lynx population could not increase fast enough to exterminate the hares unless some other factor, perhaps insufficiency of food, slowed the growth rate of the hare population. Second, lynx population peaks occasionally coincide with or precede hare population peaks rather than following them by a year or so. Third, on some islands from which lynx are absent, snowshoe hare populations fluctuate just as much as they do on the mainland. Perhaps the population cycles are caused by periodic decline in quality or quantity of plants on which the hares feed, which in turn causes a decline in the hare population (and the lynx population), and subsequently allows the plants a chance to recover from overeating by hares.

FIGURE 13–16   A snowshoe hare in its winter coat. The hares are brown in summer.

Attempts to dissect the population cycles of arctic mammals into their working parts by field observation and experimentation have met with little success. Lemmings have been studied intensively at Barrow, Alaska, where population density varies in a three- to five-year cycle by factors of 100 or more between trough and peak years. Predators undoubtedly exert a strong depressing influence on population growth because most of the increase in number of lemmings occurs during the winter beneath a protective mantle of snow. Nevertheless, the reduction of lemming population size in summer by predators is small compared to longer-term changes in population size. The only habitat characteristic that appears to parallel the lemming cycle at Barrow is the quality (not quantity) of the vegetation (Table 13-6). During one peak year, vegetation contained 22 per cent protein, but only 14 per cent during the next population trough. If lemming populations and plant quality have parallel cycles, the two may be interrelated: the total available nutrients in the tundra ecosystem may be of such small quantity that they are mostly transferred to the bodies of lemmings during peak years, thereby depressing plant growth and nutrient quality. Nutrients would then be restored to the cycle only after large numbers of lemmings die and their remains decompose.

Time Lags and Population Cycles

We expect density-dependent factors to restore population size to an equilibrium level, and yet cyclic populations never reach a single equilibrium point, but fluctuate around that point. If we can determine why density-dependent responses fail to damp population cycles, we may come closer to understanding predator–prey cycles. Experiments by Australian ecologist A. J. Nicholson on population regulation in the sheep blowfly demonstate that if the action of density-dependent factors is delayed, population cycles will occur.

TABLE 13-6   Features of a lemming population cycle, including vegetation characteristics, at Barrow, Alaska.

| | *Year* | | | |
	1960	1961	1962	1963
Relative peak density	125	0.5	1 to 10	50
Male body weight (g)	92	47	69	59
Litter size	7.6	7.0	7.3	6.7
Breeding season (days)	58	80	73	83
Green vegetation (lbs/A)	111	278	115	149
Per cent protein	22	14	17	19
Protein in plants (lbs/A)	24	40	20	28

The life cycle of the fly includes four stages: egg, larva, pupa, and adult. In laboratory cages both the larvae and the adults are fed liver. When larvae were fed 50 grams of liver each day, and adults received unlimited sugar solution to supplement their diet, the number of adults in the population fluctuated in a regular cycle from a maximum of about 4,000 per cage to a minimum of zero (at which point all the individuals in the population were either eggs, larvae, or pupae). The period of the oscillation varied between 30 and 40 days, which is the maximum lifespan of individual flies (Figure 13–17).

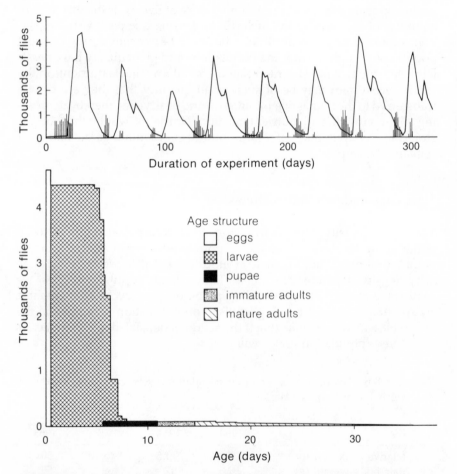

FIGURE 13–17   Fluctuations in laboratory populations of the sheep blowfly. Larvae were provided 50 grams of liver daily; adults were given unlimited supplies of food and water. The continuous line represents the number of adult blowflies in the population cage. The vertical lines represent the number of adults that eventually emerged from eggs laid on the days indicated by the lines. The average age structure of the population, representing the average survivorship curve, is shown below.

Regular fluctuations in the blowfly population were caused by a time lag in the response of fecundity and mortality to the density of adults. Large adult populations laid many eggs, but most of the larvae hatched from those eggs starved or failed to pupate owing to an inadequate amount of food; few became adults. Because large adult populations produced few progeny, and because adults live a maximum of four weeks, peak populations declined rapidly. During adult population lows, few eggs were laid and most of the larvae survived. The population then increased to peak levels again. Fluctuations in population numbers were not damped because the effects of adult population size were felt during the larval period of the *next* generation. These effects were expressed in adult populations two to four weeks after the eggs were laid when the progeny were adults of the next generation.

The time lag in the blowfly population could be eliminated by restricting the amount of food available to adults. Since adults must consume protein to produce eggs, limiting their food curtails egg production to a level determined by the availability of liver rather than by the number of adults in the population. When adult blowflies were fed only one gram of liver per cage per day, and the larvae were fed a special food preparation that the adults could not eat, the recruitment of new adults into the population was determined at the egg-laying stage by the influence of food on fecundity (nearly all the larvae survived); as a result, fluctuations in the population all but disappeared (Figure 13–18).

Behavioral Aspects of Population Regulation

Social behavior among animals ranks an individual according to its position on a scale of social dominance or according to the quality of territory it procures. Strong territorial systems or hierarchical systems of social dominance can prevent many individuals in a population from reproducing. Environmental fluctuations often affect the proportion of individuals reproducing in addition to the fecundity or survivorship of individuals that do reproduce. The attainment of sexual maturity by mammals often depends on the successful procurement of breeding territories. Within social groups, individuals vie for position in the dominance hierarchy and for the opportunity to reproduce, which depends on high social status (Figure 13–19).

By limiting the occupation of suitable habitats, territoriality can exert a strong stabilizing effect on population size. Social behavior immediately adjusts the density of organisms to year to year fluctuations in their resources, and thereby eliminates the time lag from the density-dependent response. Mammals that maintain breeding territories in the summer and hibernate through the winter — chipmunks, squirrels, bears, raccoons — do not exhibit well-defined population fluctuations. Mammals and birds that do show cyclic populations are typically nonterritorial for a large part of the year.

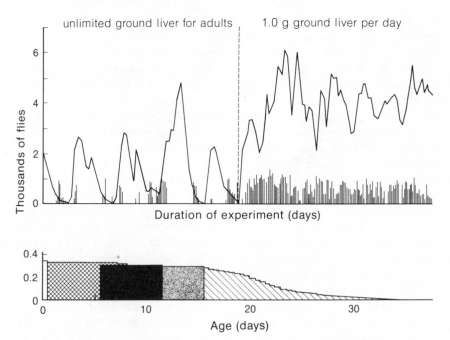

FIGURE 13–18   Effect on fluctuations in a laboratory population of sheep blowflies of limiting the amount of liver available to adults. The experiment resembled that depicted in Figure 13–17 in all other respects. Average age structure is shown for the latter half of the experiment.

Individuals excluded from holding territories in optimum habitats either establish territories in poor habitats where they cannot breed or wander through the population searching for the occasional opening resulting from the death of a territory holder. Experimental removal of territorial birds, fish, and insects has demonstrated the presence of large reserves of nonbreeding individuals that quickly occupy vacated habitats. In addition to limiting reproduction, failure to procure a territory exposes the individual to increased risk of mortality through starvation, disease, exposure, or predation.

Social Pathology

Working with artificial laboratory populations of mammals, many physiologists have observed decreased survival and extreme impairment of reproduction at high population density, even when food, water, and nesting sites are provided in excess of requirements. Social stress leads to a variety of abnormal physiological symptoms, collectively referred to as the general adaptive syndrome, which includes

FIGURE 13–19 Two bull elks fighting among a herd in Wyoming. The outcome of the encounter will affect each adversary's social position and his
prospects of reproducing.

curtailment of growth and reproduction, delay of sexual maturity, increased mortality of embryos, inadequate lactation, and increased susceptibility to disease.

When house mouse populations are confined with unlimited food
and nesting boxes, mice continue to reproduce regardless of the density
of the population, but the survival of embryos and young drops so low
that population growth ceases and population size may even decline
(Figure 13–20). As density increases, individuals fight more frequently
and the number of wounded or diseased adults, mostly males, rises.

Social pathology has been suggested as a mechanism of population
regulation in natural populations of mammals and as a cause of population cycles, but concepts of social stress developed for confined populations do not apply widely in nature. First, few natural populations
achieve the densities that can be maintained in the laboratory. The
densities of natural populations are kept low by emigration of socially
subordinate individuals from regions of high population density. Second, where resources can support dense populations, evolutionary responses in social behavior will favor those individuals capable of tolerat-
ing extreme crowding. We see this in many colonial organisms that
reproduce normally at densities far beyond levels that would be tolerated by species which maintain large breeding territories. That repro-

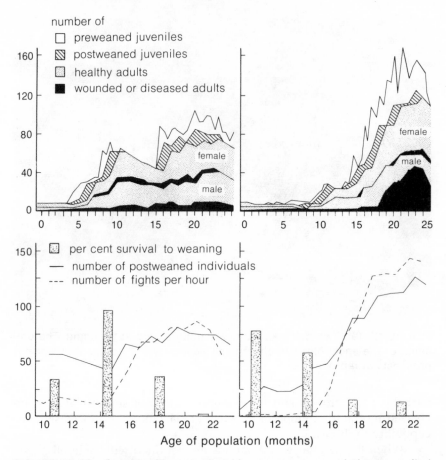

FIGURE 13–20 Growth of two confined house mouse populations supplied with unlimited food, showing the age structure, level of aggression, survival of young to weaning, and the incidence of diseased or wounded adults.

ductive behavior can evolve to function normally at high population densities can best be appreciated by a visit to a colony of prairie dogs or sea birds, where young are successfully raised amidst utter bedlam (Figure 13–21).

We have examined the balance of forces that regulate population size. Some forces are density-dependent. The available resources in its environment limit the inherent capacity of a population to grow and tend to keep the population at a level for which the resources are adequate. If the availability of resources changes, the size of the population changes in an appropriate manner. Other forces — cold, drought, and the like — are density-independent. They may limit the growth of a population without regard to its density or to its resources.

FIGURE 13–21   A dense breeding colony of royal terns on the coast of North Carolina. The social behavior of the terns has evolved to allow tolerance of high densities.

Changes in population size are compensated by homeostatic population responses, including growth and dispersal. The response of a population to its resources can lead to cyclic fluctuations in population size if the influence of the response is delayed by time lags. Such cyclic fluctuations have great influence on the total stability of the ecosystem and on the efficiency of energy and nutrient flow.

Inseparable from the notion of density-dependent forces and responses of populations to them is the fact that individuals compete with one another for resources. In the next chapter, we shall examine the influence of competition, whether between individuals of the same or of different species, on population processes and on the coexistence of populations of different species within the community.

14

Competition

Energy flows through the community from one link in the food chain to the next, from one trophic level to the next. We might infer that predator–prey relationships dominate the trophic structure of the community. We should not forget, however, that many kinds of organisms occupy each trophic level, each vying with the others to fill its need for energy and nutrients. Organisms that potentially may use the same resources are called competitors. Competition between species with similar resource requirements determines the organization of each trophic level, and therefore helps to regulate ecosystem structure and function.

We may define *competition* as the use of a resource (food, water, light, space) by an organism which thereby reduces the availability of the resource to others. If competing individuals belong to the same species, their interaction is called *intraspecific* (or within-species) competition; if the individuals belong to different species their interaction is called *interspecific* (or between-species) competition. In either case, a resource consumed by one individual can no longer be used by another. When a fox captures a rabbit, there is one less rabbit in the prey population for other foxes, or for bobcats, hawks, and others that also prey on rabbits.

The slowing of population growth as a population approaches the carrying capacity of its environment results from competition between individuals in the population. When a population is small, intraspecific competition is weak and resources are abundant. At high population densities, intense intraspecific competition reduces resources below the level that will sustain further population growth, and thus regulates the size of the population.

Both intraspecific and interspecific competition depress population growth rate; a resource is depleted, no matter who consumes it. Populations are usually less dense in the presence of a competing species than when they exist alone. Each trophic level survives on a limited supply of nutrients and energy. The more species, on each trophic level, the less common each must be. The organization of each trophic level — including the number of species and their relative abundance — is determined by competitive interactions among its members. We shall examine competition between species in this chapter to assess its narrow role in regulating population size and its broad consequences for community structure.

Coexistence and Competitive Exclusion

Field observations and laboratory experiments convey opposite impressions of nature. We frequently observe many ecologically similar species coexisting in nature, clearly using many of the same resources. Closely related species of trees grow in the same habitat, all needing sunlight, water, and soil nutrients. Coastal estuaries and inland marshes harbor a variety of fish-eating birds, including egrets, herons, terns, kingfishers, and grebes (Figure 14-1). As many as six kinds of warblers (small insectivorous birds), all with similar morphology and feeding habits, may be found in one locality in forest habitats of the northeastern United

FIGURE 14–1   Three closely related species of egrets (common, reddish, and snowy) feeding in the Aransas National Wildlife Refuge, Texas.

States. As many as eight species of large predatory snails frequent the shallow waters of Florida's Gulf Coast.

By contrast, closely related species rarely coexist in the laboratory. If two species are forced to live off the same resource, one inevitably persists and the other dies out. Reconciling these observations has been a major task for ecologists over the past half-century, since the Russian biologist G. F. Gause first tried to make two similar species coexist in the laboratory.

Gause established laboratory cultures of closely related species of protozoa, of the genus *Paramecium*, in the same nutritive medium. The species flourished when separate, but in mixed cultures only one species survived (Figure 14–2). Similar experiments with fruit flies, mice, flour beetles, and annual plants always produce the same result: one species persists and the other dies out, usually after 30 to 70 generations.

The results of laboratory competition experiments led to the formulation of the *competitive exclusion principle,* also called Gause's principle, which states that two species cannot coexist on the same limiting resource. The word "limiting" is included in the statement of the principle

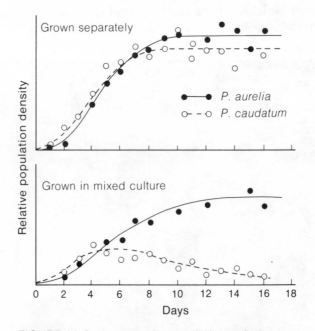

FIGURE 14–2 Increase in populations of two species of *Paramecium* when grown in separate cultures (above) and when grown together (below). Although both species thrive when grown separately, *P. caudatum* cannot survive when grown with *P. aurelia*.

because only resources that limit population growth can provide the basis for competition. Nonlimiting resources, such as atmospheric oxygen, are superabundant compared to the needs of organisms and their use by one organism does not make them less available to others. How can the principle of competitive exclusion, verified by laboratory experiments, be reconciled with observations of similar species coexisting in nature?

Species Diversification

Species coexist because their resource requirements differ. Whenever ecologists have examined groups of similar species in the same habitat they have found small but significant differences in size or foraging behavior that enable the species to use slightly different resources and avoid intense competition. Imagine that we are watching the great cormorant and the shag cormorant (large, dark-colored, diving sea birds) feed in the same area and apparently in the same manner. They both swim on the surface of the water and dive for food. The cormorants seem to be competing for food, but the appearance is deceptive. Had we followed the cormorants beneath the water and observed their feeding habits directly, we should have found that one feeds primarily at the bottom and the other at intermediate depths. Because the species feed in different parts of the habitat, their diets do not overlap greatly and they do not compete intensely for food. This fact was not appreciated until studies of their stomach contents revealed that more than 80 per cent of the diet of the great cormorant consisted of sand eels and herring, which swim well above the bottom, whereas the shag cormorant ate bottom-living forms, particularly flatfish and shrimp.

Four species of honeybees common in English fields avoid competitive exclusion through morphological and behavioral specializations to slightly different food resources. All four species gather nectar and pollen from flowers, but each has a different tongue length and visits flowers of correspondingly different size. The species with the longest tongue feeds only at a few kinds of plants, those with long tubular flowers. The other species have more broadly overlapping flower preferences but two of them feed in shrubby habitats, the other in open fields. The two shrub-habitat bees further avoid competition by appearing at slightly different times of the year.

A most interesting example of specialization in eight closely related species of worms, parasitic in the lower digestive tracts of European tortoises, demonstrates the generality of species diversification. All eight species of worms feed on unassimilated remains of food, but each apparently prefers a slightly different consistency and composition associated with a specific region of the colon. The species are concentrated within different regions along the length of the colon and some also exhibit a distinct preference for the peripheral area next to the lining of the colon (Figure 14–3). Species that occupy the same region of the colon may have different food preferences. One species consumes mostly gut bacteria, another feeds on the gut contents more generally.

Population Effects of Competition

The species we find coexisting in nature are the successful ones. What of the species that failed? Fossils of extinct species prove that populations have died out. Was their demise caused by superior competitors?

Competitive exclusion is a transient phenomenon. The evidence of

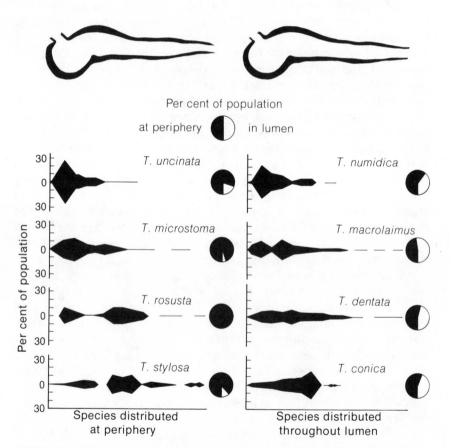

FIGURE 14–3   Distribution of parasitic worms of the genus *Tachygonetria* in the colon of the European tortoise. Species in the left-hand group were concentrated toward the edge of the colon; species in the right-hand group were more evenly distributed throughout the contents of the colon.

exclusion having taken place is lost when the poorer competitor is eliminated. We can observe competitive exclusion in the laboratory because we can mix populations according to our whims and follow the course of their interaction. The closest natural analogy to the laboratory experiment is the accidental or intended introduction of species by man. When new immigrants are superior competitors to resident species, the immigrants can reduce or eliminate local populations. The explosion of rabbit populations following their introduction to Australia (see Chapter 10) worsened range conditions and thereby reduced populations of many native marsupial herbivores, such as kangaroos and wallabies. (Rabbit control programs were aimed more at preserving the range for another introduced competitor — the sheep — than preserving the native fauna.)

Competition between introduced species has produced some vivid demonstrations of the competitive exclusion principle. When many species of parasites are intentionally introduced to control a weed or insect pest, the control species are brought together in the same locality to prey on, or parasitize, the same resource. We should not be surprised that competitive exclusion has occurred frequently under these conditions. Between 1947 and 1952, the Hawaii Agriculture Department released 32 potential parasites to combat several species of fruit pests, including the Mediterranean fruit fly. Thirteen of the species became established, but only three kinds of braconid wasps proved to be important parasites of fruit flies. Populations of these species, all closely related members of the genus *Opius,* successively replaced each other from early 1949 to 1951, after which only *Opius oophilus* was commonly found to parasitize fruit flies (Figure 14-4). As each parasite population was replaced by a more successful species, the level of parasitism of fruit

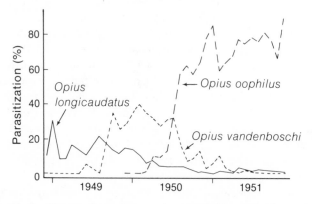

FIGURE 14-4 Successive change in predominance of three species of wasps of the genus *Opius,* parasitic on the Oriental fruit fly.

flies also increased, indicating the superior competitive ability of the newly established population.

A similar pattern of competitive replacement involving wasps that are parasites of scale insects, which are pests of citrus groves, has been thoroughly documented in southern California. With the failure, owing to the evolution of resistance by pests, of chemical pesticides to provide adequate, long-lasting control, agricultural biologists turned to the importation of insect parasites and predators. Yellow scales have infested California citrus groves since oranges and lemons were first planted there. In the late 1800's, the red scale was accidently introduced and has replaced yellow scale almost completely, perhaps itself a case of competitive exclusion. Of the many species introduced in an effort to control citrus scale, tiny parasitic wasps of the genus *Aphytis* (from the Greek *aphyo,* to suck) have been most successful. One species, *A. chrysomphali* (literally the golden-navel sucker), was accidently introduced from the Mediterranean region and became established by 1900.

The life cycle of *Aphytis* begins when adults lay their eggs under the scaly covering of hosts. The newly hatched wasp larva uses its mandibles to pierce the body wall of the scale and proceeds to consume nearly all the body contents. After the wasp pupates and emerges as an adult, it continues to feed on scales while producing eggs. Each female can raise 25 to 30 progeny under laboratory conditions and the development period is so short (14 to 18 days egg to adult at 80°F) that populations may produce 8 to 9 generations per year in the long growing season of southern California.

In spite of its tremendous population growth potential, *A. chrysomphali* did not effectively control scale insects, particularly not in the dry interior valleys. In 1948, a close relative from southern China, *A. lingnanensis,* was introduced as a control agent. This species increased rapidly and almost completely replaced *A. chrysomphali* within a decade (Figure 14–5). When both species were grown in the laboratory, *A. lingnanensis* was found to have the higher net reproductive rate whether the two species were placed in separate or in mixed population cages.

Although *A. lingnanensis* had replaced *A. chrysomphali* throughout most of southern California, it did not provide effective biological control of scale insects in the interior valleys because cold winter temperatures periodically reduced parasite populations. Wasp development slows to a standstill at temperatures below 16°C (60°F) and adults cannot tolerate temperatures below 10°C (50°F). Pupae are more resistant to cold, but average winter pupal mortality was 42 per cent, with extremes of about 80 per cent in the cold interior valleys.

In 1957, a third species of wasp, *A. melinus,* was introduced from areas in India and Pakistan where temperatures range from below freezing in winter to above 40°C in summer. As expected, *A. melinus* spread rapidly throughout the interior valleys of southern California, where

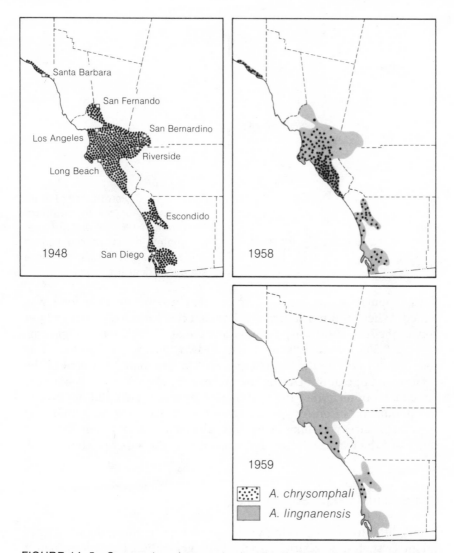

FIGURE 14-5 Successive changes in the distribution of *Aphytis chrysomphali* and *A. lingnanensis*, wasp parasites of citrus scale, in southern California. *A. lingnanensis* was first released in 1948 and rapidly replaced *A. chrysomphali* throughout the region.

temperatures resemble the wasp's native habitat, but did not become established in coastal areas (Figure 14-6).

Competition between *A. lingnanensis* and *A. melinus* was studied under controlled laboratory conditions at 27°C (80°F) and 50 per cent relative humidity. The wasps were provided with oleander scale grown on lemons as hosts, and new scale-infested lemons were added every 17

FIGURE 14–6 Distribution of three species of *Aphytis* in southern California in 1961. *A. melinus* predominates in the interior valleys while *A. lingnanensis* is more abundant near the coast.

days. Populations of wasps were counted just before new food was added. When cultured separately, both species parasitized about 40 per cent of the scales and wasp populations averaged 6,400 for *A. lingnanensis* and 8,300 for *A. melinus*. When grown together, *A. melinus* was reduced in four months from 50 per cent to less than 2 per cent of the total wasp population. During the process of competitive exclusion, the combined population of the two species (average 7,400) did not exceed the population of either species grown separately. We may infer that population size was limited by host availability and under the conditions of the experiment, *A. lingnanensis* was the superior competitor.

Competition in Plant Populations

Plants differ from animals in two ways that bear on the study of competition. First, few terrestrial species have generation times less than a year. Plant ecologists therefore often cannot continue experiments for a long enough period to demonstrate competitive exclusion. Second, plant growth, as well as survivorship, is greatly affected by the variety of conditions under which the plant may live. In particular, plants grow slowly when crowded and do not attain their full stature, even though they may produce seed. In contrast, animal populations usually respond to crowding with increased mortality and reduced fecundity, not stunted growth.

When horseweed seed is sown at a density of 100,000 seeds per square meter (equivalent to about 10 seeds in the area of your thumb nail), the young plants compete vigorously. As the seedlings grow, many die and the density of surviving seedlings decreases (Figure 14–7). At the same time, the growth of surviving plants exceeds the decline of the population, and the total weight of the planting increases. Over the entire growing season, the hundred-fold decrease in population density is more than balanced by a thousand-fold increase in plant weight.

The relationship between plant weight and density, shown in Figure 14–7, is often called a *self-thinning curve*. When obtained from very dense plantings, self-thinning curves reflect the maximum capacity of the soil to support plant growth. Combinations of density and size lying outside the self-thinning line never occur because they are beyond the carrying capacity of the environment. If horseweed were sown at a density of 10,000 seeds per square meter, rapid growth with little early mortality would carry the population to the self-thinning line as shown in Figure 14–7. Sown at densities of 1,000 per square meter, few of the horseweed plants would die before reaching maturity.

Each species has a characteristic thinning curve under particular conditions of soil, light, temperature, and moisture. Regardless of the density of the initial planting, the eventual harvest of mature plant biomass is surprisingly uniform. The self-thinning curve does not, however, tell the full story of within-species competition. When sown at low initial density, most individuals grow vigorously. At high density, only a few individuals reach large size. Other survivors do poorly because

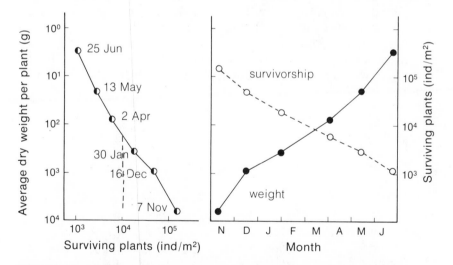

FIGURE 14–7 Progressive change in plant weight and population density in an experimental planting of horseweed (*Erigeron canadensis*) sown at a density of 100,000 seeds per square meter (10/cm²). The relationship between plant density and plant weight as the season progressed is shown at left, where the self-thinning curve of a planting with a density of 10,000 seeds per square meter is shown by the dashed line.

they are crowded by individuals that obtained an initial growth advantage owing to a favorable site for germination (Figure 14-8).

Because of the flexible growth response of plants and their long generation times, botanists usually assess competition between plant populations by comparing total plant weight or number of seeds produced in an experimental plot. These indices were used to measure competition between two closely related species of oats, *Avena barbata* and *A. fatua*. Plants were grown in watered pots at six densities: 8, 16, 32, 64, 128, and 256 individuals per pot. When grown separately, density had relatively little effect on germination, establishment of seedlings, or survival to maturity. Growth responses were, however, markedly different. At high densities, individual plants attained smaller average weight and height and produced fewer seeds.

To measure the influence of interspecific competition on plant growth, pots were sown at each of several densities with different proportions of the two species. In experiments with seed densities of 128 per pot, for example, a group of pots would be planted with 128 *barbata* and 0 *fatua*, 112 *barbata* and 16 *fatua* (total 128), 64 *barbata* and 64 *fatua* (total 128), and so on, each totalling 128 seeds. For each species, the pots created conditions ranging from pure intraspecific competition (128 *vs.* 0) to strong interspecific competition (16 *vs.* 112).

When grown separately at densities of 128 seeds per pot, both species showed survival rates of about 96 individuals per pot (75 per cent). If the survival of a *fatua* individual was affected by competition with *barbata* individuals no differently than from competition with other *fatua* individuals, we would expect 75 per cent of *fatua* seeds to survive regardless of the ratio of the two species in the pot. Thus 12, 48, and 84 individuals would survive from initial plantings of 16, 64, and 112 seeds.

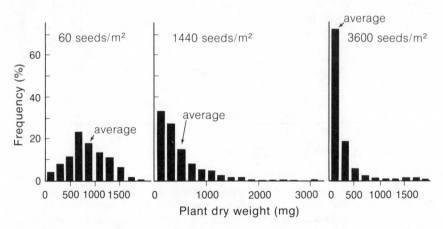

FIGURE 14-8   The distribution of dry weights of individuals in populations of flax plants sown at different densities.

The relationship between the initial planting and the final outcome is often depicted on a graph called a *replacement series diagram* (Figure 14-9). The left-hand figure represents the expected number of *fatua* surviving to maturity if interspecific competition and intraspecific competition had equal influence on survival. A straight line drawn between 96 individuals on the right-hand axis (128 *fatua* seeds per pot) and 0 in the left-hand axis (0 *fatua* seeds per pot) represents 75 per cent survivorship of *fatua* seeds. If interspecific competition depressed the survival of *fatua* seedlings, relative to the effect of intraspecific competition, the outcomes of the experiment in the mixed species pots would fall below the line of equal competitive effect. If, on the other hand, *fatua* grew better under conditions of strong interspecific competition than it did under strong self-inhibition, the points would fall above the line of equal competitive effect.

Survival of both *fatua* and *barbata* exhibited no differential response to intraspecific and interspecific competition. Weight and seed production did, however, display differences (Figure 14-10). *A. fatua* produced more seeds per plant when grown in competition with *barbata*. On the other hand, *A. barbata* produced fewer seeds per plant when grown with *fatua,* than in pure culture. Thus *fatua* competes well against *barbata* under the conditions of the experiment and we would expect *fatua* eventually to exclude *barbata*.

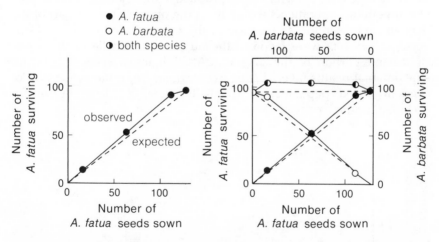

FIGURE 14–9 Replacement series diagrams representing the outcome of competition between *Avena fatua* and *A. barbata* grown at a density of 128 seeds per pot. The left-hand diagram shows the expected response of *fatua* survival if the effects of intra- and interspecific competition did not differ. Actual results fall closely to this expectation. At right, the results of *A. fatua* and *A. barbata* are plotted on the same diagram. The expected total survivorship of both species is added to the diagram.

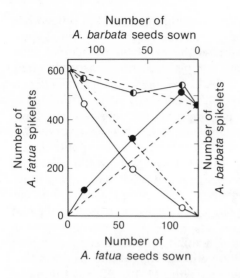

FIGURE 14–10 Replacement series diagram for seed production (measured as number of flower spikelets per pot) in *Avena fatua* and *A. barbata*.

We may visualize the outcome of competition more readily on a ratio diagram (Figure 14–11) in which the ratio of seeds produced by one species to seeds produced by the other is compared to the initial ratio of seeds planted. If the ratio of the species of seed produced from a pot is identical to the ratio of seeds sown, neither species has a competitive advantage. This situation is represented by the diagonal line in Figure 14–11. If *fatua* were relatively more productive than *barbata,* the outcome of competition experiments would lie in the upper left-hand part of the diagram, representing an increase in the ratio of *fatua* to *barbata.* Conversely, if *barbata* were favored, the outcome would lie in the lower right-hand portion of the diagram. At high initial seed density, all experimental results favored *fatua* over *barbata.* Since the proportion of *fatua* in mixed populations continually increased, *fatua* eventually excluded *barbata.*

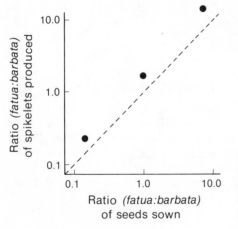

FIGURE 14–11 Ratio diagram of competition between *Avena fatua* and *A. barbata*, representing the outcome of the experiment depicted in Figure 14–10.

Ecological Replacement of Species

When closely related species are thrown into direct competition in the laboratory, which species turns out to be the superior competitor depends upon the conditions of the experiment. When two species of flour beetle, *Tribolium castaneum* and *T. confusum,* are grown together in wheat flour, *castaneum* excludes *confusum* under cool, dry conditions, but itself is excluded when colonies are kept warm and moist (Figure 14-12). When grown separately, both species reach their highest densities in the moister conditions. Thus a species may be a superior competitor under sub-optimum conditions; the outcome of competition is not determined by the performance of a species in the absence of competition, rather it is determined by the relative productivity of species grown together. The point is further emphasized by the observation that the species which was superior in competition at each combination of temperature and humidity was not necessarily the most productive species when grown alone under the same combination. Competition involved the special interaction between species not often evident when observing each species alone.

In the complex mosaic of conditions in the natural environment, we

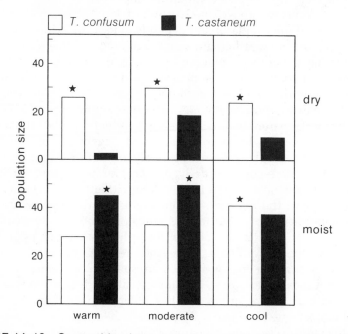

FIGURE 14–12   Competition between two species of flour beetles at different temperatures and relative humidity. Each section of the diagram shows the population density of the two species when grown separately (vertical bars); the superior competitor under each set of conditions is indicated by the star.

would expect species to be superior competitors in one place and to be inferior elsewhere. As we travel from one locality to the next, changing environments are followed by a successive replacement of species, each superior to its competitors at a particular point along our route. We have seen, for example, that the ability of the parasitic wasp *Aphytis melinus* to tolerate winter cold enabled it to displace the closely related *A. lingnanensis* in the interior valleys of southern California. Whereas *lingnanensis* was formerly widespread throughout the region, its range is presently restricted by a superior competitor to coastal regions.

If we were to climb to the summit of the White Mountains in southern California, we would pass through bands of vegetation types restricted to the narrow range of altitude where conditions give them superior competitive ability. Near timberline, a common species of fleabane (herbaceous perennial plants), *Erigeron clokeyi,* is abruptly re-placed by the closely related *E. pygmeus.* Near the upper limit of the distribution of each species, low temperatures slow development so much that in many years flowering does not occur before the first frost. At the lower limits of their distribution, both species are adversely affected by the summer drought conditions characertistic of lower al-titudes. Because *clokeyi* is more tolerant of drought and *pygmeus* is more tolerant of cold, the relative competitive superiority of *clokeyi* at lower elevations is reversed above an elevation of 11,000 to 12,000 feet. The exact altitude of their replacement depends in part upon the underlying bedrock. Sandstone weathers into a dark-colored soil that absorbs sun-light and retains warmth. Dolomite forms a light-colored, relatively cool soil. The sandstone causes temperature and moisture in a locality to resemble conditions at a lower elevation. Hence *clokeyi* replaces *pygmeus* at a higher elevation on sandstone than on dolomite.

Underlying parent bedrock has also influenced the outcome of competition between bristlecone pine and sagebrush in the White Mountains. Sagebrush is found primarily on south-facing and east-facing slopes where soils are derived from granite or sandstone; pine predominates on soils derived from dolomite (Figure 14–13). The com-petitive ability of the two species depends upon the relative moisture and phosphorus contents of the soil. Sandstone and granite weather to form soils with high levels of available phosphate, but poor water retention qualities. In addition, these soils are dark-colored and surface temperatures may be 5°C warmer than soils on dolomite. Heat, of course, increases water stress. Dolomite weathers into a soil with rela-tively good moisture retention, but low phosphate content.

Laboratory measurements of photosynthesis showed that sage-brush tolerates lower soil moisture than pine. Sagebrush seedlings reached peak photosynthetic rates at about five per cent soil moisture, which is three per cent lower than the point at which photosynthesis in pine seedlings begins to shut down.

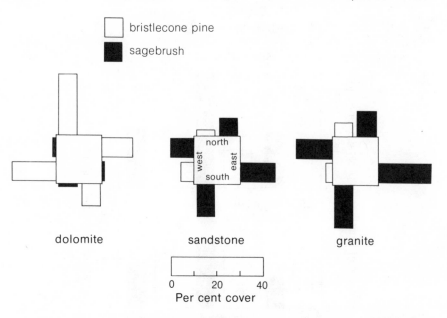

FIGURE 14–13 The percentage vegetation cover of bristlecone pine and sagebrush between 9,500 and 11,500 feet elevation in the White Mountains of southern California with respect to exposure and underlying bedrock.

By growing romaine lettuce in pots of each type of soil in the laboratory, dolomitic soils were shown to provide poor nutrition for plants. Of the total production on granite soil, lettuce produced only 36 per cent as much on sandstone soil and 18 per cent as much on dolomite soil. Production on dolomite was boosted by adding phosophorus, but not nitrogen or potassium, to the soil. Dolomite soils have a low phosphorus content partly because the abundant calcium and magnesium carbonate in the rock produce a soil with a slightly basic reaction. As we have seen (Chapter 9), phosphorus availability is greatest in slightly acid soils, such as the soils produced by granite and sandstone.

The influence of the mineral content of soil on competition between pine and sage was determined by measuring the growth of seedlings of each species, grown separately and together. One-inch tall seedlings were planted in pots with soil taken from localities with dolomite, sandstone, and granite bedrock. The pots were placed out-of-doors and supplemented with distilled water to avoid drought stress. The seedlings were harvested, dried, and weighed after six months. The experiment demonstrated that the presence of a sagebrush seedling reduced the growth of pine seedlings by less than 30 per cent, and not at all on dolomite soils, whereas pine seedlings reduced the growth of sage by between 24 and 40 per cent, depending on the soil type (Figure 14–14).

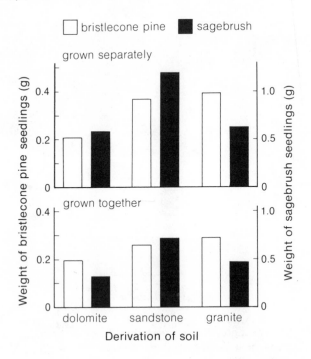

FIGURE 14–14   Relative growth of bristlecone pine and sagebrush seedlings planted separately and together in soil derived from different parent materials.

The superiority of pine seedlings on poor soils can be related to a close association between the pine roots and fungi. Certain species of fungi form what are known as *mycorrhizal* associations in which the fungal mycelium extends from the cells of the pine roots into the soil. The mycorrhizal fungi acidify the soil, thereby increasing the availability of phosphorus, and decompose organic materials, thereby releasing nutrients to the pine. In return, the fungi obtain sugars from the pine to sustain their metabolism.

Mutual exclusion of two species of barnacles at different heights in the intertidal zone of the coast of Scotland provides an excellent example of how competitive ability varies with local conditions. Adult *Chthamalus* are found higher in the intertidal zone than *Balanus*. The line of demarcation between the vertical distributions of adults of the two species is sharp, even though the vertical distributions of newly settled larvae overlap broadly. *Chthamalus* is not restricted to the zone above *Balanus* by physiological tolerance limits; if *Balanus* is removed from rock surfaces in the lower part of the intertidal zone, *Chthamalus* thrives there. The two species normally compete for space where the larvae grow up together (Figure 14–15). The heavier-shelled *Balanus* grows more rapidly

FIGURE 14–15   Competition for space among barnacles on the Maine coast. Above their optimum range in the intertidal zone, the barnacles are sparse and there are bare patches for the young to settle on (left). Lower in the intertidal (right) the barnacles are so densely crowded that there is no room for population increase; the young barnacles are forced to settle on older individuals.

than *Chthamalus,* and as individuals expand, the shells of *Balanus* literally pry the shells of *Chthamalus* off the rock. Hence rapid growth gives *Balanus* a distinct competitive edge in the lower parts of the intertidal zone. At higher levels, *Chthamalus* achieves competitive superiority because it resists desiccation better than *Balanus.*

Ecological Compression and Ecological Release

Exclusion is an extreme consequence of competition. Many of the species that coexist in any one place share a portion of their resources but do not overlap enough to cause the elimination of species. Nonetheless, each species depresses the populations of other species by an amount related to their ecological similarity. If a species is removed from a habitat, the populations of ecologically similar species increase in response to the additional resources made available. This response is often referred to as *ecological release,* several examples of which are presented below.

An experimental demonstration of the influence of interspecific competition for light and water on plant growth comes from the tropical forests of Surinam, where forest ecologists set out to determine whether the growth of commercially valuable trees could be improved by removing species of little economic importance. The foresters poisoned 79 per cent of undesirable trees with girths greater than 30 centimeters in one area and greater than 15 centimeters in another, leaving the desirable species untouched. The increase in girth of the desirable trees was then measured over a year's time in experimental plots and in control plots which had not been selectively thinned (Figure 14–16). Removing trees greater than 30 centimeters in girth increased the penetration of light to the forest floor by six times. Removing smaller trees (15 to 30 cm girth) further increased light penetration by about one-third. Additional light stimulated the growth of the trees left on the experimental plot; improvement was greatest among small individuals, which are normally most shaded by competing species. Although removal of small trees did not increase light levels as much as removal of large trees, it produced a striking response in growth rate, particularly among the remaining large trees. The improved growth could not have been caused by increased light because many of the trees that responded were much taller than the trees which were poisoned. Therefore, the added growth stimulus probably resulted from reduced competition for either water or mineral nutrients in the soil.

The depressing effect of intraspecific competition on growth of trees has been demonstrated in many forest thinning experiments in temperate-zone forests. The additional growth of young longleaf pine

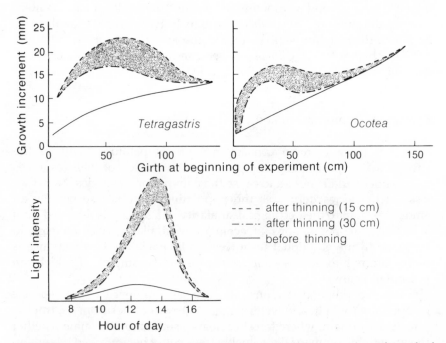

FIGURE 14–16 Effect on the increase in girth of two species of tropical trees, *Ocotea* and *Tetragastris*, achieved by removing competing trees greater than 15 centimeters or 30 centimeters in girth. Increase in light intensity which resulted from thinning is shown at left. Shaded area represents the difference between plots with 15-centimeter girth and 30-centimeter girth trees removed.

trees in response to selective thinning of trees over 15 inches in diameter is shown in Figure 14-17. Each core of wood was obtained by boring into a tree trunk, from the bark to the center, with a long, tubular device called an increment borer. The core of wood removed by the borer tube gives a record of annual ring growth without cutting down the entire tree. These cores show a very rapid increase in growth rate, particularly in summer (dark wood), during the 18 years between the time the forest was logged and the time the cores were taken.

A study of competition between shrubs and grasses in the dry foothills of the Sierra Nevada Mountains in California demonstrated that the type of vegetation that gained an initial foothold effectively excluded the other type. Rapid-growing, deep-rooted annual grasses reduced soil moisture below the point that shrub seedlings could survive. On the other hand, because of their tall growth form, shrubs shade out grasses and other herbaceous vegetation. Competition of grasses on shrub seedlings was demonstrated by removing grasses from experimental plots a little more than six feet on a side (0.001-acre area), and following the growth and survival of shrub seedlings. One plot was completely denuded of grass and other herbaceous vegetation and in another the grass was clipped to a height of one-half inch each week. A third plot was left undisturbed as a control. Sets of three such plots were established at ten localities within the study area.

At the beginning of the study in early spring, numerous seedlings of wedgeleaf ceanothus, chaparral whitethorn, and manzanita sprouted

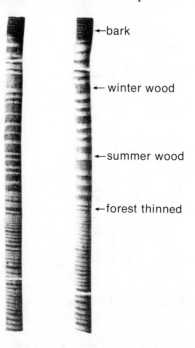

←bark

← winter wood

←summer wood

←forest thinned

FIGURE 14–17 Cores of two longleaf pines obtained near Birmingham, Alabama, showing the effect of removing large trees on subsequent growth.

in all the plots. After the last rains of the season in April, the soil contained 13 to 14 per cent moisture at all depths. As the summer dry season progressed, soil moisture levels in control plots decreased quickly, reaching the wilting point in the top foot of soil by mid-May (Figure 14–18). Because grass roots penetrate to a depth of four to five feet, grasses can obtain moisture throughout the summer. The roots of

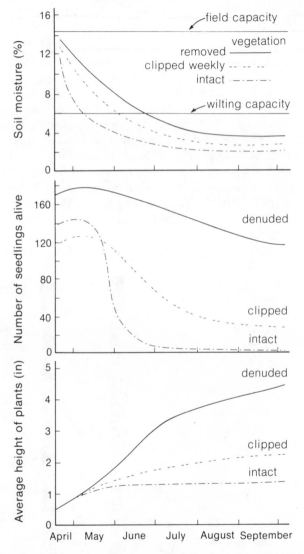

FIGURE 14–18 Seasonal progression of moisture content in the uppermost foot of the soil profile, survivorship of brush seedlings, and average height of surviving brush seedlings in plots where herbaceous vegetation was removed, clipped weekly, and left intact.

shrub seedlings do not grow fast enough to reach below the moisture-depleted upper layers of soil and die rapidly after soil moisture drops below the wilting point. In clipped and denuded plots, the soil retained its moisture long enough for the seedlings to become firmly established. In mid-July, by which time most of the grasses had matured, set seed, and begun to die back, soil moisture profiles on clipped and control plots had reached their greatest difference (Figure 14–19). To the depth of penetration of grass roots, soil moisture was reduced below the wilting point, showing the influence of transpiration on soil moisture. Shrub roots had penetrated no farther than a foot in control plots and most of the seedlings had died. In denuded plots, soil moisture was sufficient for plant growth in all but the top foot of soil, and bush seedling roots had penetrated more than three feet into the soil.

Once annual grasses become established in the dry foothills at low elevations, they reseed the area each year and exclude bush seedlings.

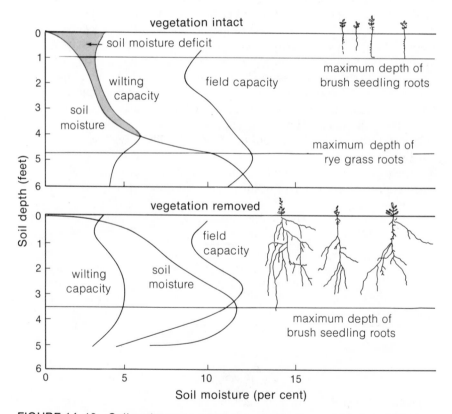

FIGURE 14–19   Soil moisture in relation to depth on control plots and plots denuded of herbaceous vegetation in early July. Depletion of soil moisture in control plots extends to the depth of grass roots, showing the influence of plant transpiration on soil moisture. Silhouettes of brush seedlings show the extent of their root systems.

At higher elevations, moisture levels in the soil are high enough to support bush seedling growth all summer, regardless of whether grasses and other herbs are present, and shrubs and small trees are the dominant vegetation components. At lower elevations, bushes sometimes become established in moist gullies or in the shade of rocks, and young bush seedlings grow at the edge of mature plants. Shrubby vegetation can invade grasslands in this way. If an area is denuded by fire, whichever type of vegetation reseeds the area in the greatest numbers will likely persist. Range management biologists have taken advantage of this fact to restore rangeland for cattle by controlled burning of bushland followed by heavy reseeding with annual grasses.

Ecological release following removal of competitors and, conversely, ecological compression following introduction of competitors maintain ecosystem function regardless of the specific composition of the community, or number of species in the community. Nowhere is this more clearly demonstrated than on oceanic islands. Distance from mainland sources of colonization has kept the number of species in most island ecosystems far below comparable mainland ecosystems. In response to the low diversity of competitors, island populations expand through habitats from which they would be excluded on the mainland, and they attain greater abundance in each habitat. For example, censuses of land-bird species in small patches of wet tropical habitats of comparable area and variety revealed 135 species in Panama, 108 in Trinidad (a large island near the coast of Venezuela), 56 in Jamaica, and 33 in St. Lucia (a small island in the Lesser Antilles). Ecological release of bird populations on the small islands compensated for the reduction in number of species: the total populations of all species of birds did not vary significantly between island mainland areas (Table 14–1).

TABLE 14–1   Relative abundance and habitat distribution of birds in four tropical localities.*

Locality	Number of species observed	Habitats per species	Relative abundance per species per habitat (density)	Relative abundance per species	Relative abundance of all species
Panama	135	2.01	2.95	5.93	800
Trinidad	108	2.35	3.31	7.78	840
Jamaica	56	3.43	4.97	17.05	955
St. Lucia	33	4.15	5.77	23.95	790

*Based on 10 counting periods in each of 9 habitats in each locality. The relative abundance of each species in each habitat is the number of counting periods in which the species was seen (maximum 10); this times number of habitats gives relative abundance per species; this times number of species gives relative abundance of all species together.

The size of song sparrow territories on small islands with different numbers of competing bird species demonstrates the principle of ecological release for a single species. Where many species of birds are present, the types of food resources available to song sparrows are a relatively small fraction of the total, the rest having been eaten by competitors. Sparrows must defend large territories to include enough food resources to breed successfully; as territory size increases, population density decreases. Where few competitors occur, the sparrows utilize a wider variety of food types; smaller territories can satisfy food requirements, and population density thus increases (Figure 14–20).

How Competition Occurs

We have seen many examples in this chapter of competition through the mutual use of limited resources. Such *indirect competition* need never bring competitors face to face. They do battle by seige and starvation rather than by direct attack.

One occasionally finds cases in which individuals disputing some resource do attack each other directly. *Direct competition* usually occurs over space. Defense of territories by birds and other animals is a conspicuous example of competition by interference, although most conflicts are restricted to individuals of the same species. It would be impossible to defend individual prey living within an area, and so competition is transferred from defense of the resource itself — food or water — to defense of an area containing it.

Space is the direct object of competition between barnacles;

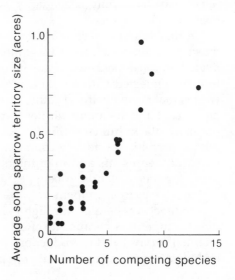

FIGURE 14–20 Territory size of song sparrows in relation to numbers of species of small land birds on small islands off the coast of Washington.

some species physically remove individuals of other species by prying them loose. One could construe the shading of competitors by plants as a form of direct competition, and possibly as severe a blow as a plant could deal. But not to be outdone by the behavior of which animals are capable, some species of plants inhibit the growth of other species by excreting or otherwise introducing toxic chemicals into the soil. The decaying leaves of walnut trees release toxic substances into the soil which inhibit seedling growth in many other species of trees. Such toxic restraints are frequently referred to as *allelopathy*. Perhaps the best-known case of chemical interference involves several species of sage of the genus *Salvia*. Clumps of *Salvia* are usually surrounded by bare areas separating the sage from neighboring grassy areas (Figure 14–21). Sage roots extend to the edge of the bare strip but not into the bare area beyond. Thus it seems unlikely that a toxic substance was exuded into the soil by the roots. The leaves of these species produce volatile terpenes (a class of organic compounds that includes camphor) which apparently affect nearby plants directly through the atmosphere. Heavy rainfall washes the toxic compounds out of the atmosphere, reducing the prohibitory effect of sage on other species. This may explain the general absence of direct chemical competition in wet climates.

Although terpenes produced by *Salvia* have been shown to suppress the growth of other species, *Salvia* may have evolved the ability to produce volatile chemicals as a means of attracting insects to pollinate its flowers. We are all acquainted with the taste of sage honey. The close association of the honeybee (*Apis mellifera*) and sage plants is underscored by the scientific names of two species of sage, *S. apiana* and *S. mellifera*. Furthermore, one of the terpene compounds produced by sages — cineole — acts as a powerful attractant to many kinds of bee. *Salvia* may have only secondarily adapted attractants to function as toxins to other plants.

A final example of competition, which is difficult to classify as either direct or indirect, involves the mutually depressing effects of closely related internal parasites. Parasitologists have long known that if a human is infected with a mild case of schistosomiasis caused by blood worms that normally infect cattle, he is unlikely to be severely affected if parasitized by the more virulent species of schistosome worm that normally attacks humans. Similar examples of cross-immunity to closely related parasitic species are common. It has been suggested that cross-immunity represents a form of interspecific competition between parasites. A parasite stands to benefit if it can stimulate its host's immune system to reject its competitors.

Indirect competition through exploitation of resources differs from direct competition because its effects are expressed slowly through differential survival and reproduction. Direct interference can immediately exclude a competing individual or population from a resource. Although

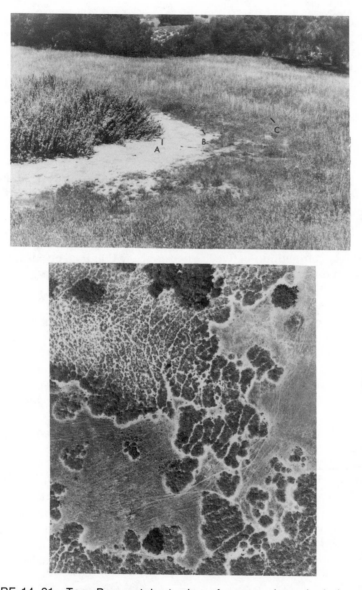

FIGURE 14–21   Top: Bare patch at edge of a sage clump includes a two-meter wide strip with no plants (A-B) and a wider area of inhibited grassland (B-C) lacking wild oat and bromegrass, which are found with other species to the right of (C) in unaffected grassland. Bottom: Aerial view of sage and California sagebrush invading annual grassland in the Santa Inez Valley of California.

interference reduces competition immediately, it also exacts a high price in terms of time, effort, and, potentially, survival. In evolving a strategy of competition, these costs must be weighed against the intensity of competition and the practicality of defending resources or resource substitutes. Interference competition occurs most frequently between individuals of the same species, hence between individuals that are ecologically most similar.

Coexistence and Resource Partitioning

Up to this point, competition has been discussed in terms of exclusion and success, elimination and persistence, superiority and inferiority. Although these terms describe what has *happened* in communities, coexistence describes what is found. For decades, ecologists have pondered the conditions necessary for species to coexist. Mathematical analyses of competition between species show that if a species limits its own numbers more than it limits the population of a second species, and vice versa, the two species can coexist. These conditions are fulfilled when each species uses somewhat different resources than the other. Consider the simple case of two species subsisting on three equally abundant resources. Species 1 consumes only resources A and B. Species 2 consumes only B and C. Because individuals of species 1 compete with others of their species for both A and B, but compete with individuals of species 2 only for resource B, intraspecific competition is more intense than interspecific competition and the two species will coexist.

This kind of coexistence can be seen in the replacement series diagram developed by botanists to evaluate the effects of competition (see page 277). If both species are ordinarily inhibited more by intraspecific competition than by interspecific competition, the production of each species should lie above the line of expected production when intraspecific and interspecific competition are equivalent. Because individuals of both species are more productive when mixed than when separate, the total production of mixed species populations exceeds that of single species populations maintained at the same density (Figure 14-22).

Coexistence depends on avoiding competition. Any number of species can coexist so long as each exploits a portion of the resources in the habitat more efficiently than all the other species. To coexist with other species in a community, each species must excel in some way; otherwise, it will be excluded by superior competitors.

Species avoid ecological overlap by partitioning the available resources according to their size and form, their chemical composition, their place of occurrence, and their seasonal availability. Tolerance of extreme physical conditions allows some species to feed where others would succumb. Species that would compete intensely if together usu-

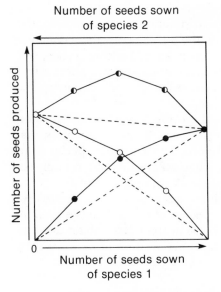

Number of seeds sown
of species 2

Number of seeds produced

0

Number of seeds sown
of species 1

FIGURE 14–22   Replacement series diagram of competition between two species with different ecological requirements. Mixed species plantings are more productive than separate plantings of either species maintained at the same density.

ally do not occur in the same locality or habitat. Resource partitioning within habitats often involves similar spatial separation of species according to their size and behavior. For example, each of five species of warblers that breed in spruce forests in Maine feed in different parts of the trees and use somewhat different foraging techniques as they search for insects among the branches and foliage (Figure 14–23). Four

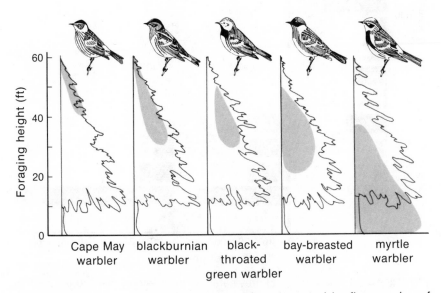

FIGURE 14–23   Location and method of foraging used by five species of warblers (genus *Dendroica*) in spruce forests in Maine.

species of lizards of the genus *Anolis* coexist on the island of Bimini in the Bahamas, where they forage in different parts of the habitat. *Anolis sagrei*, a large, brown lizard, is the only species that commonly ventures to the ground to feed. The American chameleon, *A. carolinensis*, hunts for its prey among leaves. The remaining two species feed along branches, but the larger *A. distichus* is found more commonly than *A. angusticeps* on branches of large diameter.

Food specialization is often based on prey size. Among predatory species such as hawks, large organisms usually eat larger prey than small organisms (Figure 14-24). Among species which partition resources solely by prey size, the body length of one predator is rarely less than 1.3 times the length of the next smallest predator. This difference apparently represents the smallest difference in resulting prey size that prevents competitive exclusion. The absence of size differences between some similar species of predators living in the same habitat does not necessarily imply intense competition; other avenues of resource partitioning may be used.

FIGURE 14-24 Size range of prey eaten by three widely distributed species of hawks. All three species belong to the accipiter group, which is adapted to pursue small birds and mammals.

The morphology and behavior of predatory snails along the Gulf Coast of Florida emphasize the variety of ways ecological overlap can be avoided. The small murex snail, for example, cannot pry or chip open the shell of a large clam or oyster. Instead, it uses the filelike teeth (radula) at the tip of its proboscis to drill a neat hole in the shell of its prey, through which it scrapes away the flesh. Because *Murex* has a poorly developed foot, it can neither pursue other predatory snails nor dig clams out of the sand or mud. Its diet is therefore restricted to such prey as the rock cockle, which rests exposed on the sea floor. The large conches of the genus *Busycon* use the heavy edges of their shells to chip at the edges of large clam shells or to wedge them open. *Busycon contrarium* opens the thick-shelled *Chione* by grasping the prey and aligning the shell margin at a 45-degree to 90-degree angle to the lip of its own shell. Using its shell as a hammer, the conch then chips away enough of the clam's shell to insert its proboscis or to wedge the clam open. Conches pry open thin-shelled clams and species that do not close completely simply by using the edge of their shell as a wedge and pressing hard on the margin between the two halves of the clam's shell. *Busycon spiratum* has a thinner shell than the larger species, *B. contrarium*. It usually preys upon clams that have thin shells or do not close completely. The horse conch and tulip shell attack other predatory snails smaller than themselves. The smaller *Fasciolaria hunteria* has a particularly long proboscis which is adapted to attack tube-building worms in their burrows beneath the mud. These differences in size and feeding technique enable the snails to avoid intense competition and permit their coexistence.

Evolutionary Divergence

Resources are not partitioned among species haphazardly. Competitive exclusion eliminates cases of extreme ecological similarity. Suppose we were to pick species at random from several localities and introduce them to an island whose climate resembled their place of origin. By chance, some of the species would have nearly identical resource requirements and compete intensely. The inferior competitors among these species would soon disappear from the community, leaving only those species that are ecologically distinct. But we would also notice that some of the species would begin to diverge in appearance due to evolutionary changes in each population. When similar species coexist, interspecific competition promotes the evolution of ecological divergence to reduce resource overlap.

Consider two hypothetical species which use many of the same resources. Suppose that the size range of prey eaten by each species overlaps that eaten by the other, as shown in the upper diagram of Figure 14–25. Where the species overlap most, they compete most in-

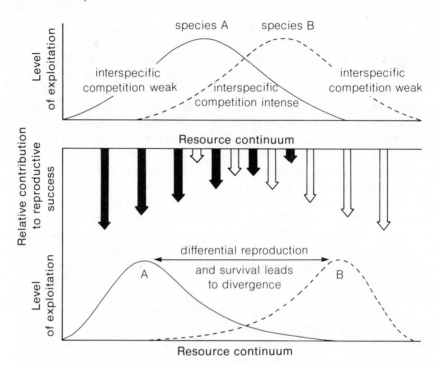

FIGURE 14–25   Diagram showing evolutionary divergence between populations caused by the influence of interspecific competition on the survival and reproduction of individuals that are adapted to eat prey of different sizes.

tensely. Individuals of species A that eat prey smaller than the average have a greater evolutionary fitness than the population as a whole because their productivity is influenced less by competition. The same is true of individuals of species B that consume prey larger than average. Differential reproduction and survival caused by interspecific competition lead to the evolution ecological divergence of close competitors. If three species feed off the same range of prey size, the middle-sized predator will become adapted to eat prey intermediate in size between the prey consumed by the largest and smallest species (see Figure 14–24).

In most cases of resource partitioning we cannot determine whether the degree of ecological overlap between two species has been adjusted by evolutionary divergence. Evolution usually proceeds too slowly to be observed directly. We may, however, infer evolutionary divergence if a species displays appearance or behavior that differs in the presence of a competitor from appearance or behavior in the absence of a competitor. This phenomenon, referred to as *character displacement*, provides convincing evidence that competition can promote evolutionary divergence. When the ranges of two species do not overlap completely, interspecific competition influences evolution in the zone of overlap but not elsewhere. This principle is demonstrated in Figure 14–26 by illustrating the beak size of ground finches in the Galapagos Islands. The islands of Abingdon and Bindloe have three species of *Geospiza*, which partition seed resources according to size. With its large beak, *Geospiza magnirostris* is adapted to husk large seeds which smaller finches could not break open. With its small beak, *G. fuliginosa* can husk small seeds more efficiently than larger species. *G. fortis* feeds on seeds of an intermediate size range. The largest species does not occur on Charles Island nor on Chatham Island. Individuals of *Geospiza fortis* on these islands tend to have slightly heavier beaks, on the average, than on Abingdon or Bindloe. On Daphe Island, where *fortis* occurs in the absence of *fuliginosa*, its beak is intermediate in size between the two species on Charles and Chatham Islands. On Crossman Island, *fuliginosa* occurs in the absence of *fortis*, and *its* beak is intermediate in size there. Because

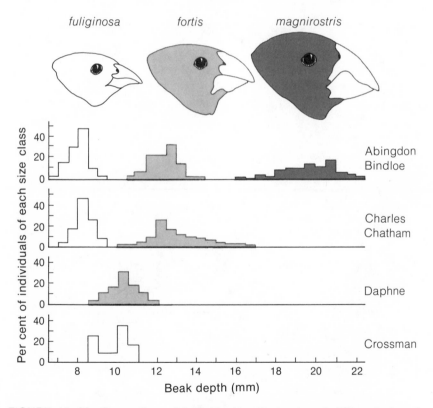

FIGURE 14–26 Proportion of individuals with beaks of different sizes in populations of ground finches (*Geospiza*) on several of the Galapagos Islands.

habitats on the islands are similar in most respects, the simplest explanation for these patterns is that where the species occur together, competition has caused character divergence.

The evolutionary and ecological results of interspecific competition seen in this chapter organize the species on each trophic level according to the manner in which they exploit resources. Competitive exclusion and evolutionary divergence ensure, by reducing interspecific competition, that resources are utilized efficiently and that the overall flux of energy and nutrients in the ecosystem is maintained at the highest level possible with the array of populations present in the community.

15

Predation

Predation, in its broadest sense of food consumption, is the prime mover of energy and materials in the ecosystem. To the extent that death is caused by predation, the efficiency of predators in finding and capturing their prey determines the rate at which energy flows from one trophic level to the next. As a building block of community structure and a source of community stability, predation differs from competition in one important respect: whereas competitors exert a mutual influence on each other, predation is one-sided. It is true that predator and prey affect each other, but changes in their relationship that benefit one, hurt the other. Competitive exclusion and evolution of ecological divergence stabilize the structure of trophic levels by minimizing the interaction between species. Conversely, adaptations that benefit either predator or prey need not increase the inherent stability of the community.

Understanding predation — including relationships between predator and prey, parasite and host, herbivore and plant — is a key to understanding the internal workings of the community. We may ask many questions about the way in which predators affect their prey and the consequences of their relationship to the community as a whole. For example, to what extent do predators stabilize prey populations or, alternatively, cause them to fluctuate? Do predators limit prey populations below the carrying capacity of the environment for the prey? If predators are so efficient that they can substantially reduce prey populations, how do they keep from overeating their prey? How do predators influence competition between prey species? Do predators act to maximize their returns? That is, do they "manage" prey populations? Our present knowledge cannot provide definitive answers to these questions, partly because no one answer applies to all predators. As we shall see in this chapter, diverse predator–prey relationships have different properties and offer correspondingly different insights into community function.

The Impact of Predators on Prey Populations

We should distinguish two types of predators. One feeds mainly on the surplus of its prey, capturing the sick, the old, the vulnerable young, and the displaced, socially subordinate, but leaving the reproductively fit — the well-spring of the prey population — untouched. The other feeds so efficiently on all classes of individuals that it can seriously impair the growth potential of the prey population.

The prey themselves and the habitat in which they live determine which type of predation they suffer. Populations of short-lived organisms with high reproductive rates are frequently held under control by their predators. Individuals in such prey species have adopted a strategy of maximizing their production of offspring at the risk of increasing their vulnerability to predators. But what choice has an aphid? If it is to suck the juices from the veins of a sycamore leaf, it must sit on a flat surface exposed to every passerby (Figure 15-1). The tiny algae of the phytoplankton have nowhere to hide. Their survival depends purely

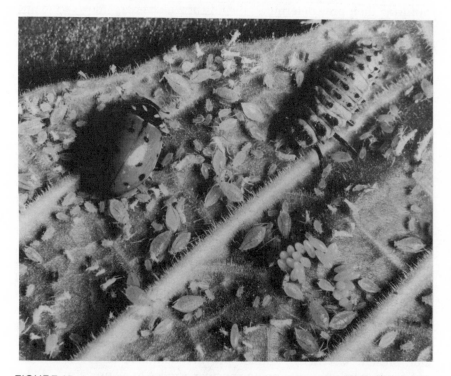

FIGURE 15–1   Adult and larvae ladybeetles (family *Coccinellidae*) feeding on aphids in a laboratory culture. The flightless aphids are easy prey for the predatory beetles. Note the abundant hairs on the veins of the leaf. These help to deter the aphids from penetrating the plant and sucking its juices.

on chance. Animals that, because of their own food supply, have low reproductive rates must invest much more heavily in avoiding predators if they are to shift the balance between predator and prey in their favor. Prey species are aided toward this goal when their habitat contains hiding places to which they can escape.

Instances in which predators have been shown to depress prey populations below the carrying capacity of the environment are widely scattered. Several such studies are recounted here in detail because of the fundamental consequence of predation for community ecology.

The cyclamen mite is a pest of strawberry crops in California. Populations of the mites are usually kept under control by a species of predatory mite of the genus *Typhlodromus*. Cyclamen mites typically invade a strawberry crop shortly after it is planted, but their populations do not reach damaging levels until the second year. Predatory mites usually invade fields during the second year and rapidly subdue the cyclamen mite populations, which rarely reach damaging levels a second time.

Greenhouse experiments have demonstrated the role of predation in keeping the cyclamen mites in check. One group of strawberry plants was stocked with both predator and prey mites; a second group was kept predator-free by regular applications of parathion, an insecticide that kills the predatory species but does not affect the cyclamen mite. Throughout the study, populations of cyclamen mites remained low in plots shared with *Typhlodromus*, but their infestation maintained damaging proportions on predator-free plants (Figure 15–2). In field plantings of strawberries, the cyclamen mites also reached damaging levels where predators were eliminated by parathion, but they were effectively controlled in untreated plots (a good example of an insecticide having the wrong effect). When cyclamen mite populations began to increase in an untreated planting, the predator populations quickly responded to reduce the outbreak. On the average, cyclamen mites were about 25 times more abundant in the absence of predators than in their presence.

The effectiveness of *Typhlodromus* as a predator owes to several factors in addition to its voracious appetite. Its capacity for population increase is of the same order as that of its prey. Female cyclamen mites lay three eggs per day over the four or five days of their reproductive life span; female *Typhlodromus* lay two or three eggs per day for eight to ten days. But even its high reproductive rate does not tell the whole success story of *Typhlodromus*. Seasonal synchrony of reproductive activities with the growth of prey populations, ability to survive at low prey densities, and strong dispersal powers all contribute to its efficiency. During the winter, when cyclamen mite populations are reduced to a few individuals hidden in the crevices and folds of leaves in the crown of the strawberry plants, the predatory mites subsist on the honeydew produced by aphids and white flies, and they do not reproduce except

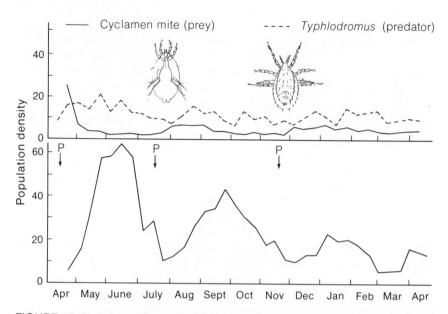

FIGURE 15–2   Infestation of strawberry plots by cyclamen mites in the presence of the predatory mite, *Typhlodromus* (above), and in its absence (below). Prey populations are expressed as numbers of mites per leaf; predator levels are based on the number of leaflets in 36 which had one or more *Typhlodromus*. Periodic parathion treatments (indicated by P's) kept the predatory mites from reinvading the predator-free plots.

when they are feeding on other mites. Whenever predators are suspected of controlling prey populations, one usually finds a high reproductive capacity compared to the prey, combined with strong dispersal powers and the ability to switch to alternate food resources when primary prey are unavailable.

Inadequate dispersal is perhaps the only factor that keeps the cactus moth from completely exterminating its principal food source, the prickly pear cactus. When prickly pear was introduced to Australia, it spread rapidly through the island continent, covering thousands of acres of valuable pasture and rangeland. After several unsuccessful attempts to eradicate the prickly pear, the cactus moth was introduced from South America. The caterpillar of the cactus moth feeds on the growing shoots of the prickly pear and quickly destroys the plant — by literally nipping it in the bud. The cactus moth exerted such effective control that within a few years, the prickly pear became a pest of the past (Figure 15–3). The cactus moth has not, however, eradicated the prickly pear because the cactus manages to disperse to predator-free areas, thereby keeping one jump ahead of the moth and maintaining a low-level equilibrium in a continually shifting mosaic of isolated patches. Indeed, one would probably not guess that the cactus moth

FIGURE 15–3 Photographs of a pasture in Queenland, Australia, two months before and three years after introduction of the cactus moth to control prickly pear cactus.

keeps the prickly pear at its present low population levels; the moths are actually scarce in the remaining stands of cactus in Australia today. The same moth is probably responsible for controlling prickly pear in some areas of Central and South America, but its decisive role may have gone unnoticed unless the appropriate experiment was performed in Australia.

The mite–mite and moth–cactus interactions prove that a predator can keep prey populations far below the capacity of the environment to support them. Yet in both situations, the prey originally occurred at unnaturally high levels in managed environments and the predators were introduced by man. These studies do not, therefore, help us to determine whether natural populations are ever controlled by predators.

Experiments on the effect of sea urchins on populations of algae have demonstrated predator control in some natural marine ecosystems. The simplest experiments consist of removing sea urchins, which feed on attached algae, and following the subsequent growth of their algae prey. When urchins are kept out of tidepools and subtidal rock surfaces, the biomass of algae quickly increases, indicating that predation reduces algal populations below the level that the environment can support. Different kinds of algae also appear after predator removal. Large brown algae flourish and begin to replace the coralline algae (whose hard, shell-like coverings deter grazers) and small green algae (whose short life cycles and high reproductive rates enable algal population growth to keep ahead of grazing pressure by sea urchins). In subtidal plots kept free of predators, brown kelps formed thick forests below the ocean's surface and shaded out most small species.

Changes in kelp beds off the coast of southern California provide a striking demonstration of the role of sea urchin predation on algal populations. The most important brown algae in the kelp beds is *Macrocystis*. Its life cycle begins when an embryo, having developed from a single cell floating in the plankton, settles to the bottom at a depth of 8 to 25 meters. As the young plant grows, it is secured to the bottom by a holdfast; its fronds extend towards the surface, buoyed up by gas-filled floats (Figure 15–4). Mature kelps form dense stands and greatly influence the overall economy of the coastal marine ecosystem. Kelps not only are major primary producers, but also provide refuges for fish populations. (Kelp beds figure in the human economy of southern California as well, because they are commercially valuable as a source of iodine and fertilizer. The 200 or so square kilometers of kelp beds between Santa Barbara and San Diego yield a harvest of 100,000 to 150,000 tons annually at a price of $20 per ton.)

Kelps are long-lived compared to other marine plants. Once a bed has become established it will last between one and ten years, depending on the locality and exposure to waves. Death occurs mostly by

FIGURE 15–4   The kelp *Macrocystis*, showing holdfast and gas-filled floats on fronds.

storms, high water temperature, and grazing by fish and sea urchins. Once a bed has been devastated, by whatever cause, it normally becomes re-established within a few years. The coastline thus witnesses the regular disappearance and persistent reappearance of kelp beds as a normal course of events in the marine community.

Beginning in 1940, kelp beds all but disappeared from areas near Los Angeles and San Diego. Disposal of sewage was suggested as a cause because deterioration of the beds started near sewage outlets and spread outward. Areas in which kelp beds had disappeared were practically devoid of all types of algae, and were instead swarming with dense populations of immature sea urchins. Normally when a kelp bed disappears, the urchins disappear with it, giving newly settled plants a chance to regenerate. Pollutants did not affect mature kelp plants: a few beds persisted within polluted areas in sheltered spots free of wave damage. The kelp beds disappeared because no young kelp grew where normal causes of mortality had removed mature plants.

Sea urchins apparently can obtain nourishment from suspended and dissolved organic matter in sea water. Sewage maintains an urchin population in the absence of adequate algal food in the same way that honeydew from aphids maintains predatory mites when their normal prey, the cyclamen mite, is unavailable. When young *Macrocystis* plants reinvade a devastated kelp bed, urchins are there to meet them and quickly devour newly settled plants.

When urchin populations were controlled by dumping quicklime (calcium oxide) into devastated areas, and these areas were reseeded with young *Macrocystis* plants, the kelp beds quickly returned to their former state. The proof of the pudding came when, in 1963, San Diego stopped dumping sewage directly into the ocean. Reseeding then quickly brought kelp beds back without the need to destroy the urchins. In fact, these beds now contain fewer, but larger, urchins whose grazing does not seriously depress algal growth.

Nature would not, however, leave us content with so simple a system. We must add yet another component to the dynamics of the kelp community — the sea otter. Once abundant along the west coast of the United States and Canada, sea otters were hunted to the verge of extinction during the nineteenth century by Russian and American fur traders. Under close protection and surveillance by the California Fish and Game Commission, the otters have staged a successful comeback. Wherever otter populations have reached their former densities, kelp beds also flourish. Needless to say, otters eat urchins. And a final note: otters also eat abalone, a valuable commercial shellfish in California, and so while they are encouraged by the kelp industry, they are illegally persecuted by abalone fishermen.

Predator–Prey Cycles

Populations of predators and prey often vary in what appear to be closely linked cycles. The periodic fluctuations of the snowshoe hare, followed closely by fluctuations of lynx, one of the hare's major predators, is a classic example (see Figure 13–15). Because the cycles persist for long periods, they represent a stable interaction between predator and prey populations. As we saw in Chapter 13, ecologists have not determined whether predation by lynx on hares actually causes their population cycles. Hare populations may fluctuate in conjunction with their own food supply, with lynx populations passively following the trend of their prey.

Predators have been shown to cause population cycles in laboratory experiments in which prey are provided a constant, abundant source of food. In such situations, the cycles are enhanced by slow response of predator populations to changes in prey density.

When azuki bean weevils are maintained in cultures with predatory braconid wasps, the populations of predator and prey often fluctuate out of phase with each other in regular cycles of population change. In the life cycles of the wasp, eggs are laid on beetle larvae (Figure 15–5), which the wasp larvae proceed to consume after hatching. Abundance of prey therefore influences the number of adult wasps in the *following* generation, after the wasp larvae have metamorphosed into adults. This

FIGURE 15–5 A braconid wasp laying an egg in a cotton boll worm. When the egg hatches, the wasp larva will consume the boll worm. Various species of these tiny wasps attack many kinds of insect larvae and are frequently major factors in pest population control.

built-in time lag enhances the population fluctuation (see page 259). The diagram in Figure 15–6 follows the course of wasp and weevil populations in a laboratory culture for 30 generations (about one and a half years). When the predator and prey populations are both low, the prey increase rapidly. As the prey become abundant, the population of wasps also increases. Eventually, the predators overeat their prey and the prey population declines. The wasp is never so efficient that all weevil larvae are attacked, hence a small but persistent reserve of weevils always remains to initiate a new cycle of prey population growth when the predators become scarce.

Whether predation will cause population cycles depends on the time lag in the response of predators to prey abundance. The time lag is short when prey-capture efficiency and population growth response of the predator are great. The wasp *Heterospilus* readily finds weevil larvae in laboratory culture, but its reproductive rate is too low to keep up with the initial population growth rate of its prey. This difference in reproductive rate places a lag of two to four generations between prey and predator populations and promotes population cycling. In contrast,

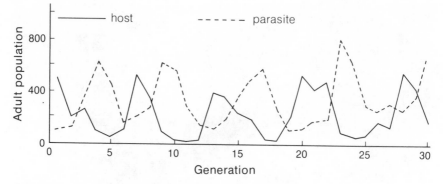

FIGURE 15–6   Population fluctuations in a predator–prey system involving the azuki beanweevil and a braconid wasp predator.

Neocatolaccus has a reproductive potential six times greater than that of its close relative, *Heterospilus.* When *Neocatolaccus* and the azuki bean weevil are raised together, the quick population response of *Neocatolaccus* obliterates population cycles.

With an extremely efficient predator, prey populations are often eaten to extinction, and their predators soon follow. This type of predator–prey interaction can become stable only if some of the prey can find refuges in which they can escape predators. G. F. Gause demonstrated this principle with his early studies on protozoa. He employed *Paramecium* as prey and another ciliated protozoan, *Didinium,* as predator. In one experiment, predator and prey individuals were introduced to a nutritive medium in a plain test tube. In that simple environment, the predators readily found all the prey; when they had consumed the prey population, the predators died from starvation. In a second experiment, Gause added some structure to the environment by placing glass wool at the bottom of the test tube, within which the *Paramecium* could escape predation. In this case, the *Didinium* population starved after consuming all readily available prey, but the *Paramecium* population was restored by individuals concealed from predators in the glass wool.

Gause finally achieved recurring oscillations in the predator and prey populations by periodically adding small numbers of predators — restocking the pond, so to speak. The repeated addition of individuals to the culture corresponds, in natural predator–prey interactions, to repopulation by colonists from other areas of a locality in which extinction of either predator or prey has occurred. This is reminiscent of the interaction between the cactus moth and prickly pear cactus (page 302), in which the cactus escapes complete annihilation by dispersing to predator-free areas.

C. B. Huffaker, a University of California biologist who pioneered the biological control of crop pests, attempted to produce just such a mosaic environment in the laboratory. The six-spotted mite was prey; another mite, *Typhlodromus occidentalis*, was predator; oranges provided the prey's food. The experimental populations were set up on trays in which the number, exposed surface area, and dispersion of the oranges could be varied (Figure 15–7). Each tray had 40 possible positions for oranges, arranged in four rows of ten each; where oranges were not placed, rubber balls of about the same size were substituted. The exposed surface area of the oranges was varied by covering different amounts of the oranges with paper; the edges of the paper were sealed with wax to keep the mites from crawling underneath. In most experiments, Huffaker first established the prey population with 20 females per tray, then introduced 2 female predators 11 days later. Both species reproduce parthenogenetically (without sex).

When six-spotted mites were introduced to the trays alone, their populations levelled off at between 5,500 and 8,000 mites per orange area (Figure 15–8, left-hand diagram). The predators introduced to the system increased rapidly and soon wiped out the prey population. Their own extinction followed shortly (Figure 15–8, right-hand diagram). Although predators always eliminated the six-spotted mites, the position of the exposed areas of oranges influenced the course of extinction. When the orange areas were in adjacent positions, minimizing dispersal

FIGURE 15–7 (Above) One of Huffaker's experimental trays with four oranges, half exposed, distributed at random among the 40 positions in the tray. Other positions are occupied by rubber balls. (Right) An orange wrapped with paper and edges sealed with wax. The exposed area is divided into numbered sections to facilitate counting the mites.

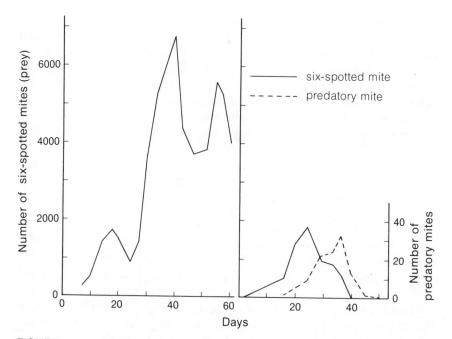

FIGURE 15–8 Number of six-spotted mites per orange area when raised alone (left) and in the presence of the predatory mite, *Typhlodromus*. The food was arranged in 20 small orange areas alternating with 20 foodless positions.

distance between food sources, the prey reached maximum populations of only 113 to 650 individuals and were driven to extinction within 23 to 32 days after the prey were introduced to the trays. When the same amount of exposed orange area was randomly dispersed throughout the 40-position tray, the prey reached maximum populations of 2,000 to 4,000 individuals and persisted for 36 days. These experiments demonstrated that the survival of the prey population can be prolonged by providing remote areas of suitable habitat. The slow dispersal of predators to these areas delays the extinction of their prey.

Huffaker reasoned that if predator dispersal could be slowed further, the two species might coexist. To accomplish this, Huffaker increased the complexity of the environment and introduced barriers to dispersal. The number of possible food positions was increased to 120 and the equivalent area of 6 oranges was dispersed over all 120 positions. A mazelike pattern of Vaseline barriers was placed among the food positions to slow the dispersal of the predators. *Typhlodromus* must walk to get where it is going, but the six-spotted mite spins a parachutelike silk line that it can use to float on wind currents. To take advantage of this behavior, Huffaker placed vertical wooden pegs throughout the trays. The mites used the top of the pegs as jumping-off points in their wanderings. This arrangement finally produced a series of three predator–prey cycles over eight months (Figure 15-9). The distribution of the predators and prey throughout the trays continually shifted as the prey, exterminated in a feeding area, recolonized the next, one jump ahead of the predators.

Thus we see that a spatial mosaic of suitable habitats allows predator–prey interactions to achieve stability. But, as we saw in Gause's experiment with protozoa, predator and prey can coexist in a locality if some prey can take refuge in hiding places. And if the environment is so complex that predators cannot easily find scarce prey, stability will again be achieved.

Rules of Predator–Prey Stability

Experimental studies in conjunction with theoretical mathematical analyses of population dynamics have revealed four factors that enhance the stability of predator–prey interactions: (1) predatory inefficiency (or prey escape), (2) external ecological restrictions on either

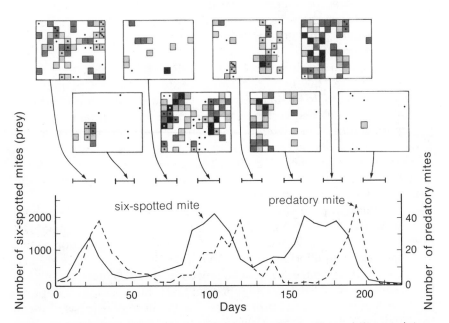

FIGURE 15-9 Population cycles of the six-spotted mite and the predatory mite, *Typhlodromus*, in a laboratory situation. The boxes above show the relative density and positions of the mites in the trays. Shading indicates the relative density of six-spotted mites; circles indicate presence of predatory mites.

population, (3) alternative food sources for the predator, and (4) reduced time lags in predator population response.

A predator–prey system will achieve, or oscillate around, one of two possible equilibrium points. The first is set largely by the carrying capacity of the environment for the prey in the absence of predation. In this case, the predator exerts a minor influence on the prey population, which is ultimately limited by food or some other resource. The second equilibrium point is set at a much lower prey population level by the ability of prey to find refuges or hiding places. In this case, the predator reduces prey populations far below the carrying capacity determined by food, to a level determined by habitat complexity. The equilibrium point achieved by a particular pair of predator and prey species is decided by the hunting efficiency of the predator relative to the growth potential of the prey population. Efficient predators drive the prey population to its lower equilibrium point. Inefficient predators remove the vulnerable surplus of a prey population to a level near the carrying capacity of the environment. Predatory inefficiency increases the stability of a predator–prey system around its upper equilibrium point by placing the burden of population regulation on density-dependent responses of the prey to its food resources. Near the lower equilibrium point, inefficiency increases because prey become fewer and farther apart and a larger proportion of them have access to good hiding places. In the case of the six-spotted mite, predator inefficiency eventually kept the prey from becoming extinct.

The upper equilibrium point of a predator–prey interaction is set by external limits to the prey population. External control of predator populations, owing to limited nest sites, water, alternative food, reduces the impact of predators, and increases the control of external factors, on prey populations. Cold winter temperatures severely reduce populations of *Aphytis lingnanensis,* a wasp predator of citrus scale, in the interior valleys of southern California (see page 272). Although the wasp controls citrus scale populations in the mild climate near the Pacific coast, it is ineffective in the interior.

The ability of predators to use alternative food sources when their prey are scarce or inactive greatly hastens their response to increases in prey populations and thereby helps to keep the prey population at the lower of its two equilibrium points. (We have seen this effect in studies on cyclamen mites, page 301, and kelp beds, page 304, where alternative foods always kept predator populations at levels high enough to depress prey population growth.) Furthermore, use of alternative foods reduces the likelihood of a predator eating its principal prey to extinction.

Decreasing the lag time in the response of a predator population to prey population growth dampens population oscillations and increases the overall stability of the predator–prey interaction at either equilibrium point. Because predator response greatly influences the stability of predator–prey systems, we shall examine further two avenues of response to changes in prey population size.

The Functional Response

The relationship of an individual predator's rate of food consumption to prey density has been labelled the *functional response* by Canadian ecologist C. S. Holling. When tested in the presence of increasing levels of prey density in the laboratory, praying mantises increase their rate of food consumption in proportion to prey density, at first, but at high prey density, their feeding rate eventually levels off (Figure 15–10, curve B). Two factors dictate that the functional response of the individual should reach a plateau. First, as the predator captures more prey, the time spent handling and eating the prey cuts into hunting time. Eventually the two reach a balance and prey capture rate levels off. Second, predators become satiated — continually stuffed — and cannot feed any faster than they can digest and assimilate their food.

Hunger clearly influences a predator's motivation to hunt. The distance over which a praying mantis will strike at a fly depends on the time since its last feeding, which, since one cannot ask a mantis how hungry it is, serves as a reasonable index of hunger. Holling also experimented with deer mice to determine how motivation affects the functional response of a vertebrate predator. Cocoons of the pine sawfly (see page 254) were buried at several densities in sand on the floor of large cages; dog biscuits were provided as an alternative food. Palatability of the cocoons influenced the functional response: deer mice dug up many more fresh cocoons than cocoons collected during the previous year and stored prior to the experiment. Deer mice apparently are not fond of stale cocoons. When fresh cocoons were buried deeper in the sand, the mice spent less time digging for them. The type of alternative food also influenced the functional response. With cocoons at a density of 15 per square foot, mice consumed 200 cocoons per day when dog biscuits were provided. When sunflower seeds were added to the menu, consumption of cocoons dropped to just over 100 per day. Deer mice evidently are not fond of dog biscuits, either.

Holling determined the functional responses of three small mammals to the density of pine sawfly cocoons in the relatively natural habitat of pine plantations in Ontario, Canada. The eggs of the sawfly, which are laid in live pine needles in the fall, hatch early in the spring. The larvae feed on the needles of the pine. In early June, full-grown larvae fall to the ground and crawl into the litter, where they spin cocoons. The adults do not emerge to lay eggs until September. For

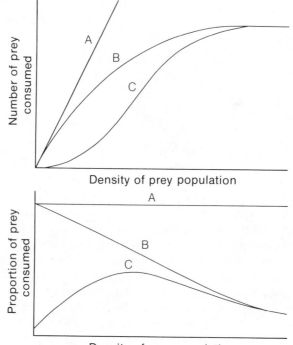

FIGURE 15–10 The functional response of predators to increasing prey density: (A) predator consumes a constant proportion of the prey population regardless of its density; (B) predation rate decreases as predator satiation sets an upper limit to food consumption; (C) predator response lags at low prey density owing to low hunting efficiency or absence of search image. The upper diagram portrays the functional response in terms of the *number* of prey consumed; the lower diagram, in terms of the *proportion* of the prey population consumed.

three months during the summer, the forest floor is scattered with varying numbers of cocoons from a few thousand to more than a million per acre, depending on the level of sawfly infestation. Small mammalian predators, particularly the common shrew, the deer mouse, and the short-tailed shrew, open the cocoons in ways characteristic of each mammal, enabling investigators to tally instances of predation separately for each species. Densities of the predators and prey can be assessed by trapping the rodents and sampling the forest litter for cocoons.

The functional response curve of each predator was unique. The short-tailed shrew, the least common species in the study area, increased its consumption of sawfly pupae in response to their density much more rapidly than either the common shrew or deer mouse (Figure 15–11). Because the short-tailed shrew increases its consumption of cocoons in response to small increases in their availability it appears to take full advantage of a highly variable food resource.

The functional response of many predators increases more slowly at low prey densities than at higher prey densities (see Figure 15–10, curve C). Two factors can cause this lag. First, hunting efficiency is decreased at low density because the few prey have the best hiding places for escape. Second, vertebrate predators are thought to adopt their hunting behavior and prey recognition to the most worthwhile prey — usually an abundant, oft-encountered species. A preconception of what a given prey looks like and where it is found is called a *search image*. We use search images all the time. A lost object is easier found if we know its shape, size, and color. Predators base search images on prior experience. The more abundant the prey, the more often it is found and the better prepared the predator to find it. Conversely, search-image formation works against a predator's being able to locate uncommon prey, hence the lag in the predator's functional response at low prey density.

The Numerical Response

Individual predators can increase prey consumption only to the point of satiation. Predator response to increasing prey density above that point can be achieved only through an increase in the number of predators,

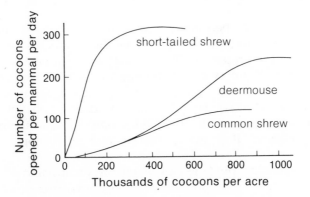

FIGURE 15–11 Functional response of three mammalian predators to the density of pine sawfly cocoons in the litter of pine plantations.

either by immigration or population growth, which together constitute the *numerical response.* Many predators congregate wherever their prey become abundant. Bay-breasted warblers specialize in feeding on periodic local outbreaks of the spruce budworm, during which the warbler population may locally reach 120 pairs per 100 acres, compared to about 10 pairs during years when there is no budworm outbreak. Mammal populations in pine plantations in Ontario varied greatly in relation to the availability of sawfly cocoons in the forest-floor litter. Common shrews increased from about 3 per acre at low sawfly levels to 24 per acre at medium and high prey densities; the numerical response of the deer mouse population was less marked; the short-tailed shrew population showed no response to prey density.

Three predatory birds, the pomarine jaeger, the snowy owl, and the short-eared owl, each respond in a different manner to varying densities of lemmings on the arctic tundra (Table 15-1). Lemming populations exhibit great fluctuations; high and low points in a population cycle may differ by a factor of one hundred. At Barrow, Alaska, during the summer of 1951, when lemmings were scarce, none of the predatory birds bred; short-eared owls did not even appear in the area. During the following summer, one of moderate lemming density, both the jaeger and snowy owl bred, but short-eared owls again were absent. In 1953, a peak year for lemmings, all three species of bird predators bred. Jaegers were four times more abundant in 1953 than in 1952. In contrast to the jaegers, the density of snowy owls did not increase. Instead, each pair of birds raised more young. Most snowy owls laid two to four eggs during the year of moderate lemming abundance, and up to a dozen during the peak year.

A Model of Predator–Prey Equilibrium

The diverse relations between predator and prey populations can be summarized in a diagram that compares the productivity of the prey population to the proportion of prey that are removed by predators, both as a function of prey density (Figure 15-12). The two curves in the diagram represent the net addition of new prey to the population (in excess of deaths due to causes other than predation), either by reproduction or immigration (collectively called *recruitment*), and the removal of prey by predators. Both are expressed as a proportion of the prey population.

Recruitment and predation are analogous to birth and death. When recruitment exceeds predation, the prey population grows; when predation exceeds recruitment, prey numbers decline. Points at which the recruitment and predation curves cross are population equilibria for the prey.

TABLE 15–1 Response of predatory birds to different densities of the brown lemming near Barrow, Alaska.

	1951	1952	1953
Brown lemming (ind/acre)	1 to 5	15 to 20	70 to 80
Pomarine jaeger	Uncommon, no breeding	Breeding pairs 4/mi²	Breeding pairs 18/mi²
Snowy owl	Scarce, no breeding	Breeding pairs 0.2 to 0.5/mi² many nonbreeders	Breeding pairs 0.2 to 0.5/mi² few nonbreeders
Short-eared owl	Absent	One record	Breeding pairs 3 to 4/mi²

The recruitment curve of the prey population declines with increasing prey density, owing to intraspecific competition for resources, and falls to zero when the prey attain the carrying capacity of the environment. In the absence of predators, prey populations are regulated at this point by the balance between their biotic potential and resource limitations of the environment.

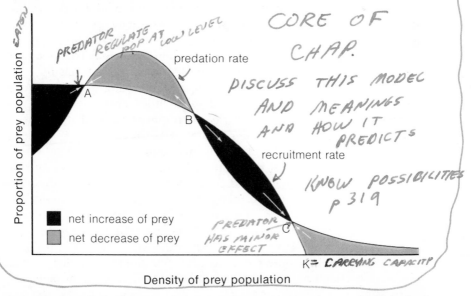

FIGURE 15–12 Predation and recruitment rates in a hypothetical predator–prey system. When predation exceeds recruitment, prey populations decrease, and vice versa (as shown by arrows). Points A and C are stable equilibria for the prey population; the lower point (A) represents population control by predators; the upper point (C) represents population control by food and other resources.

The shape of the predation curve is determined by the functional and numerical responses of the predators. Escape ability of the prey and the failure of predators to form search images limit the proportion of prey captured at low prey density. At moderate prey density, functional and numerical responses increase the effectiveness of the predators as a whole and the proportion of prey removed increases. These responses are eventually satiated at high prey density, and although the number of prey consumed may continue to rise slowly, the proportion of individuals removed from the prey population decreases.

The recruitment and predation curves in Figure 15-12 are drawn to produce three equilibrium points for the prey population. The highest and lowest points represent stable equilibria around which populations are regulated, the middle equilibrium is an unstable transient point through which a population passes from one stable point to the other. The lower equilibrium point (A) corresponds to the situation in which predators regulate a prey population substantially below the carrying capacity of the environment. The upper equilibrium point (C) corresponds to the situation in which a prey population is regulated primarily by availability of food and other resources; predation exerts a minor depressing influence on population size.

Predators maintain a shaky hold on prey populations at point A. If a heavy frost or an introduced disease reduced the predator population long enough to allow the prey population to slip above point B, the prey would continue to increase to the higher stable equilibrium point (C), regardless of whether the predator population recovered. To the farmer, this means a crop pest, normally controlled at harmless levels by predators and parasites, suddenly becoming a menacing epidemic. After such an outbreak, predators could exert little control over the pest population until some quirk of the environment brought its numbers below point B, back within the realm of predator control.

The effectiveness of predators in maintaining prey populations at low densities depends on the relationship between predation and recruitment curves. The higher and broader the predation curve, the more effective predators are as control agents. Functional and numerical responses enhance predator effectiveness. Several species of predators attacking the same prey can control the prey population better than any one of them alone (Figure 15-13). Well-planned biological control programs take advantage of this principle in attempting to establish several species of predator and parasite, each with different predator tactics, to control pest populations.

Using the predation-recruitment rate diagram in Figure 15-12, we can examine the consequences of different levels of predation for prey population control (Figure 15-14). Inefficient predators cannot regulate prey populations at low density; they depress prey numbers slightly, but the prey population remains near the equilibrium level set by re-

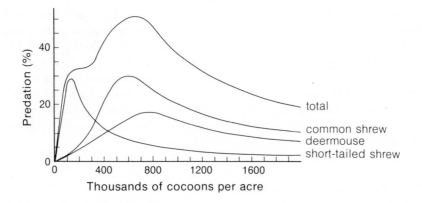

FIGURE 15–13 The combined functional and numerical responses of small mammals to density of sawfly cocoons showing the relative influence of several species of predator on prey populations. The total curve corresponds to the predation curve in Figure 15–10.

sources (upper-left diagram, point C). Increased predation efficiency at low prey density can result in predator control at point A (upper-right diagram). If functional and numerical responses are sufficient to maintain high densities of predators, or if prey are limited relative to predation by a low carrying capacity, predators may effectively control prey

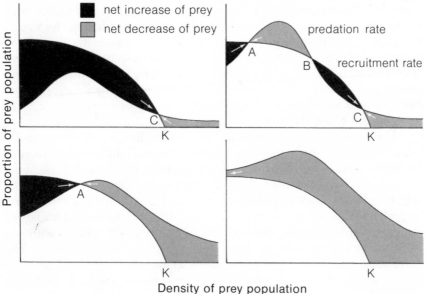

FIGURE 15–14 Predation and recruitment curves at different intensities of predation.

under all circumstances, and equilibrium point C disappears (lower-left diagram). We could also envision predation being so intense at all levels of prey density that the prey are eaten to extinction (lower-right diagram, no equilibrium point). We would expect this situation only in simple laboratory habitats, or when predators maintained themselves at high population levels by feeding off some alternative prey. Indeed, many ecologists have suggested that biological control of pests would be enhanced by providing parasites and predators of the pest with innocuous alternative prey.

Prudent Predators and Managed Prey

A predator that eats its prey down to a low level takes the food out of its own mouth. Predators can capture prey more easily, and therefore be more productive, when prey are numerous. The ability of a prey population to support predators varies with its density. A small prey population can support correspondingly few predators because, while each individual's reproductive potential may be high, the total recruitment rate of a small population is low. Prey populations near their carrying capacity are also unproductive because although numerous, each individual's reproductive potential is severely limited as a result of intraspecific competition for resources. The total recruitment rate of every prey population reaches a maximum at some density below the carrying capacity. Because predators can remove a number of individual prey equivalent to the annual recruitment rate without reducing the size of the prey population, the prey population that yields the maximum recruitment also will support the greatest number of predators. Ranchers and game management biologists are clearly concerned with maintaining populations of beef cattle, deer, and geese at their most productive levels to maximize man's ability to harvest these species without reducing their populations.

The achievement of optimum yield can be illustrated with populations of guppies maintained at different densities in aquaria. Recruitment (the number of immature fish produced in three weeks) reached a peak of 33 when there were 30 adult guppies per tank, and dropped to about 7 when adult populations exceeded 100 individuals per tank (Figure 15-15). In the absence of predators, the natural mortality of guppies would stabilize the population at about 120 adults, with the recruitment rate of 7 every three weeks just balancing mortality. The maximum sustainable yield would be achieved when predators removed about 40 per cent of the adult population every three weeks (about 2 per cent of the population per day).

Maximum potential daily yields in laboratory populations have been calculated as 3 per cent for flour beetles, 13 per cent for unicellular

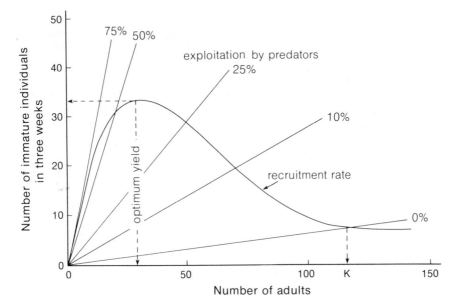

FIGURE 15–15 Recruitment curve and hypothetical exploitation rates for aquarium populations of guppies. In the absence of predation (0 per cent exploitation curve) the natural mortality of adult guppies would stabilize the population at about 120 individuals. The maximum exploitation rate possible is about 50 per cent per three weeks, at which point an adult population of about 30 and a yield of 33 would be maintained. A 75 per cent exploitation rate is more than the population can bear.

algae, 23 per cent for water fleas (*Daphnia*) and 99 per cent for the sheep blowfly. Few estimates of maximum potential yield are available for natural populations of large animals. The population of ring-necked pheasants on Protection Island, Washington, could withstand removal at a rate of 1 to 3 per cent per day. Wolves kill about 25 per cent of the moose population of Isle Royale each year (0.07 per cent per day), and perhaps 37 per cent of the white-tailed deer population of Algonquin Park in Ontario (0.1 per cent per day). The population of ring seals on Baffin Island can withstand exploitation by Eskimos at a rate of 7 per cent per year (0.02 per cent per day).

Annual recruitment rates of adults into natural populations may be used as a crude index to the rate at which predators remove individuals from populations. These rates vary widely, from 83 to 120 per cent for plaice and haddock, 60 to 80 per cent for several species of salamanders in Virginia, 40 to 60 per cent for many of our backyard birds, to less than 5 per cent for some large mammals and sea birds. Predators are not responsible for the loss of all the adults which the recruits replace, but they also capture many young before they reach reproductive age.

Could levels of exploitation observed in nature actually be maximum yields? Or would competition between individual predators cause overexploitation of their prey? Territorial animals, which exclude competitors from their feeding areas, could indeed space themselves with respect to their prey to achieve maximum yields. When the feeding areas of predators overlap, however, intraspecific competition dictates that each predator maximizes its immediate harvest at the expense of long-term yields. Man behaves no differently. Intelligently managed ranches, with fences to exclude "competing" livestock, can achieve maximum yields. Alas, in highly competitive situations — fishing in international waters, to name one — man has proved to be pathetically short-sighted and imprudent. Fishing and hunting practices that would attain long-term yields give way to practices that maximize today's harvest. For example, after World War II, the North Sea, between England and Norway, was fished so intensively that reducing fishing effort by 15 per cent and increasing the mesh size of nets to let more fish through actually would have *increased* the total catch by 10 to 20 per cent. Overexploitation of whale populations has similarly led to the near extinction of some species and has virtually doomed the whaling industry.

Evolution of the Predator–Prey Equilibrium

Regardless of whether predators obtain maximum yields from their prey, the predator–prey equilibrium represents a balance between adaptations of the predator to find and capture prey and adaptations of the prey to escape predation. Selection constantly seeks to improve the adaptations of both predator and prey, achieving an evolutionary balance analogous to the ecological balance between predator and prey populations.

We shall turn once again to laboratory populations of insects for evidence of evolutionary adjustment between predators and their prey. In an ingenious experiment with a host–parasite system, David Pimentel and his co-workers at Cornell University demonstrated that the relationship between populations of houseflies and a wasp parasite could be changed by artificial selection. In one cage housing flies and wasps, the fly population was held constant by replenishing losses from a stock population that had no contact with the wasp. Any flies that escaped parasitism by wasps during the larval stage were removed from the cage to keep selection from acting on the fly population. That is, any genetic traits that enhanced resistance to wasp parasitism were not permitted to be transmitted to progeny in the cage. Pimentel also held the level of flies constant in a second cage, but allowed emerging flies to remain and perpetuate genetic factors that might have enhanced their resistance to wasps. After three years, adaptations that fostered resistance to wasp parasitism had accumulated in the second fly population. As a result, the reproductive rate of wasps had dropped from 135 to 39 progeny per female, and their adult life span had decreased from 7 to 4 days. No changes occurred in the first population. The average level of the parasite population was lower in the second cage (1,900 wasps) than in the first (3,700 wasps).

Pimentel then established larger, 30-cell cages in which the numbers of flies and parasites were allowed to vary (Figure 15-16). Flies were neither added nor removed from the cages. One 30-cell cage was started with flies and wasps that had no previous contact with each other; another was started with flies and wasps taken from the second system described above, in which the flies and wasps had had prolonged contact. In the first population cage, wasps were efficient parasites but their population growth response lagged behind that of the flies, causing the system to shift back and forth between upper and lower equilib-

FIGURE 15–16 A 16-cell cage, used to study competition between populations of flies. Note the vials with larval food in each cage and passageways connecting the cells. The dark objects concentrated in the upper-right hand cells are fly pupae. (Right) Fly pupa and parasitic wasp.

rium points. In the second system, the wasp population never depressed the fly population far below the carrying capacity of the population cage (Figure 15-17). The startling difference in the results from the two 30-cell cages reinforced the conclusion drawn from earlier experiments that the flies had evolved resistance to the wasp parasites.

If flies respond to wasps by evolving adaptations to escape parasitism, and if wasps respond to flies by evolving adaptations to exploit them, the two must reach an evolutionary equilibrium on some middle ground. This argument is given strength by the relatively constant ratios of predators to prey in widely diverse circumstances. For example, the ratio of wolf populations to their prey fall within a fairly narrow range of 1 pound of wolf to each 150 to 300 pounds of prey, regardless of the principle prey or the locality (Table 15-2). Similar population ratios of predators to large ungulates on African savannas and in fossil remains of extinct mammals in Alaska (approximately 1:100), and of pomarine jaegers to lemmings on the arctic tundra (1:90 by weight) give further substance to the notion of constancy among predator and prey interactions.

Parasites and Host Populations   *NOT CENTRAL*

Parasites and predators use different strategies to exploit their prey or host populations. The death of a prey organism is the objective of predators, but the death of a host often causes the death of its parasite. Although predators evolve adaptations to maximize their ability to capture prey, selection acts strongly on parasites to adjust their food consumption to a level that the host can withstand. Parasite and host together evolve a balance: adaptations of the parasite reduce its virulence, and adaptations of the host (immunity and resistance) reduce the parasite's hazard to health. The balance achieved between parasite and host is often upset when a parasite is accidentally transferred to a new host, frequently man, his livestock, or his crops, and sweeps through the population in a devastating epidemic.

Little is known about the incidence of infection or the effects of parasites on natural populations. Factors that limit populations of parasites extend far beyond the skin of their hosts. Most parasites undergo complicated life cycles involving changes in form and stages of dispersal that take the parasite into the hostile exterior environment. For an organism adapted to living within the tissues of a host, the environment through which its progeny travel to reach another host, often by way of intermediate species, must be forbidding.

Infections of many parasites vary seasonally. The occurrence of the harvest mite, *Thrombicula*, an external parasite of mice and voles in England, reaches a peak in the early fall and declines through the winter

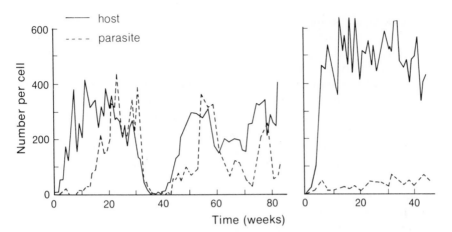

FIGURE 15–17 Populations of the house fly and a parasitic wasp in 30-cell laboratory cages. Control (left-hand graph): flies had no previous exposure to the wasp. Experimental (right-hand graph): flies and wasps derived from populations that had been in contact for over 1,000 days.

and spring, finally disappearing completely until the seasonal cycle is begun once more in late summer. But the pattern of abundance also varies from host to host. The harvest mite is far more persistent on the bank vole (*Clethrionomys*), than on either field mice (*Apodemus*) or meadow mice (*Microtus*) both of which are completely free of the mites by late fall.

Parasite populations also fluctuate from year to year. For example, incidence of trypanosome parasites in populations of red squirrels and eastern chipmunks at Trout Lake, Manitoba, varied considerably during a three-year study. (Trypanosomes are parasitic protozoa that infect the bloodstream. In man, trypanosomes cause several fatal illnesses, of which

TABLE 15–2 The ratio of populations of wolves to their prey.

Locality	Principle prey	Prey weight (kg)	Density of wolves (ind/100 mi²)	Predator–prey ratio based on Numbers	Biomass
Jasper National Park	Elk	300	1	1:100	1:250
	Mule deer	70			
Canadian tundra	Caribou	200	1.7	1:84	1:186
Wisconsin	White-tailed deer	60	3	1:300	1:300
Algonquin Park	White-tailed deer	60	10	1:150	1:150
Isle Royale	Moose	350	10	1:30	1:75

sleeping sickness has received the most attention. The parasites are usually carried from host to host by insects such as tsetse flies.) Squirrels and chipmunks apparently are infected by different, host-specific strains of the trypanosome because the occurrence of trypanosomiasis in the two hosts fluctuated independently (4, 37, and 15 per cent in the squirrel population and 42, 26, and 12 per cent in the chipmunk population, in 1961, 1962, and 1963). The incidence of the disease varied because adults that had previously been exposed to the disease became immune, and the number of juveniles in the population, all of which are susceptible, varied from year to year. In the eastern chipmunk, the incidence of the trypanosome in juveniles was more than four times that in adults (48 versus 11 per cent over three years), and the per cent of juveniles in the population during the summer months decreased from 68 per cent in 1961 to 29 per cent in 1963. This drop, combined with the reduction of disease incidence in adults from 19 to 4 per cent over the three years, accounted for the decrease of trypanosomiasis in the population.

Infection by trypanosomiasis did not reduce the survival of squirrels and chipmunks. Endemic occurrences of disease organisms usually cause few detrimental effects in healthy organisms. Parasites can, however, intensify the harm caused by such stressful conditions as cold or lack of food, and thus can increase the incidence of death under these conditions. Infection by *Trypanosoma duttoni* was found to reduce the tolerance of laboratory mice to stressful conditions. Groups of mice were given either full or half rations of food in either warm (19 to 22°C) or cold (3 to 8°C) environments. With full rations, none of the mice died over a 19-day period, regardless of parasitism or the temperature at which they were kept. The nonparasitized mice did, however, gain twice as much weight as the parasitized mice (14 versus 7 per cent increase). On half rations, all groups of mice had shorter average survival times in cold than in warm environments, as one would expect, but parasitism by *Trypanosoma* significantly reduced survival in both environments.

Rapid development of immunity characterizes epidemic outbreaks of most diseases. Perhaps the most famous epidemics of all times have been the outbreaks of the Black Plague in human populations. The bubonic plague organism, a bacterium, is endemic in many wild populations of rodents. The disease can be spread to man by rodent fleas, particularly the rat flea. Several times in history the natural balance between the bacteria, rodents, and fleas has been upset so badly that the plague spread to the human population, in which it caused epidemic disease. To produce a major plague epidemic, rat populations must be so great that hordes of rats, searching for food in houses, come into close contact with humans. Furthermore, rat fleas must heavily infest rats before they will abandon their preferred hosts for humans. These conditions have occurred infrequently, even in the crowded, garbage-ridden conditions prevalent in Medieval Europe. But once the plague takes hold in a human population, its course runs swiftly and surely.

A typical epidemic of Black Plague initially spreads rapidly, infects a large part of the population, and takes a high toll in human lives. But as susceptible individuals either die or become immune to the disease organism, the number of lethal cases and the mortality rate drop almost as rapidly as the disease first strikes. This is illustrated by data for a localized outbreak of the Plague in India between 1953 and 1959:

Year	Contracted cases	% lethal
1953	20,539	70.5
1954	6,670	84.5
1955	705	23.1
1956	331	20.5
1957	44	0
1958	26	0
1959	37	0

The Great Plague of the fourteenth century originated in 1346 during the Siege of Caffa, a small military post on the Crimean Straits. From there, the epidemic spread to Italy and the south of France by 1347, and during the next year it had reached all of Europe. The Plague did not disappear entirely until 1357. The Plague reappeared in Europe three more times during the fourteenth century, in 1361, in 1371, and in 1382, but with lower incidence and mortality rate:

Year	Approximate % of population afflicted	Resultant deaths
1348	67	Almost all
1361	50	Almost all
1371	10	Many survived
1382	5	Almost all survived

The Plague visited London three times during the seventeenth century, but unlike the epidemic waves that struck Europe during the fourteenth century, the effects were not attenuated during successive outbreaks:

Year	Population of London	Plague deaths	Deaths as a % of total population
1603	250,000	33,347	13
1625	320,000	41,313	13
1665	460,000	68,596	15

These outbreaks differed from those of the fourteenth century in one significant way: successive epidemics during the fourteenth century were separated by 13, 10 and 11 years; during the seventeenth century, intervals were 22 and 40 years. The longer interval increased the severity of successive epidemics in two ways. First, many more individuals lost their immunity over a 20-year to 40-year period than over a decade. Second, the proportion of the population born since a plague epidemic, and therefore not immune, was much larger after intervals of 22 and 40 years than after the shorter intervals during the fourteenth century. Because more humans were susceptible, the plague organism spread more rapidly through the population.

The ecology of a disease vector (an organism that carries and transmits parasite organisms) frequently limits the spread of parasite populations. For example, when avian malaria and bird pox were introduced to the Hawaiian Islands, where neither disease had previously occurred in the avifauna, highly susceptible local populations of birds were quickly destroyed and several species became extinct. But malaria and bird pox organisms are carried from host to host by a species of mosquito that does not venture above an elevation of 600 meters, and so birds that lived at higher altitudes completely escaped the diseases.

Herbivores and Plant Populations

The influence of herbivores on plant populations differs with animals that consume whole plants (usually seeds and seedlings) from the influence of those that graze on vegetation. The first are predators. Grazers and browsers are akin to parasites. We can infer that herbivores have an important role in the lives of plants from the elaborate morphological and physiological defenses of plants against attack. Plant toxins like hypericin, digitalis, curare, strychnine, and nicotine are a fair match for most herbivores. Thick bark, spines, thorns, and stinging hairs are also strong deterrents.

Herbivores can stabilize plant succession on disturbed sites and prevent the encroachment of new forms into an area. For example, after the decline of rabbit populations in Australia following the introduction of myxamatosis virus (see page 228), the native pine *Callitiris* regenerated extensively in New South Wales. Grazing by rabbits evidently had prevented the growth of pine seedlings. Even though cactus moths consume only a small portion of the net production of the prickly pear, it successfully controls the introduced cactus population in Australia. In fact, herbivorous insects are frequently used to control imported weeds. Klamath weed, a European species toxic to livestock, and the source of the drug hypericin, accidentally became established in northern California in the early 1900's. By 1944, the weed had spread over 2,000,000 acres of rangeland in 30 counties. Biological control specialists borrowed an herbivorous beetle (*Chrysolina*) from an Australian control program. (Everything seems to become a pest in Australia.) Ten years after the first beetles were released, the Klamath weed was all but obliterated as a range pest. Its abundance was estimated to have reduced by more than 99 per cent.

The impact of herbivores in plant communities, measured by the total net primary production consumed, is least in forests, intermediate in grasslands, and greatest in aquatic environments. Grazing herbivores consume between 2 and 10 per cent of the net production of forests. The rest enters detritus pathways in the community. Seed predators are much more efficient, consuming anywhere between 10 and 100 per cent of their food supply. Although they consume relatively little of the total biomass of the plant, seed predators attack at a vital stage in the life cycle and can influence plant populations greatly. Ecologists are just beginning to study their role in natural communities.

Grazing herbivores, particularly large mammals, consume 30 to 60 per cent of grassland vegetation. The story of their influence on plant production is particularly well told by the results of exclosure experiments. A study in California employed wire fences to exclude voles (mouselike rodents) from small areas of grassland. Seed production and composition of the standing crop of plants were followed for two years after the experiment began and were compared with unfenced control plots. The results, summarized by the bar diagram in Figure 15–18, show that grazing by voles in the unfenced plots reduced the abundance

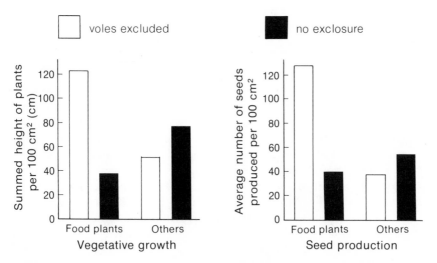

FIGURE 15–18   Species composition and seed production in grassland plots fenced to exclude voles and in unfenced control plots. The bar graphs present results of the experiment after two years. Food plants are mostly annual grasses; nonfood plants include perennial grasses and herbs.

and seed production of food plants (mostly annual grasses) but did not affect perennial grasses and herbs absent from the vole's diet. Furthermore, competition from annual grasses in the fenced plots apparently depressed the growth of the nonfood species, suggesting a large role for the vole in determining the structure of the plant community.

Aquatic herbivores consume most of the net production of aquatic plants. The effectiveness of marine snails as algal grazers is reflected in their abundance and annual energy flux. Representative figures for gross production (total energy assimilation) are 118 kcal/m²/yr for the fresh-water limpet, *Ferrissia*, 290 kcal/m²/yr for the periwinkle, *Littorina*, and 750 kcal/m²/yr for the turban shell, *Tegula*. Comparable figures for terrestrial grazers range between 7 kcal/m²/yr (field mouse) and 28 kcal/m²/yr (grasshoppers).

Although herbivores rarely consume more than 10 per cent of forest vegetation, occasional outbreaks of tent caterpillars, gypsy moths, and other insects can completely defoliate or otherwise eradicate entire forests (Figure 15-19).

Long-term studies of growth and survival of trees after defoliation by tent caterpillars and other insects demonstrate that there may be a considerable lag between an infestation and the expression of its effects. A spruce budworm infestation on balsam fir caused varying defoliation, mortality, and subsequent growth retardation. In one area of light infestation, the defoliation exceeded 50 per cent during only three of nine years, reaching a maximum of 80 per cent in 1947 (Figure 15-20). No mortality was recorded in this area until 1951, but growth remained below one-half the normal rate for several years after the peak of the infestation. In an area of heavy infestation, defoliation exceeded 50 per cent for five years and reached 100 per cent in 1947, during the peak of the budworm outbreak. Growth was greatly suppressed and all trees in the area had died by 1951.

A Minnesota study of defoliation of quaking aspen by tent caterpillars showed that aspen usually survived defoliation and the increased intensity of insect and disease attack which inevitably followed. During the year of defoliation, however, growth was reduced by almost 90 per cent, and by about 15 per cent during the following year. From studies of growth rings measured a decade later, foresters noted that among trees whose growth was normally suppressed by competition from dominant trees, mortality was independent of the history of defoliation and varied between 40 and 60 per cent since the tent caterpillar epidemic. Conversely, defoliation had a pronounced effect on survival of dominant trees. Only 2 per cent of the trees that were subjected to a single year of light defoliation died; at the other extreme, trees that had been badly defoliated three years in a row suffered almost 30 per cent mortality.

FIGURE 15–19   (Above) A stand of Englemann spruce in Colorado killed by an epidemic of the spruce beetle. (Below) A mixed forest of conifers and broad-leaved species in Nova Scotia. The broad-leaved species were killed by a massive outbreak of defoliating insects. Young spruce trees are rapidly growing up to take their place.

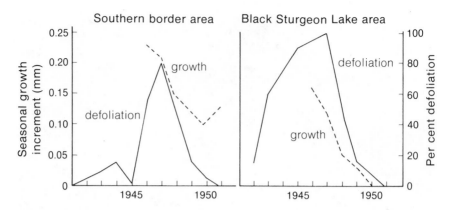

FIGURE 15–20   Defoliation of balsam fir trees in two areas during an infestation of the spruce budworm. In the area that was more heavily infested (right) all of the trees died, whereas the other area (left) had suffered no mortality by the end of the study period.

The interactions between predator and prey populations are so varied and complex as to defy summary. The examples we have looked at in this chapter lead us to the conclusion that predators play an important role in the population processes of all species, even to the extent that efficient predators can regulate prey population density below the carrying capacity of the environment. Through their influence on prey populations, predators affect both the evolution of prey characteristics and the ability of prey to compete with other species utilizing the same resources.

The extreme consequence of predation, so often observed in the laboratory, is the annihilation of the prey population. Species become extinct in nature too, although the role of predators in their extinction remains unclear, except in the most obvious cases: the extermination of ground-nesting island birds by introduced rats, the extermination of game species or pests by human hunting pressures, the extermination of specialized species by destruction of their habitat. Although extinction plays a minor role in the day-to-day working of the ecosystem, a study of extinction, the subject of the next chapter, allows us to perceive in their most exaggerated form the many adjustments of populations to the influence of others.

16

Extinction

In 1810, the American ornithologist Alexander Wilson observed an immense flock of passenger pigeons in the Ohio River Valley. For days, the column of birds, perhaps a mile wide, passed overhead in numbers to darken the sky. Wilson estimated the size of the flock at more than 2,000,000,000 birds. The last passenger pigeon died in the Cincinnati Zoological Garden just over a century later, on September 1, 1914.

The demise of the passenger pigeon was caused by two unfortunate circumstances. First, the pigeons roosted and nested in huge assemblies, sometimes numbering several hundred million birds within a few hundred square miles. Second, roasted or stewed, they tasted very good. Pigeoners, as professional hunters were called, gathered in large groups to trap and slaughter the pigeons. Nesting trees were felled to collect the squabs. As farmers cleared the forests and railroads made vast areas within the pigeons' range accessible to eastern, big-city markets, the persecution increased. By the mid-1800's, increased killing disrupted breeding, and the pigeon population began to decline. By 1870, large breeding congregations were found only in the Great Lake States, at the northern edge of the pigeon's former range. The last nest was found in 1894, and the bird was last seen in the wild in 1899.

With its extinction in 1914, the passenger pigeon joined a lengthening list of species that have vanished from the Earth. Since naturalists began describing the forms of plant and animal life, 53 birds, 77 mammals, and a host of other animals and plants have disappeared, most of them at the hand of man. By exploiting species for food or hides or destroying them as pests, by subjecting them to depredation by domesticated animals and hangers-on like rats, and by destroying their habitats, we have managed to reduce the numbers or productivity of many populations below their self-sustaining level.

More recently, our collective conscience, embodied in the Audubon Society, World Wildlife Fund, International Committee for Bird Preservation, and numerous conservation societies, has sought to reverse the trend of increasing human impact on nature by preserving species on the verge of extinction. Whooping cranes, ivory-billed woodpeckers, California condors, timber wolves, and grizzly bears are some of the more spectacular species now protected by legislation and the establishment of wildlife preserves.

This is well and good, but, although the salvation of a few endangered species may ease our conscience of our wanton past, present conservation efforts will probably prove to be a fragile dam before the flood of human population growth and technological development. Still, the extirpation of our less fortunate cohabitants of this planet provides vivid lessons concerning population regulation and the composition of biological communities. In this chapter, we shall examine the occurrence and causes of extinction, both where man is accountable and where natural changes in communities have brought the destruction of species.

Recent Extinctions of Birds and Mammals

The geographical distribution of extinction and its occurrence within diverse groups tell us that species do not disappear at random. Some are much more susceptible than others. For example, of 53 species of birds that have become extinct in the last 300 years, only 3 (the crested sheldrake of eastern Asia, and the passenger pigeon and Carolina parakeet, both of eastern North America) were found on major continents. The remaining 50 species have disappeared from islands the size of New Zealand and smaller. Island populations are particularly vulnerable because of their small size and isolation. Many island species have adapted to habitats without predators and with few competitors (see page 288) and are unable to cope with newcomers introduced from comparable mainland habitats, or with man himself. Other island species that are restricted to forest habitats suffer when land is cleared for sugar cane and coconut palm plantations. Small isolated populations are also thought to be vulnerable to random perturbations in the environment — prolonged drought, hurricanes, disease epidemics, and the like. Local extinction of these restricted populations can often mean extinction of an entire species. Prior to their extinction, many species of birds were regularly eaten by man, whose depredations directly led to the demise of ducks, pigeons, and rails (small henlike marsh birds). But small, inconspicuous, forest-dwelling birds have also disappeared.

In contrast to birds, mammals have become extinct as frequently on the major continents (27 species) as on outlying islands, including Australia (55 species). Most extinct continental species were large herbivores killed for food (ground sloths, sheep, deer, zebras) and large carnivores, killed because they competed with man for food or were thought to threaten man directly (bears, wolves, cats) (Table 16-1). Losses of mammals less involved with man's well-being (small rodents, bats, shrews, small marsupials) are confined to islands.

The Balance Between Origination and Extinction

The geological history recorded in the sedimentary rocks of the Earth's crust is filled with the fossils of animals and plants, both great and small, no longer present. Perhaps 95 per cent of the species that ever lived on Earth are now extinct. Throughout the course of biological history, new species have been added and others have disappeared. When a population becomes divided by some natural barrier, the isolated subpopulations follow independent evolutionary pathways, sometimes diverging to the point that, if they ever rejoin, individuals from different populations cannot interbreed, and so the subpopulations have achieved the status of full species. By this often-repeated sequence of isolation, divergence, and re-invasion, new species are continually added to biological communities.

Paleontologists cannot agree on the number of species that inhabited the Earth at any one time because the fossil record is too fragmentary to distinguish all forms. Nonetheless, the occurrence of species in higher levels of classification (genera, orders, families*) is better known; their diversification and extinction give us some idea of the progress of organic diversity. At the beginning of the Paleozoic Era, about 550,000,000 years ago, the first period of the Earth's history for which

TABLE 16–1   Extinctions of mammals on islands and continents.

		Number of	
	large predators	large herbivores	small species
Continents	10	17	0
Islands	1	12	42

* Taxonomists classify plants and animals according to a hierarchical scheme under which each group fits, usually with several others, into a larger group. The major levels of classification ascend in the order: *species, genus, family, order, class, phylum,* and *kingdom,* with many intermediate levels distinguished. For example, the common crow is classified as species *brachyrhynchos,* which, together with the raven (*corax*), fish crow (*ossifragus*), jackdaw (*monedula*), and others comprise the genus *Corvus.* (The full scientific name of the crow, *Corvus brachyrhynchos,* includes the genus and species names, always in italics, with the genus capitalized.) Together with the genera of jays, magpies, and nutcrackers, *Corvus* is placed in the family Corvidae, which, with swallows, thrushes, warblers, sparrows, wrens, and others, constitutes the order Passeriformes, the perching birds. The class Aves includes all birds, which, together with mammals, fish, reptile, and amphibians, make up the phylum Cordata in the kingdom Animalia.

fossil remains abound, between 50 and 100 families of animals lived in the seas. There were yet no land animals. The number of families has followed an irregular, but increasing, trend ever since:

Years before present	*Number of families*	
550	50–100	} Initial diversification of marine forms.
400	300	} Major diversification of aquatic invertebrates.
230	350	} Breakup of continents, major environmental changes.
220	250	
135	350	} Reptiles dominate the land.
0	900	} Major diversification of terrestrial forms.

Except for brief periods of major geological upheaval, new appearances have exceeded disappearances and the number of life forms has increased overall.

Regardless of the net changes in species number, species production and extinction cause the species composition of every large taxonomic group to change rapidly. Over the last 100,000,000 years, the number of species of bivalves (clams and oysters) has changed little, but few of the species initially present, or their direct descendents, are alive today. The same statement could be made of carnivorous mammals for the past 10,000,000 years. The average turnover rate of bivalve genera (per cent replacement by new genera) has varied between 1 and 5 per cent per million years during much of their history. Turnover rates of carnivores have varied between 22 and 31 per cent per million years. Throughout this period, production and extinction of species remained in approximate balance. When speciation rate increases, so does extinction rate (Figure 16–1).

The passing of many once-dominant forms of life (dinosaurs, trilobites, ammonites, placoderms) usually coincided with the appearance of new forms with similar ecological requirements. These groups soon diversified to fill the void left by their predecessors. The agnathe and placoderm types of fish that dominated Paleozoic seas were thus replaced by sharks and rays, and then, largely, by bony fish. The evolutionary success of these groups, measured by the appearance of new orders, families, and genera (Figure 16–2), shows that the Devonian (D) was a period of diversification for all groups, but that agnathes and placoderms soon dropped out. (A few agnathes — the hagfish and lamprey — survive to the present.) Bony fish have dominated fish evolution ever since the Mississippian Period (M), 345,000,000 years ago, except during the Jurassic (J), when sharks and rays staged a major comeback.

Changes in the Mammal Fauna of South America

Patterns of disappearance and replacement similar to that among fish have occurred many times. For example, ammonoids (large marine predators with coiled shells) were replaced by nautiloids (shelled relatives of squids and octopuses). The history of the mammal fauna of South America reveals a series of replacements by major groups brought about at first by the evolution and diversification of new forms within South America and later by the immigration of North American species by way of Central America. During much of the last 60,000,000 years, South America has been isolated from other continents; Central America was under water or appeared as a string of islands until about 5,000,000 years ago, when a solid land bridge finally joined the two continents. The early mammalian history of South America tells us of the diversification of notoungulates (primitive herbivores), marsupials, and many edentates (armadillos, sloths, and anteaters). Primates, primitive ro-

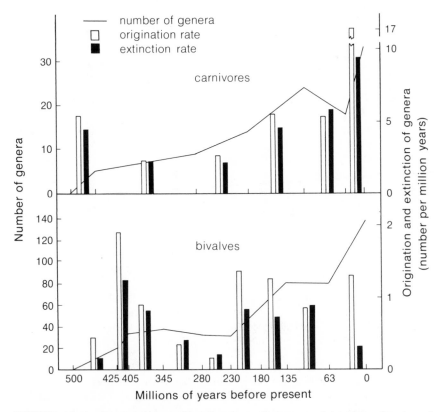

FIGURE 16-1 Production and extinction of genera of bivalves (top) and carnivores (bottom), and the net change in the total number of genera in each group.

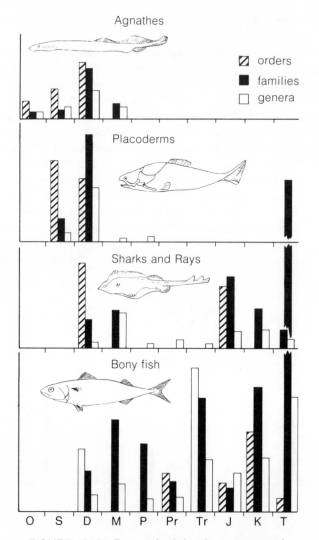

FIGURE 16–2   Rates of origination, expressed as groups per million years, for orders, families, and genera of fish. The scale for genera is reduced ten-fold compared to families and fifty-fold compared to orders. The time scale is designated by geological periods from the Ordovician (O) to the Tertiary (T).

dents, and raccoonlike forms crossed the water barrier from North America and diversified well before the modern land bridge formed.

Many changes occurred in the mammal fauna of South America during its extended isolation (Figure 16–3). Primitive ungulate herbivores, mostly notoungulates and litopterns, underwent a rapid diversification and reached their peak during the Eocene, 40,000,000 to 50,000,000 years ago. Some primitive ungulates and native marsupials evolved to fill the ecological roles of present-day rodents. This branch of the marsupial line did not persist, although marsupials comprised most of the South American carnivorous species until the immigration of North American dogs, bears, cats, and weasels. Note that on Australia, the mammal fauna is dominated by marsupials. Few other types of mammals have ever existed there (other than bats) and Australia is still sufficiently isolated to prevent the immigration of potentially superior placental mammals.

The rodentlike ungulates of South America were gradually pushed aside by the diversification of native South American rodents, represented today by capybaras, agoutis, chinchillas, and their relatives. The success of these rodents was paralleled by a decline in the larger primitive ungulates, perhaps through competitive exclusion (see page 267).

The Central American land bridge brought hordes of new immigrants from the north and caused the extinction of many South American ecological counterparts. The primitive ungulates were pushed aside by invading deer, tapirs, and camels. (Many of these are now extinct in North America. The camel family is represented in South America by llamas, vicuna, and alpaca.) Many of the early native rodents of South America vanished with the appearance of squirrels, mice and rabbits, but several unique forms (guinea pigs, capybaras) survived the onslaught. The marsupial carnivores were replaced by placental counterparts (dogs, cat, bears).

The modern mammal fauna of South America is a nearly equal mixture of distinctly South American and North American forms, yet few southern mammals invaded North America. In fact, porcupines and the now-extinct ground sloths are the only distinctly South American groups that have invaded the north temperate zone. Other southern forms, like armadillos, sloths, and some of the native rodents, have extended their ranges into Central America but have not yet moved beyond the tropical zone.

It seems odd that the native fauna of South America, adapted to the particular environmental conditions on that continent, should have been overwhelmed by invaders from another land. It is true that South American forms had not evolved in competition with northern forms. By the end of the Miocene period, 13,000,000 years ago, the native fauna had partitioned most of the possible ecological roles in the community and the major radiations of new forms were largely over. In contrast, dispersal between Asia and North America occurred regularly by way of Alaska and the Bering Land Bridge. The mammalian fauna of North America has continually been exposed to immigrants from other areas, challenging their competitive ability. We may be tempted to apply terms like "evolutionary vigor" to the mammals of North America and "evolutionary stagnation" to the mammals of South America, but evolutionary

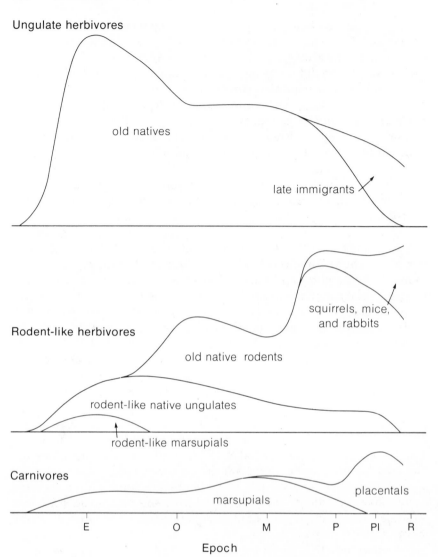

FIGURE 16-3 Diversification and reduction of several groups of mammals on the South American continent. Height of the graph in each case represents the relative number of genera. Time scale is only approximate: R = Recent, Pl = Pleistocene ($0 - 3 \times 10^6$ years before present), P = Pliocene ($3 - 13 \times 10^6$ years), M = Miocene ($13 - 25 \times 10^6$ years), O = Oligocene ($25 - 36 \times 10^6$ years), E = Eocene ($36 - 58 \times 10^6$ years). Recent exchange of species between North and South America began in the Pliocene.

biologists do not yet understand how the inherent characteristics of whole groups determine competitive ability. Why, for example, did *all* the litopterns die out? What common bonds of structure and function united these species in a fatal evolutionary pact?

The Conservatism of Evolution

Once the major radiation of a group has taken place, further change in body plan or way of life evidently becomes difficult. For example, most of the characteristics of present-day lungfish appeared during the first 50,000,000 years of lungfish evolution. The line has changed relatively little in the last 250,000,000 years. Genera of marine invertebrates and plants are recognized unchanged through hundreds of millions of years of the fossil record. These evolutionary lines have undoubtedly been buffeted by changing environments — shifts in temperature and moisture, the coming and going of predators and disease organisms — but they have retained their distinctive characteristics with a minimum of modification.

Evolutionary conservatism is well-demonstrated by the floras of the southeastern regions of Asia and North America. During the late Tertiary period, 70,000,000 years ago, vast areas of the Northern Hemisphere were covered with broad-leaved forests. The remnants of these great forests — primitive species like the rhododendron, tulip tree, magnolia, sweetgum, and hickory — are now localized in the southeastern United States and eastern Asia, where they have been isolated for more than 50,000,000 years. Yet these plants still betray their common origin in the Tertiary flora and, although many have been given different species names, most of the genera remain unchanged. Moreover, genera with southern distributions in Asia tend to have southern distributions in North America. Thus not only has form remained unchanged, but each genus has also retained a discernible portion of those physiological tolerances that determine geographical range — for more than 50,000,000 years.

The Causes of Extinction

Evolutionary conservatism in the face of environmental change could well open the door to extinction. Changes in climate, competitors, predators, and disease organisms all pose serious threats. Many changes come too quickly and too abruptly for evolution to respond. Certainly many of the results of human activities fall into this category. Natural changes in the environment take a more deliberate toll.

Climatic change by itself probably never drove a species to extinction. The physical environment sets the stage for competitive struggle between populations. As we have seen before, physical conditions influence the outcome of competition and cause species to replace each other along environmental gradients. When the climate of a whole region changes, the edge a species holds over its competitors may disappear throughout its geographical range, leading to complete extirpation unless the distribution of the species shifts to a new area with suitable climate.

In laboratory environments, predators often exterminate their prey by literally eating them out of existence. In nature, predation probably acts more often in the same manner as weather, influencing the delicate competitive balance between species. A new predator can shift that balance in favor of one species or another even though it removes only a small fraction of the total prey.

We should view extinction as a result of competition. Species may be driven to extinction by new competitors and old competitors alike; all that is required is change in the environment too rapid for the working of response mechanisms of the individual or of the population to which it belongs. The capacity for evolutionary response is an inherent population characteristic. During early stages of diversification, groups of organisms exhibit evolutionary flexibility and novelty, but these features give way to evolutionary conservatism as groups become firmly entrenched in the flora or fauna. Perhaps success is its own worst enemy.

Extinction and Species Diversity

The number of species in a community represents the balance (or equilibrium) between species production and extinction. Reduced species production and increased extinction rate both reduce the equilibrium number of species. These relationships are shown by a simple graph relating rates of species production and extinction to the number of species (Figure 16–4). On islands, species production is replaced by the rate of immigration of species from the mainland to the island. For an island with no species, the rate of immigration is initially high because each new colonist represents a new species. As diversity increases, however, more potential immigrants will have already reached the island and new species will be added more slowly. If all mainland species occurred on the island, the immigration rate would be zero.

The extinction rate increases as the number of species on the island increases, partly because there are more populations that could become extinct and partly because intensified interspecific competition increases the probability that any one species might be excluded. (Intensified competition therefore causes the extinction curve to bend upward with

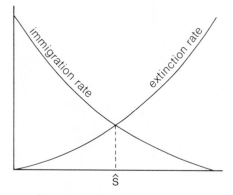

FIGURE 16–4 An equilibrium model of the number of species on an island. The immigration and extinction rates just balance at equilibrium.

Number of species on island

increasing diversity of species on the island.) Where the immigration and extinction curves cross, the number of new species balances the loss of species and diversity achieves a stable equilibrium.*

If small populations are more likely to become extinct than large populations, we would expect small islands to have fewer species than large islands that are otherwise comparable. On the diagram of the equilibrium model, the extinction rate curve for small islands would lie above that for large islands (Figure 16-5). Conversely, the rate of immigration decreases with distance of the island to mainland sources of new colonists. Hence the immigration rate curve for distant islands lies below that for close islands and, other factors being equal, species diversity should decrease with increasing distance from the mainland.

The equilibrium model has been verified for many groups of organisms on islands throughout the world. For example, number of species of birds is closely related to island area on the Sunda group of islands in the East Indies, including the Philippines, New Guinea, Borneo, and many smaller islands in the area (Figure 16-6). The Sunda islands are so close to each other and to mainland sources of colonists that no distance effect is evident. On islands to the east of New Guinea, many of them located at great distances in the Pacific Ocean, the number of bird species falls below the "saturation curve" described for species on islands of similar size close to the mainland (Figure 16-6).

The equilibrium number of species represents a steady state. The number of species may not change, but species nonetheless disappear

---

* Although ecologists widely apply the term *equilibrium* to communities whose diversity is maintained by a balance between opposing forces of extinction and species production, this condition is more properly referred to as a *steady state*. Equilibrium and steady state are similar in that the properties of a system are maintained in spite of external pressure to change, but the concepts differ in that steady state describes an open system with continual gains and losses, and equilibrium describes a closed system with its inherent properties and its substance both preserved.

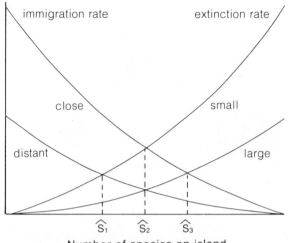

FIGURE 16–5   Relative number of species on small, distant islands ($\widehat{S}_1$) and large close islands ($\widehat{S}_3$) predicted by the equilibrium model. The expected number of species on small, close islands and large distant islands is intermediate ($\widehat{S}_2$).

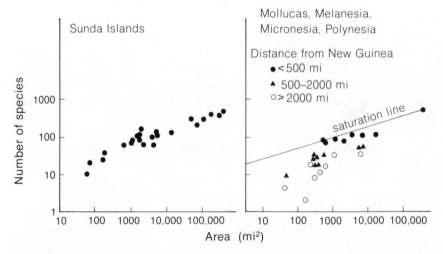

FIGURE 16–6   Species-area curves for land and fresh-water birds on the Sunda Islands (left), and several islands in the Moluccas, Melanesia, Micronesia, and Polynesia (right). The islands portrayed in the right-hand diagram show the effect of distance from the major source of colonists (New Guinea) on the size of the avifauna. The saturation line is the species-area relationship for the closely spaced Sunda Islands.

and are replaced by new colonists. The resulting turnover of species on islands can be quite high: 0.2 to 1.0 per cent per year for birds on the Channel Islands off the coast of southern California.

A Global Model of Diversity

The island diversity model can be modified to represent the balance of species production (speciation, see page 335) and extinction within large mainland areas. Rates of extinction are lower on continents than on islands because populations are larger. Furthermore, the shape of the species production curve is different because new species are added by species production within a continental land mass or ocean basin more frequently than by colonization from outside the area. In the total absence of species from an area, no new species could be produced within the area. Rate of species production increases in direct proportion to number of species present (as the number of populations able to form new species increases) but levels off at high numbers of species as the ecological opportunities for diversification become fewer (Figure 16-7). Extinction plays as large a role in the global model as it does in the island model. At the steady state, species continually disappear and are replaced by others. We should not, however, assume that species numbers have reached a global steady state. Even though the turnover rate of species far exceeds net changes in overall diversity, fossil history reveals a rising number of species, with a minor setback caused by recent glaciation, wide-spread climatic changes, and the appearance of man.

The approach of a fauna to its steady state can be illustrated by the diversity of fishes in several lakes of known age — which may be regarded as microcosms of the Earth as a whole. Lake Lanao in the

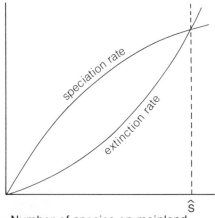

FIGURE 16–7 Equilibrium model of the number of species on a continental area, in which new species are added only by speciation; there is no immigration of species from outside the area. $\hat{S}$ represents the equilibrium number of species.

Number of species on mainland

Philippines was formed about 10,000 years ago when a lava flow dammed a large river. The lake is small (375 square kilometers), but since its formation, one species of cyprinid fish, or carp, has diversified into 18 species, 5 of which taxonomists place in 4 endemic genera known only from the lake. This initial diversification demonstrates at once the rapidity of evolution in a newly opened habitat and the importance of speciation in the diversification of a fauna.

Lake Nyasa in Africa, approximately 500,000 years old, harbors 100 species of the widespread genus of cichlid fish (sunfish family), *Haplochromis*, 70 other cichlid fish in 20 endemic genera, and several endemic species in other families of fish. Part of the high species diversity is clearly related to size rather than age. Lake Nyasa has an area of 11,430 square miles and its greatest depth is 2,000 feet. Nonetheless, the diversity and the high proportion of endemic species and genera indicates that the lake community has moved much closer to a steady state than has Lake Lanao.

The fish fauna of Lake Tanganyika, 1,500,000 years of age, suggests that Lake Nyasa has not, in fact, attained a steady-state fish fauna. Lake Tanganyika, 12,700 square miles in area and 4,700 feet deep, harbors as many cichlid species as Lake Nyasa, and, in contrast to Nyasa's 100 species that are not endemic, all but 19 species in Lake Tanganyika belong to distinctive endemic genera. Furthermore, endemic genera of other families less quick to evolve and diversify appear to fill many of the ecological roles taken by cichlids in Lake Nyasa.

The diversity and composition of the fish faunas of these lakes suggest that the number of species approaches to steady-state levels quickly by rapid speciation and diversification. After reaching an equilibrium number, the fauna then undergoes a more subtle readjustment of species composition. Groups that diversify rapidly to fill the community are replaced by species slower to evolve but perhaps inherently better suited to compete. We cannot say for certain whether global diversity is still increasing. More important to the study of ecology are differences in local diversity found in the variety of habitats on Earth. Geographical patterns of local diversity depend more on local ecological conditions than on the evolutionary history of the more numerous flora and fauna of whole continents. We shall consider the causes of these patterns in the next chapter. But before we conclude this chapter on extinction, we shall consider what might be referred to as the life history of the species: the taxon cycle.

Taxon Cycles

Species production and extinction remind us of the birth and death of individuals. In fact, the steady state between species production and extinction in a community at equilibrium corresponds closely to the steady state between birth and death in a population at its carrying capacity (see page 245). We are tempted to extend this analogy by ascribing the changes in a species between its origin and eventual extinction to phases of development, maturation, senility, and finally death. This analogy relates genetic phenomena in the population to physiological phenomena in the organism — a large jump — but the parallel is particularly evident in island birds between their initial colonization and eventual extinction. The disappearance of island populations by no means occurs at random. New immigrants pass through one or more cycles of changing competitive ability and probability of extinction. These changes are called the *taxon cycle.* A taxon is a natural biological grouping of organisms at any level in the taxomic hierarchy (e.g., species, genus, and so on) but usually refers to a species or a subpopulation of the species.

Immigrants to islands initially appear to be excellent competitors. Colonizing species are usually abundant and widespread on the mainland; these qualities make good immigrants. Most invaders of an island exhibit ecological release: their populations increase greatly and spread into habitats not occupied by the parent population on the mainland (see page 288). After immigrants become established, however, their competitive ability appears to wane; their distribution among habitats becomes restricted and local population density decreases. These trends eventually lead to extinction.

We can judge the relative ages of populations on islands by their patterns of geographical distribution and by differences in their appearance from appearance of mainland forms from which they were derived. Range maps of representative species of birds in the Lesser Antilles (Figure 16-8) demonstrate the progressive changes in distribution and differentiation of species with time. On the basis of such distribution patterns, we can assign populations to one of four stages (Table 16-2): expanding (I), differentiating (II), fragmenting (III), and endemic (IV). Species in late stages of the taxon cycle exhibit a loss of interisland movement and seasonal migration, reduced flocking behavior, and an increased tendency to occur in, or be restricted to, deep forest habitats, often montane forests. Similar habitat changes have been found in ants and beetles on islands in the Pacific.

Populations become more vulnerable to extinction as the taxon cycle progresses (Table 16-3). More endemic species (stage IV) of West Indian birds have become extinct since 1850, or are currently in grave danger of extinction, than widespread species (stages I to III). Almost a quarter of the endemic species either are extinct or are sufficiently reduced that they are presently in danger of extinction.

The decrease in competitive ability of island populations with age may be caused by evolutionary responses of an island's biota to new species. Immigrants are relatively free of parasites, predators, and efficiently specialized competitors when they colonize an island, so that

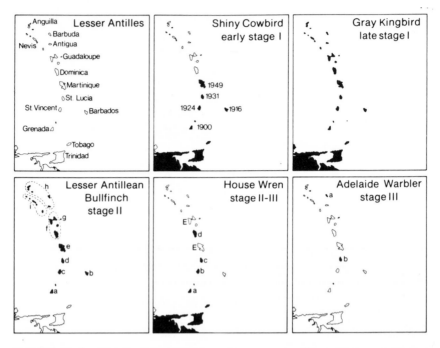

FIGURE 16–8 Distribution patterns and taxonomic differentiation of several birds in the Lesser Antilles, illustrating progressive stages of the taxon cycle. The shiny cowbird has expanded its range in the islands (dates of arrival are indicated), and the house wren has become extinct (E) on several islands during this century. Lower case letters designate subspecies.

their populations increase rapidly and become widespread. Having reached this stage, new immigrants constitute a larger part of the environment of many other species, which then evolve to exploit, avoid exploitation by, or outcompete the newcomer (*counter-adaptation*). Apparently, a large number of species, when adapting to a single, abun-

TABLE 16–2 Characteristics of distribution of species in the stages of the taxon cycle.

Stage of cycle	Distribution among islands	Differentiation between island populations
I	Expanding or widespread	Island populations similar to each other
II	Widespread over many neighboring islands	Widespread differentiation of populations on different islands
III	Range fragmented due to extinction	Widespread differentiation
IV	Endemic to one island	————

TABLE 16–3   Rate of extinction of island populations of birds in the West Indies as a function of stage of the taxon cycle.

	Taxon-cycle stage			
	I	II	III	IV
Number of recently extinct or endangered populations	0	8	12	13
Total number of island populations*	428	289	229	57
Per cent extinct or endangered	0	2.8	5.2	22.8

*A widespread species has many island populations, an endemic species only one.

dant, new population, can evolve faster than the new species can adapt to meet their evolutionary challenge. Competitive ability of the immigrants is progressively reduced by counter-adaptations of island residents until the once-abundant new species becomes rare. Species are eventually forced to extinction by subsequent arrivals from the mainland that are more efficient competitors. When a species becomes rare, other species no longer gain evolutionary advantage by adapting to it and the evolutionary pressure upon the rare species is released. If this occurs before a species' decline has proceeded too far, the species may again increase and begin a new cycle of expansion throughout the islands. This has apparently occurred many times; species distributions provide ample evidence of secondary expansions within the West Indies (for example, Figure 16–8, lower left).

The taxon cycle model re-emphasizes the primary role of competition, and the secondary roles of predators and climatic change, in extinction. Endemic island species are forced to extinction by competitively superior new immigrants. On remote islands, reached by few colonists, endemic species persist for longer periods. In fact, the inhabitants of oceanic islands include some of the unique and unusual forms of animal and plant life on Earth, protected by great distance from potentially superior mainland species. Taxon cycles probably occur on continents, only more slowly than on islands. The same principles nonetheless should pertain.

17

The Community as a Unit of Ecology

In previous chapters, we have described the structure and function of the community. We have also examined the responses of organisms and populations to environmental change and related these responses to the stability of ecosystem structure and function. Now we must turn to the most difficult and controversial question facing ecologists today: Does the community have unique regulatory properties, derived from the interactions of its constituent species, that populations separately lack? In other words, is the whole of a biological community in some way greater than the sum of its parts? While the question remains unanswered, studies of diverse communities suggest some forms an answer might take. We may suspect, on the basis of studies described in this and the two following chapters, that the question of the nature of community regulation will have many equally valid answers. Even though studies of community function may lead to general principles of organization and regulation, each community will undoubtedly demonstrate these principles in a slightly different manner, depending on its habitat and constituent species.

In this chapter, we shall discuss the community as a natural unit and trace the geographical and evolutionary bases of its organization. Then we shall examine the development of the community and its response to perturbation in the last two chapters of the book.

Definition of the Community

The term *community* has been given such a variety of meanings by ecologists that it borders on being meaningless. The term usually is restricted to the description of a group of populations that occur together, but there ends any similarity among definitions. Throughout the development of ecology as a science, the term has often been tacked on to associations of plants and animals that are spatially delimited and that are dominated by one or more prominent species or by a physical

characteristic. One speaks of an oak community, a sagebrush community, and a pond community, meaning all the plants and animals found
in the particular place dominated by its namesake. Used in this way,
"community" is unambiguous: it is spatially defined and includes all the
populations within its boundaries.

Ecologists also define communities on the basis of interactions
among associated populations. This is a functional rather than descriptive use of the term. In this and following chapters the term *association*
will be reserved for groups of populations that occur in the same area
without regard for their interactions; *community* will be used to denote
an association of interacting populations. We may find communities
difficult to delimit because interactions among populations extend beyond arbitrary spatial boundaries. Some of the oxygen molecules that a
squirrel inhales in New York could have been expired last month by a
tree in the Amazon basin of South America. Clearly, the Brazilian tree
exerts a negligible influence on the squirrel and we ought not to extend
the boundaries of the community to which the squirrel belongs to the
tropical rain forest. But biological interactions reach out in a complicated
fashion, making it difficult to set limits — other than completely arbitrary ones — to a community.

Organisms and materials move slowly across the boundaries of
most terrestrial associations of plants and animals compared to the
turnover of organisms and flux of materials and energy within the
association. Therefore, "community" and "association" are nearly identical for many terrestrial habitats, as long as the boundaries of the
habitats are not too narrowly set. Because most fresh-water and coastal
aquatic communities receive inputs from the land, their boundaries are
more difficult to establish. Two extreme cases of associations, in which
all the energy inputs come from outside, are the faunas of caves and of
unlighted depths of the oceans. Strictly speaking, the inhabitants of
these perpetually dark places depend upon primary producers in sunlit
habitats for their source of energy, but we may still speak of competition
among detritivore or carnivore members of the cave "community," if we
wish. In fact, the term community may be usefully applied to any group
of interacting populations — the herbivorous insect community, the oak
leaf community, the kelp holdfast community — as long as community
boundaries are clearly delimited.

The Community as a Functional Unit

The maintenance of community structure and functioning depends on a
complex array of interactions, directly or indirectly tying all its members
together in an intricate web. The influence of a population extends to
ecologically distant parts of the community through its competitors,

predators, and prey. Insectivorous birds do not eat trees, but they do prey on many of the insects that feed on foliage or pollinate flowers. By preying upon pollinators, birds indirectly affect the number of fruits produced, the amount of food available to animals that feed upon fruits and seedlings, the predators and parasites of those animals, and so on. The ecological and evolutionary impact of a population extends in all directions through its influence on predators, competitors, and prey, but this influence is dissipated as it passes through each successive link in the chain of interaction.

Measures of community structure and function — number of species, number of trophic levels, rates of primary production, energy flow, and nutrient cycling — reflect ecological interactions among populations and between individuals and their physical environment. The ability of the community to resist ecological perturbations no doubt reflects the homeostatic mechanisms of individuals and the growth responses of populations. But beyond these separate homeostatic capacities, we must search for properties of organization that enhance community stability and the efficiency of community function. In other words, we must ask whether the homeostatic capacity of a community transcends the summed properties of its constituent populations. We might imagine, for example, that competition and ecological release help to stabilize the function of each trophic level. If the population of one species were suppressed by climate or disease, a competitor could respond by using the first population's leftover resources and thus maintain the total production of the trophic level. We might also view predation as a destabilizing influence, magnifying population fluctuations in lower trophic levels.

Whenever we have examined homeostasis on the individual or population level, the adaptations of organisms have appeared vital. The characteristics of organisms that determine the outcome of their interaction with the physical and biological worlds are evolved properties adjusted over geological time to meet the criteria of success dictated by the environment. In our discussion of community properties we shall ask whether communities also have unique adaptations of structure and function over and above traits evolved to enhance the fitness of the individual.

The Role of Predation in Community Structure

Efficient predators can pare the size of prey populations to levels far below the capacity of the environment to support the prey, as we have seen in Chapter 15. The effect of predation extends, however, beyond the prey population. Competitors of the prey species benefit when prey numbers are reduced by predators. If owls catch voles but not rabbits, the rabbits stand to gain by the predatory activities of owls.

however,

By reducing competition between species on the trophic level below *predator* them, predators could cause an increase in the numbers of species on *susceptability* that trophic level. Conversely, removal of predators from a community *is a factor* might allow a few species, perhaps those formerly preyed upon most *in competitive* heavily, to become competitively superior to other members of their *competance.* trophic level in the absence of predators and eventually to exclude them from the community. To test whether, in the absence of natural predators, the number of prey species in a community would decline, marine ecologist Robert Paine removed predators from experimental areas of rocky shore habitat along the Pacific coast of Washington. The intertidal zone there is dominated by several kinds of herbivores that either filter phytoplankton from the water (barnacles, gooseneck barnacles, and mussels) or scrape algae off the rock surface (limpets, turban shells, and chitons). These are preyed upon by the starfish *Pisaster*. In one study area, eight meters in length and two meters in vertical extent, all starfish were removed during frequent searches of the area. (An adjacent area was left undisturbed.) Following the removal of the starfish, populations of many species of herbivores declined and seven of the fifteen species found in the experimental area at the beginning of the experiment had disappeared by its end. No changes occurred in the undisturbed area. Most of the species that disappeared from the experimental area were literally crowded out of existence by barnacles and mussels, which flourished in the absence of their starfish predators. Paine concluded that starfish were a major factor responsible for maintaining the diversity of the habitat. One must be wary, however, of inferring causative influences within a steady-state community from the results of experimental manipulation of the community. Paine's result might have been obtained if, instead of removing predators, he had lowered the tide level, increased water temperature, or placed a sewage effluent pipe nearby.

Predators are sometimes added to communities, often to control weedy or pest species (see Chapter 15). When populations of pest species are reduced by introduced predators, other species, potential competitors of the pest, often reappear in the area of predator introduction. For example, when the cactus moth was introduced to Australia to control the prickly pear cactus, domination of areas by the cactus, often to the exclusion of all other plant species, ceased, thereby opening space for other species to invade (see page 302).

Daniel Janzen has extended the idea of predator control over prey populations to a general theory explaining the high diversity of trees in the humid tropics. Whereas fewer than 150 species of trees are found in Canada and fewer than 800 species are found in all of North America, north of Mexico, many tropical countries with rain forest habitats harbor several thousand species. Janzen has suggested that herbivores, acting as plant predators by destroying growing shoots, seeds, and seedlings, might reduce the abundance of species having superior competitive ability and thus allow other, less dominant trees to persist.

Janzen's hypothesis has not been adequately tested, but it is supported by the failure, caused by insect and disease attack, of attempts to grow dense stands of particular trees in the tropics. Plantations of rubber trees, grown in their native habitat in the Amazon Basin, have met with singular lack of success owing to epidemics of plant pests. Transplanted to Malaya, however, where their natural enemies do not occur, rubber trees are successfully cultivated in large plantations.

Ecologists do not agree on the extent to which predators influence the species composition of prey trophic levels, but the controversy has not prevented some authors from taking the idea of predator control one large step further by proposing that predators might not just influence the apportioning of energy flow among competing prey species, but might even limit the total energy flux through a prey trophic level. In a controversial paper published in 1960, N. G. Hairston, F. E. Smith, and L. B. Slobodkin argued that, because herbivores appear not to eat all the food available to them (witness the abundance of green plants) and because populations of herbivores have been observed to increase greatly when predators are removed, herbivore populations, indeed the herbivore trophic level as a whole, must be limited by predators, not by the availability of plants for food. This logic has serious flaws which were first pointed out in 1966 by William Murdoch. The mere presence of food in what appears to be an unlimited supply does not necessarily exclude food as a limit to herbivore populations. Leafy vegetation is relatively unsuitable as food because it contains too little protein, too much indigestible cellulose, and, frequently, toxic doses of exotic organic poisons. Furthermore, the occurrence of large populations of herbivores could, by reasoning parallel to that of Hairston, Smith, and Slobodkin, suggest that predators are themselves limited by higher level predators rather than by the availability of their own prey. The weakness of this type of argument becomes apparent at the end of the food chain, where predators have no predators, and thus are not themselves prey.

Most ecologists still subscribe to the view that energy flux through each trophic level is limited by the availability of food on the trophic level below it. This view does not exclude the idea of predation influencing the standing crop of biomass and rate of turnover of populations on each trophic level. An experiment accidently performed in Gatun Lake, in the Panama Canal Zone, shows how a predator's influence can extend down through the trophic structure of the community, even to the primary producer level. Gatun Lake is artificial, formed in the early 1900's when the Panama Canal was built. Over the last 60 years, plants and animals from streams flowing into the former river basin became established in the lake, creating what appeared to be a stable association.

The basic features of this association could be summarized by a simple food chain: phytoplankton — zooplankton — fish. The fish kept zooplankton populations low; in the absence of abundant zooplankton, phytoplankton populations were usually dense enough to make the lake water murky, like a thin, living soup. In 1967, the peacock bass, a species of cichlid fish not actually related to the familiar bass of temperate lakes, was accidentally introduced to Gatun Lake. The bass turned out to be a voracious and efficient predator on other kinds of fish. With an added predator link, the lake's major food chain now became phytoplankton — zooplankton — planktivorous fish — peacock bass. The peacock bass reduced other fish populations, allowing zooplankton populations to increase and, in turn, leading to a marked reduction in phytoplankton populations and production. Wherever the peacock bass has extended its range within the lake basin, the waters have become less turbid and phytoplankton biomass has decreased. In this simple aquatic community, a predator exerted a profound effect on the abundance and productivity of the three trophic levels below it. Furthermore, populations of fish-eating birds, such as terns and kingfishers, have been reduced by competition from the peacock bass, while populations of mosquitos, whose larvae are normally fed upon by small fish, have increased. The peacock bass clearly created a major disturbance in the community. How the community will respond remains to be seen, but the example shows that its internal structure, the relationships within a community, may be an important component of its inherent stability.

The Community as an Evolutionary Unit

The community is an association of interacting individuals. Community function is the sum of what individuals do and thus reflects the adaptations of individuals. We may ask, however, whether attributes of a community represent more than the evolved properties of individuals. Because adaptations are evolved through natural selection to improve the fitness of the individual carrying the selected trait, we would not expect to find adaptations enhancing community function that were inconsistent with this goal. What is good for the individual prevails, regardless of whether it is good for the community. We have seen, however, that evolutionary adjustments of prey to their predators, and of competitors to each other, tend to stabilize their relationships. As a result, these adaptations improve the efficiency of ecosystem function and enhance community stability.

We may now set down, at least tentatively, a basic ecological principle: community efficiency and stability increase in direct proportion to the degree of evolutionary adjustment between associated populations. The action of this principle is shown quite clearly when foreign species are introduced to a community. In most cases they cannot successfully invade the community, and so die out, but occasionally exotic species gain a foothold and rapidly come to dominate the community. Such species can completely upset the delicate balances achieved between members of the community and, in so doing, disrupt community function. Outbreaks of introduced pests like the European pine sawfly and gypsy moth can, in fact, almost completely destroy a community by defoliating its major primary producers.

The evolutionary adjustment of populations to one another depends upon their degree of association. If two species always occurred together, their interaction would exert an important influence on the evolution of each. If, however, the ecological and geographical ranges of two species were mostly non-overlapping, each species would exert only a small portion of all the selective influences on the population of the other. A community cannot exhibit strong coevolutionary adjustment among its members if the adaptations of its species are molded primarily by relationships in other communities.

The actual degree of association between species in a community lies somewhere between two imaginable extremes: on one hand, the boundaries of a *closed community* would coincide with the boundaries of all its members. No species would occur outside the community in other associations. On the other hand, the boundaries of an *open community* would be drawn arbitrarily with respect to the geographical and ecological distributions of its component species, many of which would extend their ranges independently into other associations of species.

Ecotones

The structure of closed and open communities is depicted schematically in Figure 17–1. In the upper diagram, the distributions of species in each community are closely associated along a gradient of environmental conditions — for example, from dry to moist. Closed communities represent natural ecological units with distinct boundaries. The edges of the community, called *ecotones*, are points of rapid turnover of species. In the lower diagram, species are distributed at random with respect to each other. We may arbitrarily delimit an open community at some point, perhaps a dry forest community near the left-hand end of the moisture gradient, but some of the species included might be more characteristic of drier points along the gradient while others reach their greatest productivity in wetter sites.

The separate concepts of open and closed communities both apply to associations of species in nature. We observe distinct ecotonal boundaries between associations under two different kinds of circumstances: when there is an abrupt change in the physical environment — for

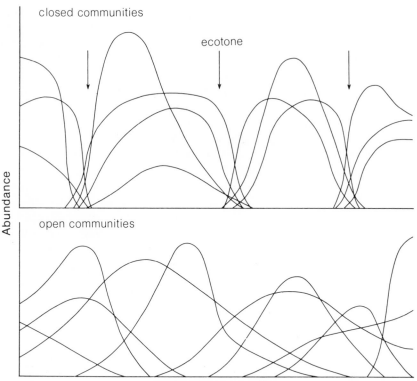

FIGURE 17–1   Hypothetical distributions of species organized into distinct associations (closed communities, above) or distributed at random along a gradient of environmental conditions (below). The species found at any point, or within any narrow region, along the lower gradient would be identified as an open community. Ecotones between communities in upper figure are indicated by arrows.

example, at the transition between aquatic and terrestrial communities, between distinct soil types, or between north-facing and south-facing slopes of mountains — or when one species or life form so dominates the environment of the community that the edge of its range signals the distributional limits of many other species.

The transition between broad-leaved and coniferous forest is usually accompanied by an abrupt change in soil acidity. At the boundary between grassland and shrubland, or between grassland and forest, sharp changes in surface temperature, soil moisture, and light intensity result in many species replacements. Sharp grass-shrub boundaries occur because when one or the other vegetation type holds a slight competitive edge, it dominates the community (see page 285). Grasses prevent shrub seedling growth by reducing the moisture content of the surface layers of the soil; shrubs prevent the growth of grass seedlings by shading them out. Fire evidently maintained a sharp boundary between prairie and forest in the midwestern United States. Perennial grasses resist fire damage that kills tree seedlings outright, but fires do not penetrate deeply into the moister forest habitats.

Sharp physical boundaries create sharp ecotones. Such boundaries occur at the interface between most terrestrial and aquatic (especially marine) communities (Figure 17-2) and where underlying geological formations cause the mineral content of soil to change abruptly. The ecotone between serpentine-derived soils and non-serpentine soils in southwestern Oregon is shown in more detail by the diagrams of soil minerals and occurrence of plant species in Figure 17-3. Levels of nickel, chromium, iron, and magnesium increase abruptly across the boundary into serpentine soils; copper and calcium contents of the soil drop off. The edge of the serpentine soil marks the boundaries of many species that are either excluded from, or restricted to, serpentine outcrops. A few species are found only within the narrow zone of transition, and others, seemingly unresponsive to variation in soil minerals, extend across the ecotone.

Gradient Analysis

Sharp physical boundaries often create abrupt changes in vegetation. It is difficult for species to be fence-sitters in such situations; they must adapt to the conditions on one side or the other. The few species

FIGURE 17-2 A sharp community boundary (ecotone) associated with an abrupt change in the physical properties of adjacent habitats. Seaweeds extend only to the high tide mark. Between the high tide mark and the spruce forest, waves wash the soil from the rocks and salt spray kills pioneering land plants, leaving the area devoid of vegetation.

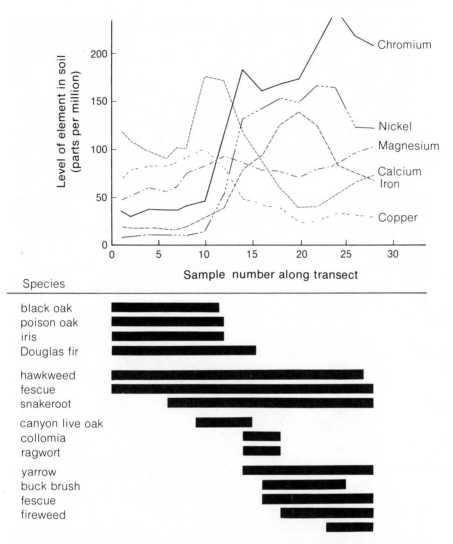

FIGURE 17–3 Changes in soil mineral content (above) and plant species (below) across the boundary between non-serpentine (left) and serpentine soils (right) in southwestern Oregon. The transect diagrammed here is somewhat atypical in that magnesium does not increase as abruptly as usual across the serpentine ecotone.

specialized to live in the ecotone — often referred to as *edge species* — are necessarily restricted in numbers by the scarcity of their habitat.

Most large-scale ecological changes are more gradual than the abrupt changes between land and water, forest and field, or geological formations. The biological communities of the Earth occupy a *continuum* of gradually changing ecological conditions, varying along one dimension from cold to warm, along another from dry to moist, along a third from seasonal to moderate, and perhaps along many other minor gradients of ecological conditions. Changes along these *gradients* occur gradually over vast geographical distances; there are no sharp physical boundaries to intercept species ranges.

When we examine the occurrence of species along gradients of moisture or temperature, we find that the local plant community must be viewed as an open system. Groups of species are not restricted to particular associations, rather each species is distributed along environmental gradients almost without regard to the occurrence of others. The environments of the eastern United States form a continuum with a north-south temperature gradient and an east-west rainfall gradient. Species of trees found in any one region, for example those native to eastern Kentucky, have different geographical ranges, suggesting a variety of evolutionary backgrounds (Figure 17–4). Some species reach their northern limits in Kentucky, some their southern limits. Few species have broadly overlapping geographical ranges, hence associa-

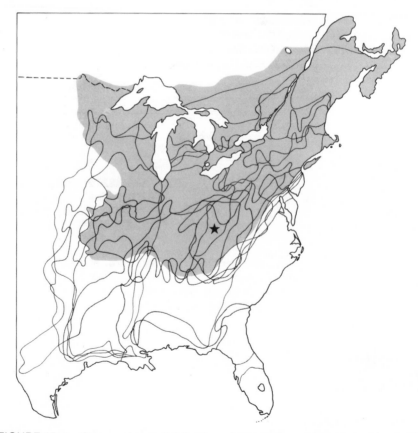

FIGURE 17–4   Geographical distribution of 12 species of trees found in plant associations in eastern Kentucky.

tions of plant species found in eastern Kentucky do not represent closed communities. Each species has a unique evolutionary history with a variable degree of association with other species in the local community.

A more detailed view of Kentucky forests would reveal that many of the tree species sort out along local gradients of conditions. Some are found along the ridge tops, others along moist river bottoms, some on poorly developed rocky soils, others on rich, organic soils. The species represented in each of these more narrowly defined associations would exhibit correspondingly closer ecological and evolutionary distributions, but the open community concept would still dominate our thinking about these associations.

Cornell University ecologist Robert Whittaker has examined plant distributions in several mountain ranges where moisture and temperature vary over short distances according to elevation, slope, and exposure. When Whittaker plotted the abundance of each species at sites at the same elevation distributed along a continuum of soil moisture, he found that the species occupied unique ranges with peaks of abundance scattered along the environmental gradient (Figure 17–5). The Oregon mountains have fewer species overall, but each species has a wider ecological distribution, on the average, than each Arizona species.

In the Great Smoky Mountains of Tennessee, dominant species of trees are widely distributed outside the plant associations that bear their

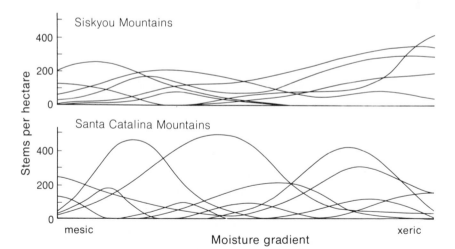

FIGURE 17–5 Distribution of species along moisture gradients at 460 to 470-meters elevation in the Siskyou Mountains of Oregon and at 1830 to 2140-meters elevation in the Santa Catalina Mountains of southern Arizona. Species in the more diverse Arizona flora occupy narrower ecological ranges; thus, in spite of the greater total number of species in the flora of the Santa Catalina Mountains, the Santa Catalina Mountains and Siskyou Mountains have similar number of species at each sampling locality.

names (Figure 17-6). For example, red oak is most abundant in relatively dry sites at high elevation, but its distribution extends into forests dominated by beech, white oak, chestnut, and even hemlock, an evergreen, coniferous species, and extends throughout the entire range of elevation in the Smoky Mountains. Beech prefers moister situations than red oak, and white oak reaches its greatest abundance in drier situations, but all three species occur together in many areas.

Species Interaction and Community Organization

Competition forces species to specialize on a narrow portion of available resources and to restrict their distributions among habitats. By preventing close ecological overlap, competition fosters the open structure of communities. The diverse distributions of plant species along gradients of temperature or moisture, shown in Figure 17-5, result from each species adapting more successfully than its competitors to a unique set of ecological conditions. Nonetheless, distributions of plants overlap more than one might expect from the principle of competitive exclusion. Overlap could be maintained in spite of strong competition by continual emigration of individuals (as seeds in the case of plants) to areas marginal for the existence of the species. Alternatively, habitats intermediate between the optima of two species may be sufficiently heterogeneous to allow both species to coexist, each restricted to tiny patches (perhaps the size of a single seedling) suitable for its establishment, mixed among patches that favor the other species. Where competition occurs by direct interference, as by interspecific territorial defense in birds, the geographical ranges of close competitors sometimes adjoin abruptly, with few individuals crossing the imaginary line of competitive balance drawn between them.

In contrast to competitors, predators and prey evolve adaptations to each other that may strengthen the unitary nature of the community. Although prey organisms might be expected to thrive where their predators are absent, no community lacks predators, and species may actually be more abundant where they have evolved adaptations to avoid local predators, than in communities with predators to which the prey have not adapted. Conversely, predators may also be more efficient where they have evolved to exploit local prey species, than in communities with prey which they have not become adapted to exploit. The mutual adaptation of species to each other, referred to as *coadaptation* even though the species may be antagonists, can enhance the association of species within communities. Many predators and parasites are specialized to attack only one species of prey. In such cases, predator and prey co-occur extensively.

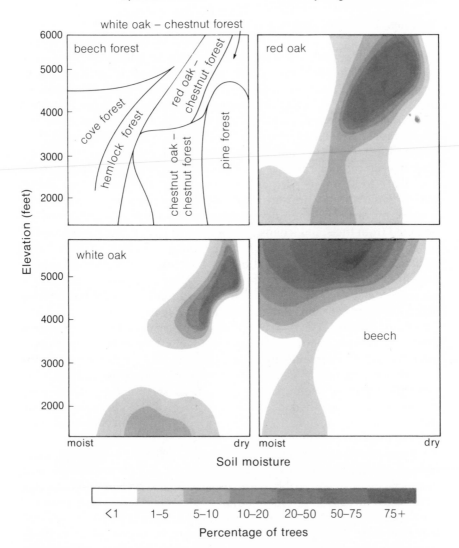

FIGURE 17–6   Distribution of red oak, white oak, and beech with respect to altitude and soil moisture in the Great Smoky Mountains of Tennessee. The approximate boundaries of the major forest associations are shown in the diagram at the upper left. Relative abundance, represented by degree of shading, corresponds to percentage of tree stems over one centimeter in diameter in samples of approximately 1,000 stems.

Mutualism

Extreme cases of association between species, in which the survival of each species depends on the other, frequently evolve out of predator-prey or host–parasite relationships. The relationship between interdependent species, in which each species benefits the other, is called *mutualism*. We have already come across cases of mutualism in the associations between legumes and the nitrogen-fixing bacteria in their root nodules (page 167), and between trees and mycorrhizal fungi (page 282). In each case, one species provides a material or service in return for a similar effort from its partner. Nitrogen-fixing bacteria provide plants with organic nitrogen in return for sugars. In return for the same substance, mycorrhizal fungi provide their plant associates with minerals obtained from the soil. Lichens are compound plants displaying another fungus–green plant association in which the green plant (in this case an alga) provides sugars produced by photosynthesis and receives minerals, sometimes dissolved out of bare rock, from the fungi. This unique association between organisms with quite different capabilities enables lichens to colonize habitats that are inhospitable to all other forms of life. Cows and termites both can digest cellulose because of their mutualistic associations with specialized microorganisms (bacteria and protozoa, respectively) in their intestinal tracts. These microorganisms have the biochemical machinery needed to break cellulose into simpler, and metabolically more useful, sugars. Of course, the microorganisms need the sugars too, but their hosts absorb some of the spill over and, in return, provide the microorganisms a constant source of food in a well-regulated environment.

Most mutualistic relationships probably evolved by way of host–parasite interactions. Nitrogen-fixing bacteria, mycorrhizal fungi, and gut microorganisms probably were disease organisms before the species coevolved mutualistic interdependency with their hosts. The arrangements between plants on one hand and the insects, birds, and bats that pollinate their flowers on the other, represent countless mutualistic associations developing out of previously exploitative interactions. Plant pollen is rich in nitrogen and other nutrients. Many animals, such as thrips, are specialized to eat the pollen and perform no useful function for the plant, but the presence of other kinds of animals flying from flower to flower to eat the nutritious pollen provided plants the opportunity to adapt their flower structure to enhance and control pollination by animals. Many plants attract pollinators with showy flowers and strong fragrances; nectar, which contains mostly sugar and very little protein, is provided as an inexpensive food incentive for pollinators to visit the flowers; flower structure has been adapted to ensure that pollinators will efficiently transfer pollen from one flower to another of the same species, but not to flowers of other species.

Plant-pollinator relationships are most highly developed in the orchid family, with its great variety of flower shapes, colors, and smells. The intricate bond between flower and pollinator is exemplified by the orchid *Stanhopea* and the euglossine bee *Eulaema*. The attraction of euglossine bees to the particular orchids they visit is unusual in that no nectar is produced by the flowers, and only male bees visit them. The flowers are extremely fragrant; each species of orchid employs a unique chemical fragrance, or combination of fragrances, so as to attract only one species of bee pollinator, preferably a species that does not visit the flower of other orchids.

When a male euglossine bee visits an orchid, he brushes parts of the flowers with specially modified forelegs, and then appears to transfer some substance to the hind leg, which is enlarged and has a storage cavity. The function of the bee's behavior is not understood, but it may be involved in mate attraction; perhaps the orchids provide the bee with a sort of perfume. For *Eulaema meriana* to pollinate *Stanhopea grandiflora*, a bee enters the flower from the side and brushes at a saclike modification on the lip of the orchid (Figure 17-7). The surface of the lip is smooth, and the bee often slips when it withdraws from the flower. (The orchid fragrances are also thought to intoxicate the bee and cause him to lose his footing on the lip of the orchid.) If the bee slips, he may brush against the column of the orchid flower, where the pollen sacs are precisely placed to stick to the hindmost part of the thorax of the bee. If a bee with an attached pollen sac slips and falls out of another flower, the sac will catch on the stigma and pollinate the flower (Figure 17-7). The placement of pollen sacs on bees differs among species of orchids and, in conjunction with specific differences in the fragrances of the orchids, ensures that bees do not accidentally fertilize the flowers of one species with the pollen of another.

Tropical ecologist Daniel Janzen has described a fascinating case of complete mutualistic interdependence between certain kinds of ants and swollen-thorn acacias in dry areas of Central America. The acacia plant provides the ants with food and nesting sites in return for protection from insect pests. Bull's-horn acacias have large hornlike thorns with a tough, woody covering and a soft pithy interior (Figure 17-8). To start a colony in an acacia, a queen ant of the species *Pseudomyrmex ferruginea* bores a hole in the base of one of the enlarged thorns and clears out some of the soft material inside to make room for her brood. As the colony grows, more and more of the thorns on the plant are filled. A colony may grow to more than a thousand workers within a year, and eventually may have tens of thousands of workers. In addition to housing the ants, the acacias provide food for the ants in nectaries at the base of their leaves, and in the form of nodules, called "Beltian bodies," at the tips of some leaves (Figure 17-8). In return, the ants kill all herbivorous insects that venture onto the acacia, dissuade large grazing mammals,

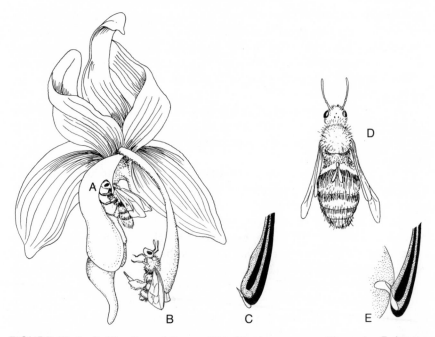

FIGURE 17–7   Pollination of the orchid, *Stanhopea grandiflora*, by *Eulaema meriana*. The bee enters from the side and brushes at the base of the orchid lip (A). If it slips (B) the bee may fall against the pollen sac, which is placed on the end of a column (C), and the pollen sac becomes stuck to the hind end of the thorax (D). If a bee with an attached pollen sac falls out of a flower, the sac may catch in the stigma, or pollen receptor organ (E), which is so placed on the column that the flower cannot be self-fertilized.

and even destroy seedling plants — potential competitors of the acacia — which sprout near the base of the acacia. The relationship between *Pseudomyrmex* and *Acacia* is *obligate:* neither the ant nor the acacia can survive without the other. Other ant–acacia associations are *facultative;* that is, the ant and the acacia can co-occur to mutual benefit but they can both exist independently as well.

The mutualistic relationship between ants and acacias has been accompanied by adaptations of both to increase the effectiveness of the association. For example, *Pseudomyrmex* is active 24 hours a day, which is unusual among ants, and thereby provides continuous protection for the acacia. In a similar evolutionary gesture, the acacia retains its leaves throughout the year, and thereby provides a continuous source of food for the ants. Most acacias lose their leaves during the dry season.

To demonstrate that acacias require ants for growth and survival, Janzen prevented ants from colonizing new acacia shoots and measured the growth of the shoots compared to others that had ants (Table 17–1).

FIGURE 17–8 Modifications of swollen-thorn acacias for mutualistic in-teractions with ants. Many of the thorns are enlarged and have a soft pith which the ants excavate for nests (left). The acacias provide food for the ants from nectaries at the base of leaves (center) and in the form of nutritious nodules at the tips of modified leaves (right).

Differences between experimental and control plants demonstrate strik-ingly the advantages to the plant of having ants. The reduced growth and survival of shoots without *Pseudomyrmex* was caused entirely by herbivorous insects; when such a shoot was grafted to a plant inhabited by an ant colony, herbivorous insects were immediately removed.

Obligate mutualism binds species into a close, long-term association

TABLE 17–1   Effect of the presence of the ant *Pseudomyrmex* on the growth and survival of bull's-horn Acacia in Oaxaca, Mexico.

	With ants	*Without ants*
Growth of suckers from stumps (cm)		
25 May to 16 June	31	6
16 June to 3 August	73	10
Size of suckers after 10 months' growth		
Average wet weight of suckers (g)	580	44
Average number of leaves	108	52
Average number of swollen thorns	104	39
Survival of stumps (per cent) from October 18 until:		
March 13	87	62
June 10	72	51
August 6	72	44

within which they behave as a single evolutionary unit. Although mutualism increases the ecological efficiency of many species, and thereby increases the overall efficiency of ecosystem function, mutualistic interactions are not common enough relative to competitive and purely exploitative interactions to unify the community in an integrated evolutionary unit. Predators promote community unity to the extent that they are specialized on particular prey, but competitive interactions within trophic levels ensure that community boundaries, however they are defined, are not respected by the plants and animals comprising a local biological association.

History and Community Structure

All communities are built upon the same framework of trophic levels but, depending upon each community's location, this framework is filled out by different kinds of species derived from different evolutionary stocks. Although most classes of plants and animals have worldwide distributions, orders, families, genera, and species are geographically more restricted. Islands lack many groups that do not have strong dispersal powers. Distributions have also been influenced by changes in the positions of continents and oceans. Forms that arose and diversified in Africa may never have dispersed as far as the New World. Owing to oceanic barriers to dispersal, Australia lacks most types of placental mammals. Only marsupials, whose evolutionary roots pre-date the breakup of the land mass that combined Australia and Africa, and bats, with strong powers of dispersal, are abundant on Australia.

Patterns of geographical distribution of taxonomic groups above the species level are illustrated by families of fresh-water fish in Figure 17-9. In most families, dispersal is limited by salt water. New streams, rivers, and lakes can be invaded only when they are joined to others. Dispersal is often a chance phenomenon determined by irregularities of geography and by the history of geographical changes. The fresh-water minnows (family Cyprinidae) are widespread throughout the world, from subarctic regions through the tropics into south temperate regions in South Africa. Minnows disperse easily and are an ecologically versatile group. They originated in the Old World, probably in central or southern Asia, and spread to the Western Hemisphere across the Bering Land Bridge. (The floor of the Bering Straits between the North Pacific Ocean and the Arctic Ocean has risen above the surface of the water several times in the last geological era.) That the distributional capability of the minnows is limited is shown by their failure to reach the Australian region, South America, and Madagascar, which has never been connected to mainland Africa.

The absence of a group from a large geographical region does not necessarily indicate limited dispersal ability; it may be the result of

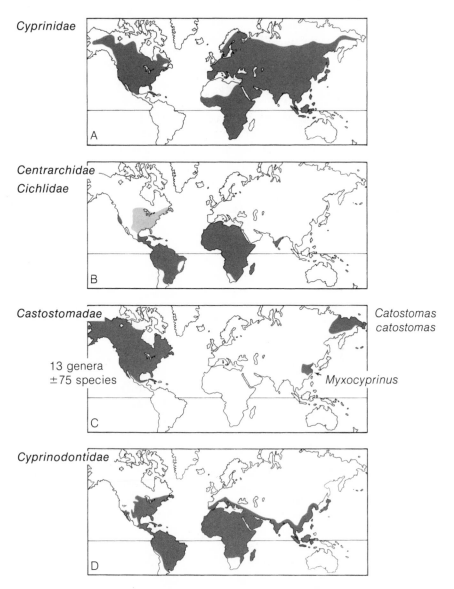

FIGURE 17–9 Geographical distributions of several families of fresh-water fish: (A) minnows (Cyprinidae); (B) cichlids (Cichlidae) and their ecological counterparts in North America, the sunfish (Centrarchidae); (C) suckers (Catostomidae); and (D) killifish (Cyprinodontidae), whose occurrence in fresh water is secondary.

competitive exclusion. Sometimes the only clue to the former presence of a group in a region is a relict population whose occurrence can be explained only by postulating previous widespread distribution of the taxonomic group to which it belongs. For example, the iguanid lizards comprise a large and successful family in the New World, but they are almost completely absent from the Old World, where their ecological roles are filled by agamid lizards. The presence of iguanids on Madagascar is a lone reminder of an extensive former distribution throughout the world.

Cichlid fish (Cichlidae) — ecological counterparts of the geographically restricted American sunfish family (Centrarchidae) — are limited primarily to tropical and subtropical fresh waters (Figure 17-9 B). Evidently they can disperse across salt water because they are found in both tropical Africa (including Madagascar) and South America, and they have recently invaded Central America and India.

The suckers (Catostomidae) probably arose in the Old World, but they are now represented there only by one ancient member of the family (Myxocyprinus). About 75 species occur in North America (Figure 17-9 C). One of these (*Catostomus catostomus*) has recently emigrated back across the Bering Straits to invade eastern Siberia, and perhaps to begin a new radiation of the family in the Old World.

Dispersal barriers differ for each group. The killifishes (*Cyprinodontidae*) are secondarily fresh-water fishes, which tolerate salt water and can cross sea barriers. Although the killifishes are limited to tropical and warm temperate regions, their worldwide distribution includes islands that lack many other kinds of fresh-water fish (West Indies, Madagascar, Bermuda, Celebes; Figure 17-9 D).

These historical accidents of distribution flavor the composition of communities like the spices found only in particular cuisines. Tropical rain forest can be identified as such in Brazil, the Congo, New Guinea, and Hawaii, but nearly all the species, most of the genera, many of the families, and some of the orders of plants and animals differ from one of these areas to the other. Standing in the foothills of the Andes in Chile, we might swear we were in the foothills of the Sierras in California, so similar do the habitats appear. Each species in one area has its ecological counterpart in the other, yet all the Chilean and Californian species differ. Ecological counterparts may even belong to different orders or classes. In the Australian deserts, lizards seem to fill many of the ecological roles played by rodents in the southwestern United States. There are no native rodents in Australia.

In spite of varied taxonomic origins, communities occupying similar environments in different continents develop similar appearance, structure, and function. Several ecological and evolutionary processes help to obscure irregularities of community structure caused by historical accidents of distribution. First, an upper limit to the number of species inhabiting a particular locality is determined by the environment of the locality and not by the total number of species inhabiting the same continent. Second, adaptive radiation (diversification) within a continent or island group can increase the number of species within regions that are impoverished owing to their distance from sources of colonists and have fewer species than their environments can support. Third, species with different evolutionary origins adapt to fill the same role in different places. We shall examine each of these processes below.

Competition and Diversity

Competitive exclusion limits the degree to which species can use the same resources and still coexist. The number of species in a community thus depends on the amount and heterogeneity of resources in the habitat. In other words, local diversity is limited by local ecology. We would therefore expect communities in similar habitats but in different regions to be composed of similar numbers of species. We know, however, that regional diversity also depends on the rate at which species are added to an area, either through internal diversification or through immigration, in addition to the rate of extinction by competitive exclusion and other causes (see page 342). The discrepancy between regional and local diversity is reconciled by the degree of habitat specialization. In regions with many species, the species occupy fewer habitats — they exhibit narrower ecological tolerance — than in regions that are depauperate in species. As a result, the species composition of communities changes rapidly from one habitat to the next (see Figure 17–5). Suppose, for example, that region A has 1,000 species and region B 5,000 species, but habitats in both regions support an average of 500 species. It follows that each species will be found in an average of 10 habitats in region A and only 2 habitats in region B.

Ecologists refer to the number of species in a locality as the *alpha diversity* of the locality. The number of species added by combining two or more localities is called *beta diversity*. Thus if two localities each have 100 species, only 50 of which are found in both places, the localities together have 150 species (50 species in common, 50 species restricted to one locality, and 50 species restricted to the other locality.) In this case, the alpha diversity is 100 species and the beta diversity 50 species. If all 150 species were found in both localities, the beta diversity would be zero. Alpha diversity is limited by the environment of each locality. Beta diversity is determined more by proximity of a region to sources of colonists and by the rate of species origination within the region. Islands tend to have few species (low beta diversity), but the tendency of these to live in many different habitats prevents a similar reduction in alpha diversity. For example, in four tropical areas in and around the Caribbean Sea, a four-fold reduction in total regional diversity between mainland and small island is paralleled by only a two-fold reduction in local (alpha) diversity (Table 17–2). The lower alpha diversity of island habitats was compensated for by denser populations of each species, bringing the number of individuals per habitat to similar levels in all four regions and thus preserving the level of ecosystem function.

Adaptive Radiation

The fossil history of the Earth records many instances of rapid diversification of groups entering new ecological realms or hitting upon novel ways of life. When plants and animals first colonized the land, terrestrial habitats presented an enormous variety of habitats to them. These habitats have since been filled by *adaptive radiations* from single lineages into numerous species having different adaptations and filling different ecological roles. Flowering plants hit upon a new and more successful way of life compared to the primitive groups of plants that preceded them. Once the basic adaptations of flowering plants were perfected in a few pioneers, their lineage began to spread and diversify, replacing competitively inferior plants by species now familiar to us.

Broad categories of life styles are often called *adaptive zones* — terrestrial primary producers, flying predators, and aquatic homeotherms might be cited as examples. The extrance of a new adaptive zone is a difficult evolutionary step, but once achieved further diversification occurs quickly.

Regions, which literally *are* zones, that are depauperate in species are the geographical equivalent to adaptive zones. Isolated island groups are separated from continents by geographical barriers rather than the evolutionary barriers that separate adaptive zones. Spectacular adaptive radiations have occurred within remote oceanic island groups. The few species that have colonized isolated archipelagos from continental regions entered an environment free of competing species and thus affording excellent opportunities for speciation and evolutionary diversification. The adaptive radiation of finches on the Galapagos Islands and honeycreepers on the Hawaiian Islands demonstrates how single populations can diversify to fill many roles (Figure 17–10). Each group of species probably originated from a single population arriving about 5,000,000 years ago at the Hawaiian Islands and perhaps only 1,000,000 years ago at the Galapagos Islands. Through speciation and

TABLE 17–2   Number of species of birds per habitat and abundance of each species in each locality in four tropical regions.*

Locality	Total number of species within region	Average number of species per locality within each region	Average number of habitats occupied by each species	Average relative local abundance of species
Panama	135	30.2	2.01	2.95
Trinidad	108	28.2	2.35	3.31
Jamaica	56	21.4	3.43	4.97
St. Lucia	33	15.2	4.15	5.77

*Based on 10 brief counting periods in each of nine habitats in each region. Not all species were observed.

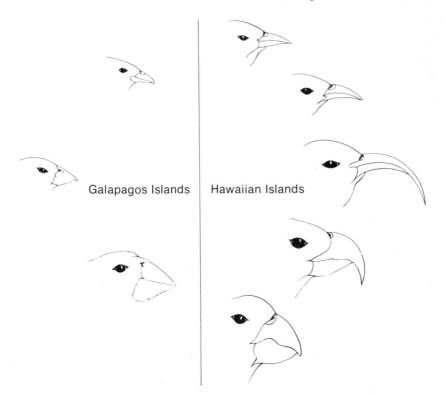

FIGURE 17–10 Representative species of Darwin's finches (Geospizinae) from the Galapagos Islands (left) and honeycreepers (Drepanididae) from the Hawaiian Islands (right) demonstrate the ranges of beak length and back depth in the two groups.

competitive displacement, each group has radiated to exploit resources in a variety of ways.

Although the finch and honeycreeper groups have evolved similar extremes in beak depth, varying from the short thin beaks of insect eaters to the heavy strong beaks of large seed eaters, only the honeycreeper group has evolved long, thin beaks, adapted for probing. Long-billed birds probably occur on the Hawaiian Islands because the moist climate supports the growth of epiphytes (mosses, lichens, orchids, bromeliads) in the recesses of which birds can probe for insects. The drier habitats of the Galapagos Islands are dominated by wind-pollinated annual plants that produce hard seeds rather than soft fruits.

Convergent Evolution

Organisms that inhabit similar environments in different geographical localities often resemble each other even though their taxonomic affinities differ (Figure 17-11). This resemblance is the result of what ecologists call *convergent evolution*. Form and function converge under the mantle of similar selective forces in the environment: the contemporaneous environment acts as a unifying principle. Trees are found virtually everywhere, and with them occur various insects that burrow through the wood of the trunks and branches. Among birds, woodpeckers have met with singular success in exploiting this food resource, which has led to their widespread distribution throughout forest habitats. But because most woodpeckers inhabit deep forests, they lack the capability of strong flight and do not disperse readily across aquatic barriers. As a result, such remote islands as the Galapagos, Hawaii, and New Zealand lack true woodpeckers. Yet in each of these island groups, unrelated birds have evolved to exploit the food resource normally utilized by woodpeckers. Moreover, each approaches this specialized task in a different way (Figure 17-11). The woodpeckers themselves combine a chisel-like bill, for hacking away bark and wood, with a long tongue, which can be extended into crevices and burrows to extract insects. On the Hawaiian Islands, honeycreepers of the genus *Heterorhynchus* have separated the excavating and probing functions between their lower and upper mandibles. They tap at wood with their short lower mandibles and probe with their elongated upper mandibles. The Galapagos woodpecker-finch digs trenches in soft wood with its stout beak and probes with a cactus spine held in the beak. The huia of New Zealand separated these functions between the sexes. The male excavated with his short stout beak and the female probed with her long

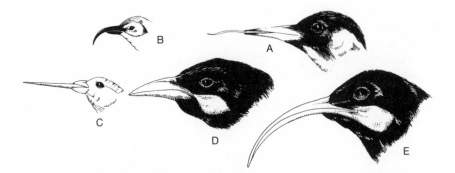

FIGURE 17-11 Birds that have become adapted to extract insects from wood in different ways: (A) the European green woodpecker excavates with its beak and probes with its long tongue; (B) the Hawaiian honeycreeper, *Heterorhynchus*, taps with its short lower mandible, probes with its long upper mandible; (C) the Galapagos woodpecker-finch trenches with its beak and probes with a cactus spine; and the New Zealand huia divides foraging roles on the basis of sex — the male (D) excavates with his short beak and the female (E) probes with her long beak.

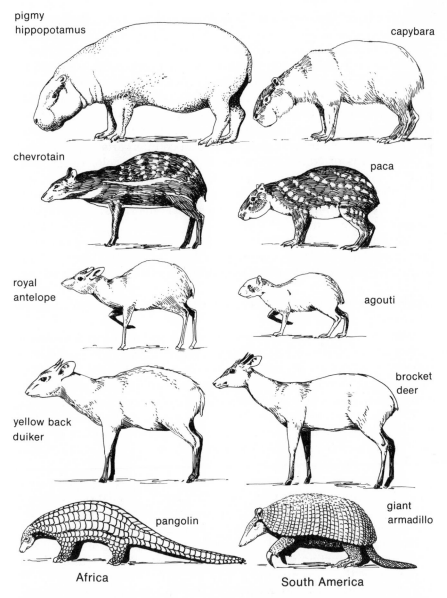

pigmy
hippopotamus

capybara

chevrotain

paca

royal
antelope

agouti

yellow back
duiker

brocket
deer

pangolin

giant
armadillo

Africa

South America

FIGURE 17–12   Morphological convergence among unrelated African (left) and Neotropical (right) rain-forest mammals. Each pair of animals is drawn to the same scale.

beak. The birds were supposed to have fed in pairs, but as the species is now extinct, this arrangement evidently did not work out satisfactorily.

Notable cases of morphological convergence have also occurred in forest-dwelling birds and mammals of tropical Africa and South America. The toucans of the New World (relatives of woodpeckers) and the hornbills of Africa (relatives of kingfishers) share many characteristics, including their large bills, fruit eating habits, nesting sites in holes in trees, and gregariousness. These two groups, with diverse evolutionary histories, have become close ecological counterparts. Similarly, the hummingbirds of North and South America have ecological counterparts in the unrelated sunbirds of Africa and Asia. Sunbirds resemble hummingbirds in their nectivorous food habits, small size, bright metallic-colored plumage, pronounced sexual dimorphism, and promiscuous mating systems.

African rain-forest mammals also have close ecological counterparts in tropical America derived from quite different groups (Figure 17-12). Ungulates, represented by the pygmy hippopotamus, chevrotain, antelope, and duiker in Africa, are poorly represented in the New World where many of their ecological counterparts are large rodents, like the capybara, paca, and agouti.

As a general principle, adaptive radiation and convergent evolution tend to make communities in similar environments more uniform in structure and function, regardless of their evolutionary derivation. Hence we can ignore the details of community composition while studying fundamental aspects of community function. According to this principle, we can attribute basic differences between the structure and functioning of communities to differences in their immediate physical environments rather than to historical accident.

In the chapters that follow, we shall examine two aspects of community function: development and regulation. Similarities between communities and individuals may suggest parallels between the changes in a community during succession and the changes in an organism during development, or between the regulation of community structure and function and the homeostatic responses of individuals. Tempting as these comparisons may be, their weakness lies in the different genetic bases of individual and community traits. The organism is a closed system with definite boundaries and a highly coevolved genotype; the community has many attributes of an open system in which the evolutionary histories of many component species are totally independent. All characteristics of the organisms are coadapted to increase the productivity of the individual. Community attributes, such as trophic structure, diversity, rates of energy and nutrient flux, ecological efficiency, and food web complexity, do not have evolutionary status of their own. They mainly depend on properties of individuals adapted to enhance individual advantage.

18

Community Development

Communities are constantly changing. Organisms die and others are born to take their places. Energy and nutrients continually pass through the community. Yet the appearance and composition of most communities do not vary. Oaks replace oaks, robins replace robins, and so on, in continual self-perpetuation. If a community is disturbed — a forest cleared for agriculture, a prairie burned, a coral reef obliterated by a hurricane — the community is slowly rebuilt. Pioneering species adapted to the disturbed habitat are successively replaced by members of the original association until the community attains its former structure and composition. The sequence of changes on a disturbed site is called *succession* and the final vegetation formation achieved is called a *climax*. These terms describe natural processes that caught the attention of early ecologists. By 1916, University of Minnesota ecologist Frederic Clements had outlined the basic features of succession, supporting his conclusions by detailed studies of change in plant communities in a wide variety of environments. In this chapter, we shall examine the course and causes of succession both from the traditional view of community development and in the light of recent studies on community succession.

Examples of Succession

The creation of any new habitat — a ploughed field, a sand dune at the edge of a lake, an elephant's dung, a temporary pond left by a heavy rain — invites a host of invading species to exploit the resources made available. Initial colonizers are followed by others slower to take advantage of the new habitat but eventually more successful than the pioneer species. In this way, the character of the community changes with time. Successional species themselves change the environment by, for example, shading the surface, contributing detritus to the soil, and altering soil moisture. These changes often act to the detriment of the species that cause them, and make the environment more suitable for other species which then exclude those responsible for the change.

The opportunity to observe succession is almost always at hand on abandoned fields of various ages (Figure 18-1). In the piedmont region of North Carolina, bare fields are quickly covered by a variety of annual plants. Within a few years, the annuals are mostly replaced by herbaceous perennials and shrubs. The shrubs are followed by pines, which eventually crowd out the earlier successional species, but pine forests are in turn invaded and then replaced by a variety of hardwood species that represent the end of the successional sequence. Change is rapid at first. Crabgrass quickly enters an abandoned field, hardly giving the ploughed furrows a chance to smooth over. Horseweed and ragweed dominate the field in the first summer after abandonment, aster in the second, and broomsedge in the third. The pace of succession falls off as slower-growing plants appear. The transition to early pine forest requires 25 years. Another century must pass before the developing hardwood forest begins to resemble the natural climax vegetation of the area.

The transition from abandoned field to mature forest is only one of several successional sequences leading to the same climax. In the eastern part of the United States and Canada, forests are the end point of several different successional series, or *seres*, each having a different beginning. The sequence of species on newly formed sand dunes at the southern end of Lake Michigan differs from the sere that develops on abandoned fields a few miles away. The sand dunes are first invaded by marram grass and bluestem grass. Plants established in soils at the edge of a dune send out rhizomes (runners) under the surface of the sand, from which new plant shoots sprout (see Figure 4-9). These grasses stabilize the dune surface and add organic detritus to the sand. Numerous annuals follow the perennial grasses onto the dunes, further contributing to the enrichment and stabilization process and gradually creating conditions suitable for the establishment of shrub species. Sand cherry, dune willow, bearberry, and juniper form shrub layers before pines become established. As in the abandoned fields in North Carolina, pines persist for only one or two generations, with little reseeding after initial

FIGURE 18–1   Stages of succession on cleared fields near Philadelphia, Pennsylvania. Top: a recently abandoned field dominated by Queen Anne's lace during the summer. Middle: shrubs and small trees begin to grow up after 10 to 15 years, yielding to a young broad-leaved forest after 25 to 50 years. (Pines do not enter the sere in this region.) Bottom: the forest approaches maturity after 100 to 200 years although signs of its earlier developmental stages persist.

establishment, giving way in the end to the beech–maple–oak–hemlock forest characteristic of the region.

Succession follows a similar course on Atlantic coastal dunes, where beach grass initially stabilizes the dune surface, followed by bayberry, beach plum, and other shrubs (Figure 18–2). Shrubs act like the snow fencing frequently used to prevent the blowout of dunes, and are called dunebuilders because they intercept blowing sand, causing it to pile up around their bases.

Primary Succession

Ecologists classify seres into two groups, according to their origin. The establishment and development of plant communities in newly formed habitats previously without plants — sand dunes, lava flows, rock bared by erosion or exposed by a receding glacier — is called *primary succession*. The return of an area to its natural vegetation following a major disturbance is called *secondary succession*. We have already followed the course of primary succession on Lake Michigan sand dunes. The sequence of species colonizing habitats exposed by receding glaciers in the Glacier Bay region of southern Alaska is quite different (Figure 18–3). The surfaces of glacial deposits are stable but the thin clay soils are deficient of nutrients, particularly nitrogen, and pioneering plants are exposed to wind and cold stress. Here the sere involves mat-forming mosses and sedges, prostrate willows, shrubby willows, alder thicket, sitka spruce, and, finally, spruce–hemlock forest. Succession is rapid, reaching the alder thicket stage within 10 to 20 years, and tall spruce forest within 100 years.

The development of vegetation on bare rock, sand, or other inorganic sediments is called *xerarch succession*. The low water retention by such habitats results in *xeric* (drought) conditions during the early stages of succession. At the opposite extreme, *hydrarch* succession begins in the open water of a shallow lake, bog, or marsh. Hydrarch succession can be initiated by any factor that reduces water depth and increases soil aeration, whether natural drainage, progressive drying up, or filling in by sediments.

The change in vegetation on bogs illustrates hydrarch succession. Bogs are formed in kettleholes or dammed streams in cool temperate and subarctic regions. Bog succession begins when aquatic plants become established at the edge of the pond (Figure 18–4). Some species of sedges (rushlike plants) form mats on the water surface extending out from the shoreline. Occasionally these mats grow completely over the pond before it is filled in by sediments, producing a more or less firm layer of vegetation over the water surface. The vegetation mat quivers underfoot, a characteristic giving rise to the name quaking bog. The

FIGURE 18–2 Initial stages of plant succession on sand dunes along the coast of Maryland. Top: beach grass on the frontal side of a dune. This grass is used widely to stabilize dune surfaces. Bottom: invasion of back dune areas by bayberry and beach plum.

FIGURE 18–3   A valley exposed by a receding glacier, visible at top center, in North Tongass National Forest, Alaska. The recently bared rock surfaces at the bottom of the valley just below the glacier have not yet been recolonized by shrubby thickets.

detritus produced by the sedge mat accumulates in layers of organic sediments on the pond bottom because the stagnant water of the pond contains little or no oxygen and thus does not support rapid microbial decomposition. Eventually these sediments become peat which is used as a soil conditioner and, sometimes, as a fuel for heating (Figure 18–5).

As the bog is filled in by sediments and detritus, bog shrubs and sphagnum moss become established along the edges, themselves add-

FIGURE 18–4 Stages of bog succession illustrated by a kettlehole bog in Massachusetts. Aquatic plants grow into the open water at right, forming a vegetation mat and building up layers of organic sediments on the bottom. Filled-in portions of the bog are then invaded by shrubs (center) and later by spruce (left).

ing to the development of a soil with progressively more terrestrial qualities. The shrubs are followed by a bog forest of black spruce and larch, which is eventually replaced by local climax species, including birch, maple, and fir, depending on the region.

In northern Michigan, the hydrarch succession that develops on bogs is only one of several seres of primary and secondary succession leading to a climax forest of spruce, fir, and birch (Figure 18–6). Following a fire, development of the climax follows a different course, passing through intermediate grass and aspen stages. If the soil is badly scorched and most of the humus is burned, a sere resembling that beginning on bare rock surface develops.

Successional States in Time and Space

The sequence of species in the sere parallels change in the physical environment as it is modified by the developing vegetation. Many of the stages in a time sequence through a sere may be found along geographical gradients in vegetation, often called *ecoclines* (Figure 18–7). For

FIGURE 18–5   A three-foot vertical section through a peat bed in a filled-in bog in Quebec, Canada. The layers represent the accumulation of organic detritus from plants that successively colonized the bog as it was filled in. The peat beds are probably several yards thick. Vegetation on the surface of the bog consists mostly of sphagnum, blueberry, and Labrador tea.

example, the xerarch succession from rock surface to forest in the eastern United States corresponds in structure, if not species composition, to the ecocline in vegetation from the nearly bare rock surfaces of the western deserts, through dry grasslands, prairie, shrubby oak woodland, to tall mixed hardwood forest that occurs along an increasing moisture gradient from west to east. The temporal sequence of primary succession in a particular place also follows an increasing gradient of moisture availability. The ecocline represents a series of stages of community development stopped at different points by lack of moisture. Each stage of the ecocline represents the local end point of succession. The species composition of a community along the ecocline differs from the corresponding stage of the complete sere in the eastern United States because species must be well-adapted to the local conditions prevalent along the ecoclinal gradient, but, again, the general structure is the same.

Hydrarch succession on a bog is mirrored in the sequence of communities present at any one time from the open water at the center of the pond to the developing forest at its edge, where sedimentation and soil development are most advanced. Concentric bands of vegetation ring the bog, progressing from an inner circle of sedge mat outward through sphagnum and bog shrubs, larch, and finally black spruce.

The Climax

Succession is traditionally viewed as leading inexorably toward an ultimate expression of plant development, the *climax community*. In fact, studies of succession have demonstrated that the many seres found within a region, each developing under a particular set of local environmental circumstances, progress towards the same climax (see Figure 18–6). These observations led to the concept of the mature community as a natural unit, even as a closed system. Frederic Clements stated the concept in 1916: "The developmental study of vegetation necessarily

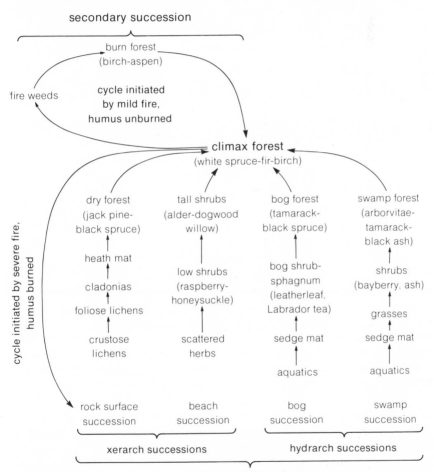

FIGURE 18–6 Trends of succession on Isle Royale, Lake Superior. Each habitat is characterized by a unique sere leading to the same climax. Secondary succession on burned sites follows different courses depending on the extent of the fire damage.

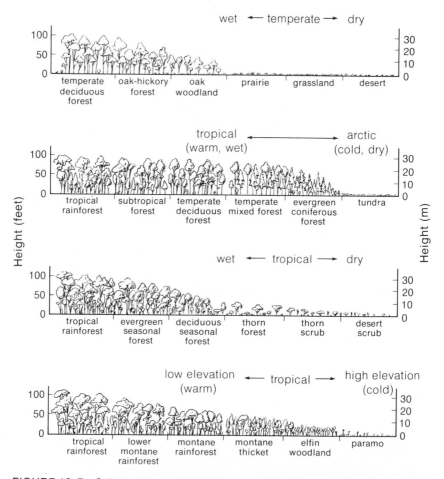

FIGURE 18–7   Schematic profiles of four ecoclines in vegetation type. Top, a wet-dry gradient from the Appalachian Mountains (left) to the southwestern United States (right); second, a warm–cold (and wet–dry) gradient from Panama to Northern Canada; third, a wet–dry gradient within the tropics; bottom, an altitudinal gradient in the tropics from the Amazonian forest into the Andes.

rests upon the assumption that the unit or climax formation is an organic entity. As an organism the formation arises, grows, matures, and dies. Its response to the habitat is shown in processes or functions and in structures which are the record as well as the result of these functions. Furthermore, each climax formation is able to reproduce itself, repeating with essential fidelity the stages of its development. The life history of a formation is a complex but definite process, comparable in its chief features with the life history of an individual plant.''

   The concept of the climax as an organism or unit has been greatly modified, to the point of outright rejection by many ecologists, with the recognition of communities as open systems whose composition varies continuously over environmental gradients. Clements recognized 14 climaxes in North America, including two types of grassland (prairie and tundra), three types of scrub (sagebrush, desert scrub, and chaparral), and nine types of forest, ranging from pine–juniper woodland to beech–oak forest. The nature of the local climax was thought to be determined solely by climate. Aberrations in community composition caused by soils, topography, fire, or animals (especially grazing) were thought to represent interrupted stages in the transition toward the local climax — immature communities.

Whereas in 1930 plant ecologists described *the* climax vegetation of much of Wisconsin as a sugar maple–basswood forest, by 1950 ecologists placed this forest type on an open continuum of climax communities extending both over broad, climatically defined regions and over local, topographically defined areas. To the south, beech increased in prominence, to the north, birch, spruce, and hemlock were added to the climax community; in drier regions bordering prairies to the west, oaks became prominent. Locally, quaking aspen, black oak, and shagbark hickory, long recognized as successional species on moist, well-drained soils, came to be accepted as climax species on drier upland sites.

Mature stands of forest in Wisconsin, representing the end points of local seres, have been ordered along a *continuum index* ranging from dry sites dominated by oak and aspen to moist sites dominated by sugar maple, ironwood, and basswood. The continuum index for Wisconsin forests is calculated from the species composition of each forest type and its value varies between arbitrarily set extremes of 300 for pure bur oak forest to 3,000 for a pure stand of sugar maple. Although increasing values of the continuum index correspond to seral stages leading to the sugar maple climax, they may also represent local climax communities determined by topographic or soil conditions. Thus the so-called climax vegetation of southern Wisconsin is actually a continuum of forest (and, in some areas, prairie) types (Figure 18-8). Some botanists prefer to retain the term *climatic climax* for the furthest point of vegetational succession within a region, relegating all other endpoints of seres to terminated stages of succession or *subclimaxes*. We run the risk of becoming entangled in semantics at this point without further elucidating the mechanism of plant succession. We would do well to follow one guideline in thinking about succession. If interrupted stages of succession are so prevalent and persistent in a region that species have adapted to the particular environmental conditions in these subclimax communities, these species should be recognized as climax forms even though they enter transitional seral stages elsewhere. The concept of climax is rooted in the self-perpetuation of an association under prevalent local conditions and in the adaptations of climax species that ensure their self-perpetuation.

The Causes of Succession

Two factors determine the positioning of species in a sere: the rate at which species invade a newly formed or disturbed habitat, and changes in the environment during succession. Some species disperse slowly, or grow slowly once they have become established, and therefore become dominant late in the sequence of associations in a sere; rapidly growing plants that produce many small seeds that are carried long distances by

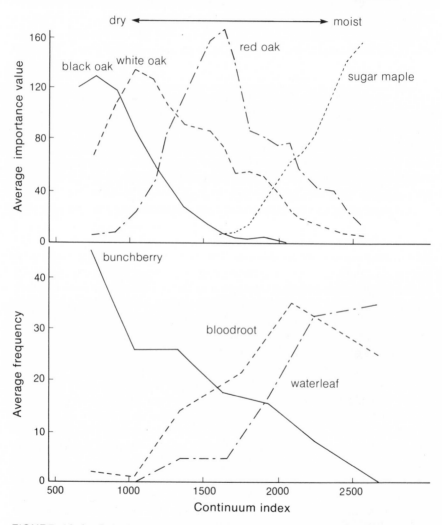

FIGURE 18–8 Relative importance of several species of trees (top) and herbs (bottom) in forest communities of southwestern Wisconsin arranged along a continuum index. Soil moisture, exchangeable calcium, and pH increase to the right on the continuum index.

the wind or by animals have an initial advantage over species that are slow to disperse, and dominate early stages of the sere. But early colonists change the environment in ways that favor the invasion of the community by species of superior competitors. Horseweed resists desiccation and rapidly colonizes abandoned farmland in the piedmont of North Carolina but, once established, horseweed plants modify the environment by shading the soil surface. Because horseweed seedlings require full sunlight, horseweed is quickly replaced by shade-tolerant species. As the community matures, the developing vegetation protects the surface layers of the soil from drying, permitting drought-intolerant species to get a foothold in the community. Progressive changes in the physical environment foster the replacement of species through seral stages.

Early stages of plant succession on old fields in the piedmont region of North Carolina demonstrate how succession is driven. The first three to four years of old-field succession are dominated by a relatively small number of species that replace each other in rapid sequence: crabgrass, horseweed, ragweed, aster, and broomsedge. The life-history cycle of each species partly determines its role in succession (Figure 18–9). Crabgrass, a rapidly growing annual, is usually the most conspicuous plant in a cleared field during the year in which the field is abandoned. Horseweed is a winter annual, whose seeds germinate in the fall. Through the winter, the plant exists as a small rosette of leaves and it blooms by the following midsummer. Because horseweed has strong dispersal powers and develops rapidly, it usually dominates one-year-old fields. Ragweed is a summer annual; seeds germinate early in the spring and the plants flower by late summer. Ragweed dominates first-year fields if the fields were ploughed under in the late fall, after

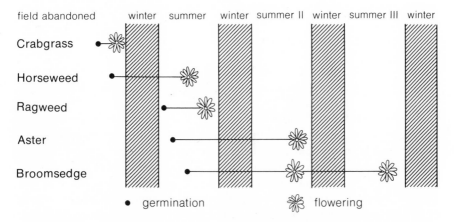

FIGURE 18–9 Schematic summary of the life histories of five early successional species of plants that colonize abandoned fields in North Carolina.

horseweed normally germinates. Aster and broomsedge are biennials that germinate in the spring and early summer, exist through the winter as small plants, and bloom for the first time in their second autumn. Broomsedge persists and flowers during the following autumn as well.

A series of experiments performed by Katherine Keaver at Duke University showed how aster replaces horseweed and how broomsedge subsequently replaces aster in early stages of plant succession. Horseweed and ragweed both disperse their seeds efficiently and, as young plants, tolerate desiccation. These characteristics allow them to invade cleared fields rapidly and produce seed before populations of competitors become established. Decaying horseweed roots stunt the growth of horseweed seedlings; this self-inhibiting effect cuts short the life of horseweed in the sere.

Aster is a relatively successful colonist of recently cleared fields but, being a slow grower, it does not become dominant until the second year. The first aster plants to colonize the field thrive in the full sunlight, but, because aster seedlings are not shade-tolerant, they shade their progeny out of existence. Furthermore, asters do not compete effectively with broomsedge for soil moisture. To determine this, Keaver cleared a circular area, one meter in radius, around several broomsedge plants and planted aster seedlings at various distances. After two months, the dry weight of asters planted 13, 38, and 63 cm from the bases of the broomsedge plants averaged 0.06, 0.20, and 0.46 grams; available soil water at these distances was 1.7, 3.5, and 6.4 grams per 100 grams of soil. Broomsedge does not, however, dominate early successional communities until the third or fourth year, in spite of its competitive edge over aster. Because their seeds do not disperse well, broomsedge plants do not increase rapidly in number until the first colonists of the field have themselves produced seeds.

Establishment of the Climax

Succession continues until the addition of new species to the sere and the exclusion of established species no longer changes the environment of the developing community. Conditions of light, temperature, moisture, and, for primary seres, soil nutrients, change quickly with the progression of different growth forms. The change from grasses to shrubs and trees on abandoned fields brings a corresponding modification of the physical environment. Conditions change more slowly, however, when the vegetation reaches the tallest growth form that the environment can support. The final biomass dimensions of the climax community are limited by climate independently of events during succession. Once forest vegetation is established, patterns of light intensity and soil moisture are not changed by the introduction of new species of trees, except in the smallest details. For example, beech and maple replace oak and hickory in northern hardwood forests because their seedlings are better competitors in the shade of the forest-floor environment, but beech and maple seedlings probably develop as well under their own parents as they do under the oak and hickory trees they replace. At this point, succession reaches a climax; the community has come into equilibrium with its physical environment.

To be sure, subtle changes in species composition usually follow the attainment of the climax growth form of a sere. For example, a site near Washington, D.C. left undisturbed for nearly 70 years developed a tall forest community dominated by oak and beech. The community had not then reached an equilibrium because the youngest individuals — the saplings in the forest understory, which eventually would replace the existing trees — included neither white nor black oak. In another century, the forest will be dominated by species with the most vigorous reproduction, namely red maple, sugar maple, and beech (Figure 18–10).

The composition and age structure of a forest in northwestern Wisconsin having had minimal human disturbance over 200 years indicates a transitory state, perhaps towards the end of a sere, between oak dominance and a basswood–maple climax. Red oak presently is the commonest large tree in the forest, but basswood and, especially, maple

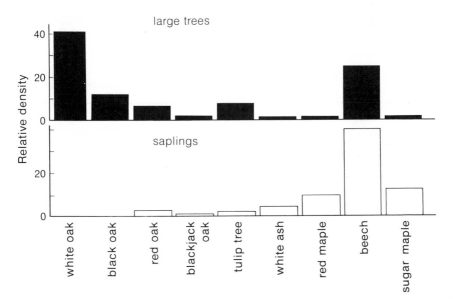

FIGURE 18–10 Composition of a forest undisturbed for 67 years near Washington, D.C. The relative predominance of beech and maple saplings in the forest understory foretells a gradual successional change in the community beyond the present oak–beech stage.

are reproducing much more vigorously (Table 18-1). The ratios of seed-lings and saplings (less than one inch diameter) to large trees (greater than ten inches diameter) are maple 186, basswood 155, red oak 18, and white oak 37. (White oak and bitternut are close to the northern edge of their ranges in northern Wisconsin and do not form a major component of the forest.)

The preponderance of red oak in the canopy of the Wisconsin forest and the evidence in the understory of successional changes yet to come indicate that the forest had been disturbed in some way, allowing seral species to enter and setting into motion the machinery of succession. The age structure of the tree populations, determined by increment borings (see page 285), suggests that fire destroyed much of the forest sometime between 1840 and 1850 (Figure 18-11). Most of the sugar maples are over 150 years old, indicating that they withstood fire dam-age. In fact, two-thirds of the sugar maple cores were so badly scarred by fire that their growth rings could not be counted accurately. Red oak and basswood both exhibited a period of rapid proliferation starting about 1850. Red oak gained its predominant position in the forest at that time and will not be excluded until present trees die and are replaced by basswood or maple seedlings.

The Pattern–Climax Theory

Clement's idea that a region had only one true climax (the *monoclimax* theory) forced botanists to recognize a hierarchy of interrupted or modified seres by attaching names like subclimax, preclimax, and post-climax. This terminology naturally gave way before the *polyclimax* viewpoint, which recognized the validity of many different vegetation types as climaxes, depending on the habitat. More recently, the de-velopment of the continuum index and gradient analysis has fostered

TABLE 18–1   Number of trees of different species in a 2,500-square meter forest area in northwestern Wisconsin. Individuals are separated into size classes.

| Species | *Diameter of trunk (inches)* | | | |
	less than 1	1 to 3	4 to 9	greater than 10
Sugar maple	3,913	2	16	21
Basswood	931	22	21	6
Red oak	781	1	34	44
White oak	75	3	9	2
Bitternut	88	4	0	0
White pine	0	0	1	2
Ironwood	1,606	40	3*	0

*Maximum size class of ironwood, an understory species.

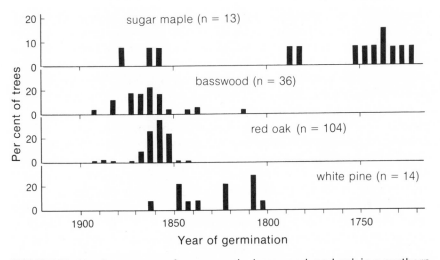

FIGURE 18–11   Age groups of sugar maple, basswood, and oak in a northern Wisconsin forest.

the broader *pattern–climax* theory of Robert Whittaker, which recognizes a regional pattern of open climax communities whose composition at any one locality depends on the particular environmental conditions at that point.

Many factors determine the climax community, among them soil nutrients, moisture, slope, and exposure. Fire is an important feature of many climax communities, favoring fire-resistant species and excluding others that otherwise would dominate. The vast southern pine forests in the gulf coast and southern Atlantic coast states are maintained by periodic fires. The pines are adapted to withstand scorching under conditions that destroy oaks and other broad-leaved species. Fire is even necessary to the life history of some species of pines that do not shed their seeds unless triggered by the heat of a fire passing through the understory below. After a fire, pine seedlings grow rapidly in the absence of competition from other understory species.

Fire also maintains the giant sequoia forests of the Sierra Nevada in California. Sequoia bark is thick and fire resistant, protecting trees from fires that kill spruce and fir. With extensive control of forest fires in sequoia groves, spruce have grown up in the understory and excluded sequoia seedlings. Needless to say, the spruce will not replace the long-lived sequoia for centuries but their presence in the understory reminds us that the sequoia groves are a fire climax community.

Any habitat that is occasionally dry enough to create a fire hazard but normally wet enough to produce and accumulate a thick layer of plant detritus is likely to be influenced by fire. Chaparral vegetation in seasonally dry habitats in California is a fire-maintained climax (see

Figure 11-11) that would be replaced by oak woodland in many areas if fire were prevented. The forest-prairie edge in the midwestern United States separates climatic climax and fire climax communities. Frequent burning eliminates seedlings of hardwood trees but the perennial grasses sprout from their roots after a fire. The forest-prairie edge occasionally shifts back and forth across the countryside, depending on the intensity of recent drought and the extent of recent fires. After prolonged wet periods the forest edge advances out onto the prairie as tree seedlings grow up and begin to shade out the grasses. Prolonged drought followed by intense fire can destroy tall forest and allow rapidly spreading prairie grasses to gain a foothold. Once prairie vegetation is established, fires become more frequent owing to the rapid buildup of flammable litter. Reinvasion by forest species then becomes more difficult. By the same token, mature forests resist fire and rarely become damaged enough to allow the encroachment of prairie grasses.

Alternating periods of flooding and drought maintain wet prairies and meadows in a quasi-equilibrium in southeastern Wisconsin. Rapid hydrarch succession on peat marshes produces wet prairie and meadow vegetation. Further encroachment of species from the local climax forest of the area is prevented by occasional flooding, alternating with fire during periods of drought. Although fire normally checks the advancement of terrestrial vegetation, its influence varies with the content of organic matter and water in the soil at the time of its occurrence. Between the turn of the century and 1930, farming, swamp drainage, and forest-clearing steadily lowered the water table and caused the peat underlying the wet prairie to dry out. By 1920, a marsh in one area was dry enough that farmers could cut hay there. Following a period of below normal rainfall, a discarded cigarette ignited the marsh in August 1930. The fire continued throughout the winter, burning the peat to an average depth of one foot and thus destroying the wet prairie vegetation, roots and all. Instead of returning to prairie and marsh vegetation, burned areas came up in aspen, evidently seeded in from a great distance inasmuch as aspen trees were not known to occur in the immediate area. By 1966, the aspen groves still remained, but their vigor was declining rapidly, with many of the trees diseased or dead. Left undisturbed, the aspen groves will eventually be invaded by hardwoods — in fact, elm had already appeared by 1940. Periodic fire, however, would preserve the aspen in a new climax because aspen resprouts from its roots following fires intense enough to kill all other hardwoods. Fire, then, plays various roles in plant succession, depending on its intensity and frequency. Although fire is usually thought of as retrogressive, returning a community to earlier seral stages, the complete destruction of prairie vegetation by an especially intense fire allowed the invasion of an area by aspen trees and moved the habitat closer to a hardwood forest climax.

Grazing pressure also can modify the climax (see Chapter 15). Grassland can be turned into shrubland by intense grazing. Herbivores kill or severely damage perennial grasses and allow shrubs and cacti unsuitable for forage to establish themselves. Most herbivores graze selectively, suppressing favored species of plants and bolstering competitors that are less desirable as food.

Transient and Cyclic Climaxes

We view succession as a series of changes leading to a climax, determined by, and in equilibrium with, the local environment. Once established, the beech–maple forest is self-perpetuating and its general appearance does not change in spite of the constant replacement of individuals within the community. Yet not all climaxes are persistent. A simple case of a *transient climax* would be the development of animal and plant communities in seasonal ponds — small bodies of water that either dry up in the summer or freeze solid in the winter and thereby regularly destroy the communities that become established each year during the growing season. Each spring the ponds are restocked either from larger, permanent bodies of water, or from spores and resting stages left by plants and animals before the habitat disappeared the previous year.

Succession recurs whenever a new environmental opportunity appears. For example, dead organisms are a resource for a wide variety of scavengers and detritus feeders. On African savannas, carcasses of large mammals are fed upon by a succession of vultures, beginning with large, aggressive species that devour the largest masses of flesh, followed by smaller species that glean smaller bits of meat from the bones, and finally by a kind of vulture that cracks open bones to feed on the marrow. Scavenging mammals, maggots, and microorganisms enter the sere at different points and assure that nothing edible remains. This succession has no climax, however, because all the scavengers disperse when the feast is concluded. Nonetheless, we may consider all the scavengers a part of a climax, which is the entire savanna community.

In simple communities, particular life-history characteristics in a few dominant species can create a *cyclic climax*. Suppose, for example, that species A can germinate only under species B, B can germinate only under C, and C only under A. This situation would create a regular cycle of species dominance in the order A, C, B, A, C, B, A . . . with the length of each stage determined by the life span of the dominant species. Stable cyclic climaxes, which are known from a variety of localities, usually follow the scheme presented above, often with one of the stages being bare earth. Wind or frost heaving sometimes drives the cycle. When heaths suffer extreme wind damage, shredded foliage and broken twigs create an opening for further damage and the process becomes self-accelerating. Soon a wide swath is opened in the vegetation; regeneration occurs only on the protected side of the damaged area while wind damage further encroaches upon the exposed vegetation. As a result, waves of damage and regeneration move through the community in the direction of the wind. If we watched the sequence of events at any one point, we would witness a healthy heath reduced to bare earth by wind damage and then regenerating in repeated cycles (Figure 18–12). Similar cycles occur where hummocks of earth form in windy regions around the bases of clumps of grasses. As the hummocks grow, the soil becomes more exposed and better drained. With these changes in soil quality, shrubby lichens take over the hummock and exclude the grasses around which the hummock formed. Shrubby lichens are worn down by wind

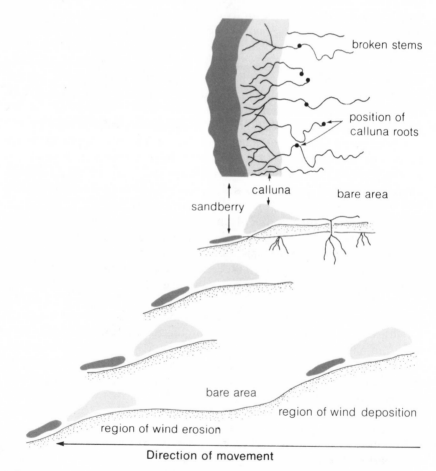

FIGURE 18–12   Sequence of wind damage and regeneration in the dwarf heaths of northern Scotland.

erosion and eventually are replaced by prostrate lichens, which resist wind erosion but, lacking roots, cannot hold the soil. Eventually the hummocks are completely worn down and grasses once more become established and renew the cycle.

Frost action and wind erosion work together in alpine meadows of the Rocky Mountains to produce a mosaic of vegetation patches. Sedges become established in wet hollows and start the process of hummock building. As peat accumulates in the developing hummock, water is absorbed by the organic detritus. Repeated cycles of freezing and thawing begin to thrust the hummock upward. This process continues until the top of the hummock protrudes above the snow surface in winter, at which point the grasses and herbaceous perennials at the crest are killed by exposure, and wind erosion begins to wear away the hummock, forming a hollow and completing the cycle.

Mosaic patterns of vegetation type are common to any climax community where the death of individuals alters the environment. Treefalls open the forest canopy and create patches of habitat that are dry, hot, and sunlit compared to the forest floor under unbroken canopy. These openings are often invaded by early seral forms, which persist until the canopy closes. Treefalls thus create a mosaic of successional stages within an otherwise uniform community. Indeed, adaptation by some species to grow in the particular conditions created by different-sized openings in the canopy could enhance the overall diversity of the climax community.

Cyclic patterns of change and mosaic patterns of distribution must be incorporated into the concept of the community climax. The climax is a dynamic state, self-perpetuating in composition, even if by regular cycles of change. Persistence is the key to stability. If a cycle persists, it is inherently as stable as an unchanging steady state.

Succession and Vegetation Structure

Succession in terrestrial habitats entails a regular progression of plant forms. Plants characteristic of early stages and late stages of succession employ different strategies of growth and reproduction. Early-stage species are opportunistic and capitalize on high dispersal ability to colonize newly created or disturbed habitats rapidly. Climax species disperse and grow more slowly, but shade tolerance as seedlings and large size as mature plants gives them a competitive edge over early successional species. Plants of climax communities are adapted to grow and prosper in the environment they create, whereas early successional species are adapted to colonize unexploited environments.

Some characteristics of early and late successional stage plants are compared in Table 18–2. To enhance their colonizing ability, early seral species produce many small seeds which are usually wind dispersed (dandelion and milkweed, for example). Their seeds are long-lived, and they can remain dormant in soils of forests and shrub habitats for years until fires or treefalls create the bare-soil conditions required for germination and growth. The seeds of most climax species, being relatively large, provide their seedlings with ample nutrients to get started in the highly competitive environment of the forest floor.

The survival of seedlings in the shaded environment of the forest floor is directly related to seed weight (Figure 18-13). The ability of seedlings to survive the shade conditions of climax habitats is inversely related to their growth rate in the direct sunlight of early successional habitats. When placed in full sunlight, early successional herbaceous species grew ten times more rapidly than shade-tolerant trees. Shade-intolerant trees, like birch and red maple, had intermediate growth rates. Shade tolerance and growth rate represent a tradeoff; each species reaches a compromise between those adaptations best-suited for its place in the sere.

The rapid growth of early successional species is due partly to the relatively large proportion of seedling biomass allocated to leaves. Leaves carry on photosynthesis and their productivity determines the net accumulation of plant tissue during growth. Hence the growth rate of a plant is influenced by the allocation of tissue to the root and the above ground parts (shoot). In the seedlings of annual herbaceous plants, the shoot typically comprises 80 to 90 per cent of the entire plant; in biennials, 70 to 80 per cent; in herbaceous perennials, 60 to 70 per cent; and in woody perennials, 20 to 60 per cent.

The allocation of a large proportion of production to shoot biomass in early successional plants leads to rapid growth and production of large crops of seeds. Because annual plants must produce seeds quickly and copiously they never attain large size. Climax species allocate a larger proportion of their production to root and stem tissue to increase

TABLE 18–2   General characteristics of plants during early and late stages of succession.

Characteristic	Early	Late
Seeds	Many	Few
Seed size	Small	Large
Dispersal	Wind, stuck to animals	Gravity, eaten by animals
Seed viability	Long, latent in soil	Short
Root/shoot ratio	Low	High
Growth rate	Rapid	Slow
Mature size	Small	Large
Shade tolerance	Low	High

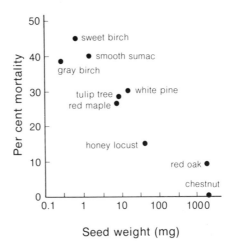

FIGURE 18–13 Relationship between seed weight and the survivorship of seedlings after three months under shaded conditions.

their competitive ability; hence they grow more slowly. The progression of successional species is therefore accompanied by a shift in the compromise between adaptation for great dispersal power and adaptation for great competitive ability.

The study of succession leads us to formulate four basic ecological principles. First, succession is one-directional; good colonizers with rapid growth and high tolerance of conditions on disturbed or newly exposed habitats are replaced by slowly growing species with great competitive ability. Second, successional species alter the environment by their structure and activities, often to their own detriment and to the benefit of other species. Third, the climax community is not a unit, rather it represents, at any given place, a point on a continuum of possible climax formations. The nature of the climax is influenced by climate, soil, topography, fire, and the activities of animals. Fourth, the climax may be a changing mosaic of successional stages maintained by wind, frost, or other sources of mortality acting locally within the community.

Successional changes are an inherent response of vegetation to disturbance, always reinstating the particular plant formation characteristic of the habitat. Succession occurs by the total replacement of populations of some species by populations of others, and thus cannot be compared to the homeostatic responses of an organism. Climax communities, nonetheless, are endowed with an inherent stability of structure and function, in part the sum of organism homeostasis and population responses, but in part the unique property of community organization. We shall examine community stability in the next chapter.

19

Community Stability

Stability is the inherent capacity of a system to tolerate or resist changes caused by outside influences. Suppose, for example, that precipitation falls 50 per cent below its long-term average, but plant production decreases by only 25 per cent and herbivore populations decrease by only 10 per cent. The relative damping of the environmental fluctuation as it is passed up through the food chain is a measure of the internal stability of the system — its capacity to resist change. In this case, stability may be derived from storage of water in the soil, physiological responses of plants to drought, and, if the drought lasts long enough, partial replacement of drought-sensitive herbs by drought-resistant species. The stability of a community depends upon the homeostatic responses of its constituent species.

We can visualize the concept of stability by considering a small ball placed in a bowl. If we nudge the ball, sending it a little way up the side of the bowl, the force of gravity quickly returns the ball to the bottom of the bowl. The steeper the sides of the bowl, the more powerful is the stabilizing force of gravity. This tendency of a system to be restored to a particular condition is referred to as a *stable equilibrium*. If the bowl itself represents the environmental factors acting on a population, the weight of the ball might correspond to the homeostatic capacities of the individuals in the population. A nudge of a given force would displace a steel ball bearing much less than it would a Ping-Pong ball, just as a cold snap has less effect on a population of bears than on a population of flies.

A ball placed on a level table represents a system having no forces tending either to maintain the position of the ball or to change it. This condition is known as a *neutral equilibrium*. A nudge applied to the ball sends it off across the table in one direction or the other and its movement is unchanged until some other outside force intercedes. Nature is rarely engineered like a level table. All systems have equilibrium points toward which they tend when displaced. Disturbances can, however, push a system beyond its capabilities of response — a ball sent careening out of the bowl; an insect population escaping predator controls and rising to outbreak levels.

When a system moves away from an equilibrium point, as when a ball rolls down the outside of an inverted bowl, the situation is referred to as an *unstable equilibrium*. The highest point on the outside surface of the inverted bowl is an equilibrium point because it is possible to balance the ball there. But the slightest disturbance sends the ball on its way.

The amount of movement a nudge of given force causes a ball in a bowl to undertake depends, then, on the weight of the ball and the configuration of the bowl. Similarly, the amount of fluctuation in a population or a community caused by an external disturbance depends on the inherent stability of the system.

Although we may define the inherent stability of a system as the ratio between variation in the environment and variation in the system itself, this definition is difficult to apply to populations or communities. Which aspect of environmental stability should we measure? Which component of the system gives the best indication of adequate function and continued persistence? Do we judge the stability of a community by the constancy of its function (production, ecological efficiency) or its structure (diversity, species turnover)? Moreover, change itself is often the best outward response to change. Hibernation and diapause represent a near total shutdown of biological activity in response to environmental change, yet dormancy allows the population to persist.

The biological significance of stability is even more elusive than its description and measurement. Constancy in the natural world is desirable to man because it enables him to predict conditions in advance and plan his activities accordingly. If weather and insect pests did not vary from year to year, farming would be simplified and a reasonably constant crop yield would be assured each year.

Virtually all human activity disturbs natural communities. Natural biological communities do not yield enough harvestable food to support dense human populations, so man selects the most "desirable" components of natural communities, alters their evolution through artificial selection to suit his purposes, and maintains these populations of crops, livestock, pulp trees, city parks, and backyards in a continually disturbed state, the populations constantly exposed to conditions they are not adapted to cope with. The price man pays for exploitation of natural resources is the price of maintaining their stability by constant management — curbing pest infestations, maintaining soil fertility, and cleaning out weeds.

Our concern with the constancy of the natural world and with the basis for stability in natural communities is understandable. This concern is shared, unconsciously, by all species. Constancy of weather, resources, predation, and competition reduces the cost of self-maintenance (homeostasis) and increases the allocation of energy and nutrients to production. Most organisms would stand to gain from a more constant world, but because their competitors and predators would also gain we cannot predict the benefit of constancy to an individual or to a species.

Constancy of the environment and the community both undoubtedly enhance production and ecological efficiency because few resources are used for homeostasis, materials do not accumulate behind bottlenecks caused by population fluctuations, and adaptations of organisms become more finely tuned to the environment. But do communities possess inherent stabilizing mechanisms in addition to the homeostatic responses of organisms and the growth responses of population? If we are to reject the notion of community adaptations (see page 355), we must also reject many contemporary ideas about community stability: that it is desirable for the community to be stable because constancy increases the efficiency of energy flow and nutrient cycling; that natural selection leads to increased complexity and diversity within the community so as to enhance the inherent stability of trophic structure and improve the ability of the community to resist perturbation; that many adaptations of organisms, such as large size, long life, low reproductive rates, and low productivity-biomass ratios, are adaptations to increase the stability of the community rather than to increase the evolutionary fitness of the organism.

If we believe that adaptations are properties of individuals and that evolutionary change is a property uniquely invested in populations, we must view the ability of the community to resist change as the sum of the individual properties of its component populations. As we shall see below, relationships between predators and prey, and between competitors, can affect the inherent stability of the community, but trophic structure does not evolve to enhance community stability.

Some Definitions

Stability and related terms are used in so many different ways by ecologists that it will be necessary to provide explicit definitions here:

> *stability* is the intrinsic ability of a system to withstand or to recover from externally caused change,
>
> *constancy* is a measure of the degree of variation of a system,
>
> *predictability* is a measure of regularity in patterns of change.

Seasonal fluctuation in the environment is predictable; day to day variation often is not.

The degree of fluctuation in the community is determined by three factors, each of which will be considered separately below: (a) the constancy and predictability of the physical environment, (b) the homeostatic mechanisms of organisms and growth responses of populations as subunits of community stability, and (c) that component of community stability uniquely contributed by the feeding and competitive relationships of populations within the community — in other words, the trophic organization of the community.

Variability in the Physical Environment

Temperature and rainfall have been measured for long periods at weather monitoring stations throughout the world. Although local climatological data do not adequately measure the environment of any particular population, they do provide an indication of the overall constancy of the environment. Three components of the environment are important to organisms: regular seasonal fluctuations, variation about seasonal norms, and the predictability of short-term variations. All habitats exhibit diurnal and seasonal changes; coastal marine environments are influenced, in addition, by lunar rhythms. Daily, lunar, and seasonal fluctuations reflect regular cycles in the physical world, as we have seen in Chapter 5. Many years of measurement would reveal the average conditions for each day of the year and each time of day. But the environment rarely exhibits average conditions. Irregularities in climate, related to changing wind patterns and random meteorological events, cause the environment to vary around its norm. For example, 100 years of weather records in Philadelphia, Pennsylvania, show that while the average July rainfall is 4.2 inches, precipitation was less than half the average in nine years and more than twice the average in four years (Figure 19-1).

Diurnal and seasonal patterns are more predictable than short-term variations because they are tied to precise physical cycles, such as the daily light-dark cycle and the seasonal change in daylength during the year. But the unreliability of weather forecasts, particularly for several days ahead, attests to the lack of predictability of short-term variations in climate; the further one is removed in time from an event of brief duration, such as a rainstorm, the less predictable it becomes. Rain can be predicted a few minutes or a few hours before a thunderstorm by the appearance of the sky. A change in wind direction during certain seasons signals the passage of a front, often accompanied by precipitation and temperature change. But it is virtually impossible to know in January, or even in May, whether June will bring drought or deluge.

Rainfall is generally most variable from year to year where it is least abundant. At a given place, dry-season precipitation is more variable and less predictable than wet-season precipitation. Similarly, tempera-

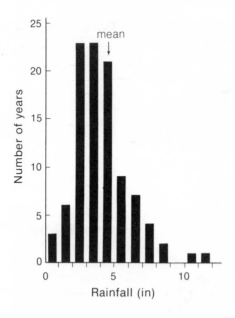

FIGURE 19-1   July precipitation for 100 years at Philadelphia, Pennsylvania.

ture variations are greatest when the average temperatures are lowest — geographically in polar regions and seasonally during winter. These patterns of variability suggest that in the tropics, the physical environment is more constant than in temperate and arctic regions; the tropics are warm the year around and the climate of most tropical areas is relatively wet. The generally observed pattern could be misleading. Dry-season rainfall in some tropical areas is less than that in the driest months in many temperate localities; rainfall during the tropical dry season can be correspondingly variable and unpredictable.

Every region, no matter how constant its environment, is subject to infrequent extremes. In a wet environment, an extreme condition encountered only once in hundreds of years may be one that differs from the normal by a factor of two. In a dry locality, an extreme condition that is encountered equally infrequently may differ from the normal by a factor of ten. The homeostatic capabilities of organisms are adapted to the range of conditions normally encountered. Regardless of the degree of fluctuation in the environment, infrequent "extreme" conditions impose a stress on organisms in any region. The homeostatic mechanisms of tropical populations might, in fact, be much more poorly developed than those of temperate and arctic populations because the environment usually varies within a narrower range.

Some types of environmental variation are so drastic that they cannot be accommodated by the homeostatic mechanisms of organisms. Such events — environmental catastrophes — include hurricanes, tornados, fires, and hard freezes (Figures 19-2 and 19-3). Although many species are adapted to prosper in the aftermath of such disasters — the weedy species that colonize disturbed habitats, and others adapted to a

FIGURE 19–2 An intense, uncontrolled fire in the Willamette National Forest, Oregon. Most of the community was destroyed. A long period of succession will follow before the community again attains its mature characteristics.

regular cycle of minor fires — natural catastrophes completely destroy the fabric of most communities. Their structure must be rebuilt gradually over long periods by succession. Many human disturbances create equally catastrophic conditions — beyond the limits of variation normally encountered by organisms — completely destroying communities and, sometimes, preventing the re-establishment of community structure.

Stability at the Individual and Population Level

The constancy of population size or organism activity reflects the interplay between environmental fluctuation and intrinsic stability. The outcome of this interaction can be seen by examining the growth rings of

FIGURE 19–3 A severe ice storm in New York damages a hardwood forest. Community function will not be completely restored for many years, during which many successional species will appear before they are crowded out by regenerating trees.

trees. A core of wood contains a record of a tree's growth rate from the sapling stage on. Because trees produce one ring each year, the rings can be dated easily and the variation in their width compared to variation in temperature and rainfall. The sensitivity of a tree to climate fluctuation depends on where it grows. In moist habitats, water may never become sufficiently limiting to affect tree growth adversely even in the driest years. Where moisture levels are marginal for a species, drought can exert a profound effect on growth. This point is illustrated by growth ring chronologies in two populations of bristlecone pine in the White Mountains of California (Figure 19–4). Trees in a moist grove with abundant winter snow accumulation exhibited little year-to-year variation in growth rate compared to stunted individuals growing on a dry, windswept, rocky ridge. Although variation in ring width at each site paralleled variation at the other, moisture deficits severely depressed growth in the moist site only three times during the 104-year chronology — 1899, 1929, and 1959.

Most of the ring-width variation in the bristlecone pine is related to the moisture level of the habitat during June, a hot month with little or

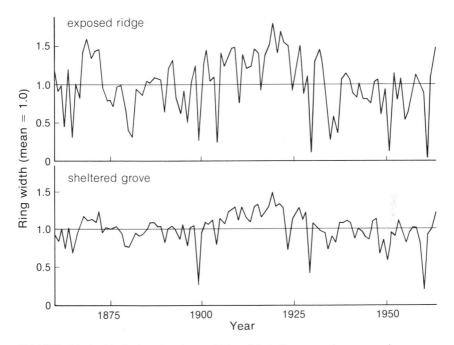

FIGURE 19–4 Variation in ring width of bristlecone pines growing on a rocky ridge (top) and in a protected grove with abundant moisture (below). The ring-width axis is scaled in such a way that the mean width at each locality equals 1.

no rainfall, and therefore with a large evapotranspiration deficit. Autumn temperatures and winter moisture also influence growth during the subsequent season. The climates of different areas vary in their effect on tree growth (Figure 19–5). Moisture stress at the beginning of the growing season is more important to the bristlecone pine in California; Douglas fir responds equally to winter and spring moisture; growth of the piñon pine is determined primarily by winter moisture, indicating that it relies heavily on accumulated ground water for growth during the arid summer months. In dry habitats in Illinois, the growth of white oak responds to conditions of moisture and temperature during the growing season but, in addition, late summer drought depresses growth during the following year, perhaps by reducing storage of food in the roots or by interfering with the formation of leaf buds.

Tree-ring data show that organisms do not completely compensate for variation in the environment, but we are left guessing about the inherent stability of a tree's growth processes. Some general principles concerning stability are apparent from our earlier study of homeostasis and population growth. Large organisms have small surface-to-volume ratios; their internal environments are therefore more independent of

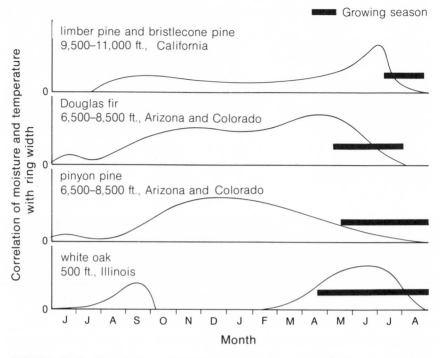

FIGURE 19–5 Correlation of tree-ring width with temperature and moisture during each of 15 months prior to and including the growing season.

the external environment than are the internal environments of small organisms. Large accumulated biomass and low turnover rate of individuals in populations also increase buffering against environmental variation (page 251). Immature organisms are generally more susceptible to environmental change than adults because they are smaller, less experienced in the case of animals, and physiologically less mature. Populations with many immature individuals tend to be less stable than populations of predominately mature individuals.

The response time of the population to fluctuations in the environment constitutes another important component of stability. Small organisms reproduce more rapidly than large organisms and their populations can respond to environmental change more rapidly. Furthermore, small organisms may choose among several avenues of response: developmental and evolutionary responses are practical ways of coping with short-term environmental change only for small organisms with short life spans (see Chapters 11 and 12). Size and biomass-to-productivity ratios have two opposing influences on stability. The number of individuals in populations of large organisms change little but respond slowly to change; the opposite is true of populations of small organisms. Which is more stable? Which is better attuned to fluctuation in the physical environment? The answers are not clear.

In the long run, populations that persist are stable, regardless of their degree of fluctuation. The ultimate measure of instability is extinction. Populations with few individuals begin at a disadvantage in comparison with larger populations and are more likely to pass into oblivion following perturbations in their environment. What factors determine the size of a population? The study of taxon cycles (page 346) indicates that the relative success of the adaptations of a species, compared to the adaptations of its predators, parasites, competitors, and prey, determines the degree of resource specialization and influences the size of the species' population. In an evolutionary struggle to achieve superior adaptations, species beset by few kinds of exploiters and competitors fare better than species that must confront the adaptations of many antagonists. For example, a grasshopper attacked by a shrew can escape by flight. If grasshoppers are preyed upon by both shrews and sparrow hawks, each predator compromises the grasshopper's escape from the other.

The number of species in communities is greatest in the humid tropics and decreases toward the poles. Populations are surely limited by a greater diversity of competing populations in the tropics than elsewhere and average population size in fact tends to be smaller. Whether tropical species are eaten by a greater variety of predators is not known. Many ecologists have asserted that species of predators are more numerous relative to species of prey in the tropics than in temperate and arctic communities. Insect communities appear to bear this out (Table 19–1). The ratio of predators and parasites to herbivores and detritivores (the predator ratio) is greater in tropical samples than in temperate zone samples. Furthermore, within the tropics both diversity and the predator ratio increase along a moisture gradient from dry

TABLE 19–1 Percentages of trophic groups in samples of insects from various localities and habitats, arranged by decreasing latitude. In general, the tropical samples contain the most parasites and the fewest detritivores.

Locality	Habitat	Herbivores	Detritivores	Predators	Parasites	Predator ratio*
Arctic Coast	Whole fauna	47	27	14	10**	0.32
Connecticut	Whole fauna	49	19	16	12	0.41
New Jersey	Whole fauna	52	19	16	10	0.37
South Carolina	Old field	68	1	19	9	0.41
Great Smoky Mountains	Average for 15 habitats	41	31	12	15	0.38
Florida Keys	Mangrove islands	41	28	22	7	0.42
Costa Rica	Average for 4 habitats	59	4	9	26	0.56

*Proportion of predators plus parasites divided by proportion of herbivores plus detritivores.
**Totals do not add to 100 per cent because some species are classified as "miscellaneous."

hillsides (ratio = 0.38) to wet lowland and river-bottom forest (ratio = 0.77). The greater predator ratio of moist tropical habitats is due primarily to an increase in the number of parasitic species, which tend to specialize their attack to a few host species.

If high diversity and complexity of community organization reduce the average population size of tropical species and increase the probability of their extinction, the composition of tropical communities may be intrinsically less stable than temperature or arctic communities. These factors could be balanced, of course, by low variability of the physical environment and a relative preponderance of mutualistic associations in the tropics. Direct measurements of the life spans of species in tropical and temperature populations would have to be based on the fossil record which is meager at best. Some groups of marine organisms are well-enough represented in the fossil record that the persistence of taxa can be compared between regions. For example, genera of planktonic foraminifers (small protozoans with calcareous shells) that lived during the Cretaceous period persisted longer in oceans north of 50° latitude than in the warm-water regions closer to the Equator (Figure 19-6). But the differences between aquatic and terrestrial environments are great enough to render this evidence slender indeed.

Diversity, Complexity, and Community Stability

If high diversity speeds individual populations to extinction thereby reducing the stability of community composition, diversity and complexity of trophic organization are widely thought to enhance the stability of community function. This principle may be stated simply: where predators eat many kinds of prey organisms, they can specialize momentarily on whichever prey species are most abundant. This switching behavior makes predators less sensitive to variation in the abundance of any one

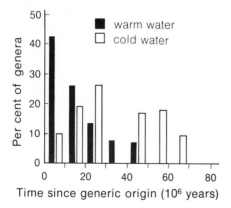

FIGURE 19–6   Life span of Cretaceous genera of planktonic foraminifera in warm-water (0 to 50°N) and cold-water (>50°N) regions.

prey species. In simpler communities, predators are restricted to eating few kinds of prey — perhaps only one — so their populations follow variations in their prey populations more closely (Figure 19-7). Studies of population cycles of fur-bearing mammals in arctic North America, where communities are relatively simple, are consistent with a direct relationship between diversity and stability (see page 257).

One can, however, take a different view of the relationship among diversity, complexity, and stability. In rigorous climates, either too dry or too cold to support diverse communities, the physical environment affects most species directly and at the same time. A drought or cold snap depresses biological activity throughout the community. Milder conditions return the community as a whole to its original state. With physical conditions exerting dominant control over fluctuations in the community, all species are linked directly to the dominant cause of the fluctuation. In the less harsh tropics, population trends are determined more by interactions with other populations than by the physical environment. Like a ripple moving across the surface of a pond, perturba-

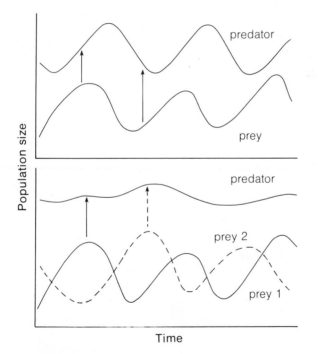

FIGURE 19-7· A diagram of population cycles of predator species feeding on either one (top) or two (bottom) kinds of prey. The predator in the two-prey system switches its feeding from one species to the other depending on which is more abundant.

tions can be passed through many links of species interaction. A population in a diverse and complex tropical community can be removed by several links from an external source of disturbance. In such circumstances, time lags in population response and time lags between the links may increase, rather than dampen, the effects of perturbation.

An example will emphasize the point. Many species of trees in tropical forests flower during the dry season when flying insect-pollinators are most abundant. Occasionally, heavy rains fall after the beginning of the dry season. Trees caught with their flowers out may not be pollinated because rain delays the activity season of the insects. If flowers are not pollinated, trees will not produce fruit during their normal fruiting period, several months later; mammals and birds, which rely on fruit when other food sources are less abundant, starve, curtail reproduction, or turn to novel sources of food; rodents and other gnawing mammals severely damage trees and kill saplings by chewing off bark to get at the nutritious growing layers underneath; herbivorous insects that normally would have eaten the tree seedlings, and predators and parasites of these insects, are affected in turn. In this way, the initial physical perturbation is maintained in the tropical community. In simpler communities, response to physical perturbation tends to be rapid, direct, and short-lived.

Biological complexity may either dampen or exaggerate perturbations to communities. Where theory is equivocal, we must turn to direct observation and experimentation. A few studies are pertinent to the diversity–stability problem. For example, ecologist Kenneth Watt examined fluctuations in populations of Canadian forest lepidoptera (moths and butterflies), comparing species that fed on few species of trees to those with broader diets. Watt's analysis (Table 19–2) showed that population size and the relative degree of fluctuation both increase as the number of tree species eaten increases. Hence, diversity of food resources does not appear to enhance the stability of butterfly and moth populations.

TABLE 19–2   Relationship of number of tree species eaten to the abundance and constancy of Canadian forest moth and butterfly populations.

Number of tree species eaten*	Relative population size	Index of variation in population size**
1.5	5	0.23
4.2	47	0.30
10.9	52	0.34
24.7	442	0.37

*Means of four groups.
**Variation relative to the size of the population.

Disturbance and Community Stability

If we want to determine the ability of a community to resist disturbance, the best approach may be to disturb the community and watch its response. We would have difficulty predicting the results of our experiment beforehand owing to the great variety of effects dependent on the nature of the disturbance and the adaptations of organisms in the community. A fire in a longleaf pine plantation in Mississippi depresses tree growth for only one year, then tree production returns to normal (Figure 19–8). When New Mexican grasslands are overgrazed, shrubs replace blue gramma and other grasses completely, and permanently alter the character of the community. Whereas grasses have extensive, shallow, fibrous root systems, which hold the topsoil firmly in place, the roots systems of many shrubs do not prevent soil erosion (Figure 19–9).

One of the most consistent effects of disturbance on the structure of communities is to reduce the total number of species while allowing some of the survivors to reach abnormally high population levels. For example, compared to natural streams, polluted streams tend to have fewer species, but some of these are extremely abundant (Figure 19–10). Strong pollution creates conditions that are lethal to most species in the community, but which are extremely favorable for a few.

Each community in a steady state has a characteristic pattern of species abundance and number of species on each trophic level. Following disturbance, communities usually regain their characteristic structure through recolonization by eliminated species, elimination of species that colonized the disturbed habitat, and adjustment of population sizes. After the insect communities of several small mangrove islands in the Florida Keys were experimentally removed by insecticides, community structure was rebuilt by immigration of species from neighboring islands. The percentage of species in each trophic class — herbivores, detritus feeders, ants (omnivores), predators and parasites — in equilib-

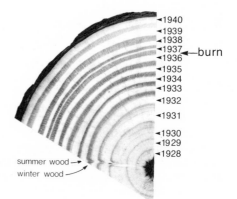

FIGURE 19–8 Cross section of a longleaf pine tree burned in January 1937 and cut in the fall of 1940 in Harrison County, Mississippi. Narrow growth rings were produced during the summer of 1937 (light wood) and the winter of 1937–1938 (dark wood).

FIGURE 19–9 Root systems of blue gramma grass (left) and snakeweed (*Gutierrezia*, right). Overgrazing on grasslands can kill grasses and allow shrubs with little forage value and poor soil-holding qualities to take over.

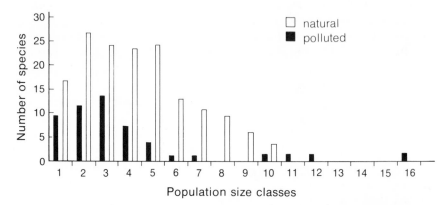

FIGURE 19–10 Distribution of abundances of diatoms in samples taken from a natural stream community and from a polluted river. Population-size classes increase from left to right by a factor of two, e.g., 1–2, 2–4, 4–8, 8–16, and so on. The largest population in the polluted sample falls into the range 32,768–65,536.

rium communities was relatively constant on the islands examined. After experimental islands were defaunated, species began to return haphazardly and community structure initially differed from the normal proportions of species in each trophic group (Figure 19-11). Rate of establishment differed from one group to another. Scavengers achieved half their final diversity in 32 days, parasites in 47 days, and predators in 51 days. Herbivores were slower to recolonize the islands (one-half of equilibrium achieved in 98 days), perhaps because they are more sedentary by nature — their food is certainly more sedentary than the food of predators. As the density of populations and number of species on the islands increased, interactions among species on the islands also increased, presumably intensified, and sped the return of the trophic structure to its equilibrium state. During the course of recolonization, the number of species on the islands overshot the equilibrium number before competitive exclusion finally sorted out the structure of the community.

Simplification and Community Stability

When the number of plant species in a habitat is reduced, by planting crops for example, the number of species on all trophic levels decreases. In these simplified communities, the abundance of some herbivore species increases to outbreak levels in the absence of effective control by predators. Herbivorous insects and disease organisms parasitic on plants are usually specialized to occur on a single host plant. Some species, adapted to feed on a particular crop, do spectacularly well in agricultural habitats, much to the farmer's dismay. On the other hand, predators require a more complex habitat and do poorly in single species stands of crops. Spruce budworms cause more damage to solid stands of balsam fir than to fir trees well spaced among other species. Infestations are worse when all the trees in the forest are the same age. Susceptibility to budworm infestation increases with age; if a forest consists entirely of old trees — often the case in managed woodlands — most of the trees become infected, easing the spread of the pest and leaving fewer unaffected trees to replace those killed by the budworm.

The relationship between community simplification and population outbreaks has been explored by entomologist David Pimentel. Insect populations on two groups of collard plants were compared. One group of collards was planted in a field that had been uncultivated for 15 years. In the old field, collards were planted nine feet apart among vegetation that contained about 300 species of plants, including five in the same family as collards (Cruciferae, the mustard family). A second group of collard plants was planted in a dense stand having no other species. The abundances of several species of insects — particularly

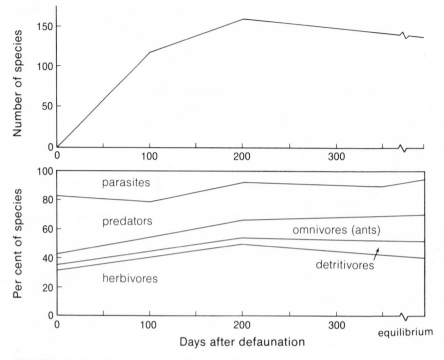

FIGURE 19–11   Return to trophic equilibrium after defaunation of insect communities on red mangrove islands near the Florida coast.

aphids and flea beetles — reached outbreak proportions in the single species plantings. The same species of insects were kept under control by predators in the old field. The number of predators and parasitic insects collected from collard plants was also higher in the single species planting than in the old field, but they appeared too late in the season to exercise effective control over herbivore populations. Also, because Pimentel collected samples only from the collard plants in both plantings, he may, in the old field, have missed the many species on other types of plants. Such species could also have preyed on herbivorous insects on the collards. This often happens. Predators on the diamondback moth, a cabbage pest, survive the winter by eating other prey species that attack hawthorn. Many of the alternate hosts of gypsy moth parasites live in the forest understory, not in the canopy; the presence of forest undergrowth is therefore indirectly important to the growth of gypsy moth populations. Unfortunately, one could point to as many examples of weed or hedgerow plants maintaining populations of crop pests as populations of their predators and parasites. The role of diversity in maintaining stability is thus cast into doubt, perhaps because crop systems are so simple that a few populations, perhaps pests, perhaps their predators, can dominate a community just by chance.

In an experiment similar to Pimentel's, arthropod populations were followed in a field that was planted with millet one year and then left to develop a natural successional community during the second year. Populations were sampled ten times each summer at regular intervals. Each trophic group of insects and other arthropods was represented by more species in the second year than in the first. Populations of predatory and parasitic species were also greater in the second year, but numbers of herbivores were reduced, presumably by the predators and parasites (Table 19-3). An interesting result of the study was that variation in the numbers of predatory insects was greater during the second year in spite of the increased diversity of prey species (see page 411). Once more, observations do not support a direct relationship between diversity and constancy.

In another study near Syracuse, New York, fertilizer was applied to abandoned hay fields of different ages. The experiment was designed to measure the effect of a perturbation — nutrient enrichment — on the composition of the community and how this perturbation was transmitted through the trophic structure. The 17-year-old field was more diverse and had greater biomass of vegetation than the 6-year-old field. Presumably the older field should have resisted perturbation to a greater degree, if increased diversity and biomass enhance stability. In fact, the older field was more sensitive to enrichment, and the effects were felt more strongly at the herbivore and carnivore levels than among the primary producers, which were affected directly by the fertilizer (Table 19-4). It is arguable whether the increased diversity and production reflected failure of internal stabilizing mechanisms or a favorable response to enrichment, further increasing stability. The point is well

TABLE 19-3 Diversity, density, and variation of arthropod populations in unharvested grain crop compared to the natural successional community which replaced it during the following year.

		Herbivorous insects	Predatory insects	Spiders	Parasitic insects	Total arthropods
Diversity*	1966	7.2	2.8	1.6	6.3	15.6
	1967	10.6	6.0	7.2	12.4	30.9
Density (ind/m²)	1966	482	64	18	24	624
	1967	156	79	38	51	355
Variation (% of mean density)**	1966	24	44	33	57	19
	1967	28	49	67	40	32

*Calculated by the equation, diversity $= (S - 1)/\log N$, where S is the number of species in the sample and N is the number of individuals.
**Coefficient of variation (%) = (standard deviation/mean) $\times$ 100.

TABLE 19–4   Percentage change in diversity and productivity of producer, herbivore, and carnivore trophic levels as a result of fertilization of 6-year-old and 17-year-old abandoned hay fields.

Trophic level	Diversity		Productivity*	
	6 years	17 years	6 years	17 years
Producers	−7	+3	+96	+70
Herbivores**	+24	+51	+31	+201
Carnivores**	+27	+74	−6	+108

*$Mg/m^2/day$.
**Diversity and productivity measured during the earlier of two growing seasons.

made, however, that in a successionally more mature community, perturbations were transferred more efficiently throughout the trophic structure.

Ecologists are terribly ignorant of stability in natural systems — the pertinent internal mechanisms of communities and how they work, the relative stability of different communities. Even recognizing stability where it occurs is difficult. Why has this single property of nature been so elusive? Stability represents the culmination of all ecological interrelationships; it is the sum of all the components and interactions that make up the community — the synthesis of all lower-order properties of community, population, and organism. To understand stability, we must understand ecological and evolutionary responses and interrelationships at all levels. The science of ecology is not yet mature enough to mold its diverse knowledge and concepts into a unified theory of stability. This goal will probably not be achieved before another decade has passed; nor should every ecologist necessarily direct his research efforts entirely, or even partly, toward understanding stability. Still, the significance of new knowledge and the importance of new ideas surely will be judged by their contribution to our understanding of the ecological synthesis — the stability of natural systems.

Conversion Factors

Length

1 meter (m) = 39.4 inches (in)
1 meter = 3.28 feet (ft)
1 kilometer (km) = 3281 feet
1 kilometer = 0.62 miles
1 micron (μ) = 10^{-6} meters
1 inch = 2.54 centimeters (cm)
1 foot = 30.5 centimeters
1 mile (mi) = 1609 meters
1 Angstrom unit (Å) = 10^{-10} meters
1 millimicron (mμ) = 10^{-9} meters

Area

1 square centimeter (cm²) = 0.155 square inches (in²)
1 square meter (m²) = 10.76 square feet (ft²)
1 hectare (ha) = 2.47 acres (A)
1 hectare = 10,000 square meters
1 hectare = 0.01 square kilometer (km²)
1 square kilometer = 0.386 square miles
1 square mile = 2.59 square kilometers
1 square inch = 6.45 square centimeters
1 square foot = 929 square centimeters
1 square yard (yd²) = 0.836 square meters
1 acre = 0.407 hectares

Time

1 year (yr) = 8760 hours (hr)
1 day = 86,400 seconds (sec)

Volume

1 cubic centimeter (cc or cm^3) = 0.061 cubic inches (in^3)
1 cubic inch = 16.4 cubic centimeters
1 liter = 1,000 cubic centimeters
1 liter = 33.8 U.S. fluid ounces (oz)
1 liter = 1.057 U.S. quarts (qt)
1 liter = 0.264 U.S. gallons (gal)
1 U.S. gallon = 3.76 liters
1 Brit. gallon = 4.55 liters
1 cubic foot (ft^3) = 28.3 liters (l)
1 milliliter (ml) = 1 cubic centimeter
1 U.S. fluid ounce = 29.57 milliliters
1 Brit. fluid ounce = 28.4 milliliters
1 quart = 0.946 liters

Velocity

1 meter per second (m/sec) = 2.24 miles per hour (mi/hr)
1 mile per hour = 0.447 meters per second
1 foot per second (ft/sec) = 1.093 kilometers per hour
1 kilometer per hour = 0.278 meters per second
1 mile per hour = 1.467 feet per second

Mass

1 gram (g) = 15.43 grains (gr)
1 kilogram (kg) = 35.3 ounces
1 kilogram = 2.205 pounds (lb)
1 metric ton (t) = 2204.6 pounds
1 ounce (oz) = 28.35 grams
1 pound = 453.6 grams
1 short ton = 907 kilograms

Energy

1 joule = 0.239 calories (cal)
1 kilowatt-hour (kWh) = 860 kilocalories
1 British thermal unit (Btu) = 252.1 calories
1 British thermal unit = 17.58 watts
1 kilocalorie (kcal) = 1,000 calories

Power

1 kilowatt (kW) = 0.239 kilocalories per second
1 kilowatt = 860 kilocalories per hour
1 horsepower (hp) = 746 watts
1 horsepower = 15,397 kilocalories per day
1 horsepower = 641.5 kilocalories per hour

Energy per unit area

1 calorie per square centimeter = 3.69 British thermal units per square foot

1 British thermal unit per square foot = 0.271 calories per square centimeter

1 calorie per square centimeter = 10 kilocalories per square meter

Power per unit area

1 kilocalorie per square meter per minute = 52.56 kilocalories per hectare per year

1 footcandle (fc) = 1.30 calories per square foot per hour at 555 mμ wavelength

1 lux (lx) = 1.30 calories per square meter per hour at 555 mμ wavelength

Miscellaneous

1 gram per square meter = 0.1 kilograms per hectare
1 gram per square meter = 8.97 pounds per acre
1 kilogram per square meter = 4.485 short tons per acre
1 metric ton per hectare = 0.446 short tons per acre

Metabolic energy equivalents

1 gram of carbohydrate = 4.2 kilocalories
1 gram of protein = 4.2 kilocalories
1 gram of fat = 9.5 kilocalories

Glossary

Acclimation. A reversible change in the morphology or physiology of an organism in response to environmental change.

Adaptive radiation. Evolutionary diversification of species derived from a common ancestor into a variety of ecological roles.

Adaptive zone. A particular type of environment requiring unique adaptations. Species in different adaptive zones usually differ by major morphological or physiological characteristics.

Allelopathy. Direct inhibition of one species by another using noxious or toxic chemicals.

Alpha diversity. See Diversity.

Ammonification. Breakdown of proteins and amino acids with ammonia as an excretory by-product.

Anaerobic. Without oxygen.

Assimilation efficiency. A percentage expressing the proportion of energy ingested that is absorbed into the bloodstream.

Association. A group of species living in the same place.

Autecology. The study of organisms in relation to their physical environment.

Autotroph. An organism that assimilates energy from either sunlight (green plants) or inorganic compounds (sulfur bacteria). See also Heterotroph.

Barren. An area with sparse vegetation owing to some physical or chemical property of the soil.

Benthic. Bottom dwelling in rivers, lakes, and oceans.

Beta diversity. See Diversity.

Biomass accumulation ratio. The ratio of weight to annual production, usually applied to vegetation.

Calcification.   Deposition of calcium and other soluble salts in soils where evaporation greatly exceeds precipitation.

Caliche.   An alkaline salt deposit on the soil surface, usually occurring in arid regions with ground water close to the surface.

Carrying capacity.   Number of individuals that the resources of a habitat can support.

Cation.   A part of a dissociated molecule carrying a positive electrical charge, usually in an aqueous solution (e.g. Ca^{++}, Na^+, NH_4^+, H^+).

Character displacement.   Divergence in the characteristics of two otherwise similar species where their ranges overlap, caused by the selective effects of competition between the species in the area of overlap.

Climatic climax.   The steady-state community characteristic of a particular climate.

Climax.   The end-point of a successional sequence; a community that has reached a steady state under a particular set of environmental conditions.

Climograph.   A diagram on which localities are represented by the annual cycle of their temperature and rainfall.

Coadaptation.   Evolution of characteristics of two or more species to their mutual advantage.

Community.   An association of interacting populations, usually defined by the nature of their interaction or the place in which they live.

Compensation point.   Depth of water at which respiration and photosynthesis balance each other; the lower limit of the euphotic zone.

Competition.   Use or defense of a resource by one individual that reduces the availability of that resource to other individuals.

Competitive exclusion principle.   The hypothesis that two or more species cannot coexist on a single resource that is scarce relative to the demand for it.

Continuum.   A gradient of environmental characteristics or of change in the composition of communities.

Continuum index.   An artificial scale of an environmental gradient based on changes in community composition.

Convergent evolution.   Development of characteristics with similar functions in unrelated species that live in the same kind of environment but in different places.

Counteradaptation.   Evolution of characteristics of two or more species to their mutual disadvantage.

Cyclic climax.   A steady-state, cyclic sequence of communities, none of which by itself is stable.

Denitrification.   The reduction by microorganisms of nitrate and nitrite to nitrogen.

Density-dependent.   Having influence on individuals in a population that varies with the degree of crowding in the population.

Density-independent.   Having influence on individuals in a population that does not vary with the degree of crowding in the population.

Detritivore.   An organism that feeds on freshly dead or partially decomposed organic matter.

Diapause.   Temporary interruption in the development of insect eggs or larvae, usually associated with a dormant period.

Direct competition.   Exclusion of individuals from resources by aggressive behavior or use of toxins by other organisms.

Diversity.   The number of species in a community or region. Alpha diversity refers to the diversity of a particular habitat, beta diversity to the species added by pooling habitats within a region.

Ecocline.   A geographical gradient of vegetation structure associated with one or more environmental variables.

Ecological efficiency.   Percentage of energy in the biomass produced by one trophic level that is incorporated into biomass produced by the next highest trophic level.

Ecological release.   Expansion of habitat and resource utilization by populations in regions of low species diversity, resulting from reduced interspecific competition.

Ecosystem.   All the interacting parts of the physical and biological worlds.

Ecotone.   A habitat created by the juxtaposition of distinctly different habitats; an edge habitat; a zone of transition between habitat types.

Ecotype.   A genetically differentiated subpopulation that is restricted to a specific habitat.

Epilimnion.   The warm, oxygen-rich surface layers of a lake or other body of water.

Euphotic zone.   Surface layer of water to the depth of light penetration at which photosynthesis balances respiration. See also Compensation point.

Eutrophic.   Referring to a body of water with abundant nutrients and high productivity.

Eutrophication.   Enrichment of bodies of water, often caused by sewage and runoff from fertilized agricultural land.

Evapotranspiration.   The sum of transpiration by plants and evaporation from water surfaces and the soil. Potential evapotranspiration is the amount of evapotranspiration that would occur, given the local temperature and humidity, if water were super abundant.

Exploitation efficiency. The percentage of potential prey or food plants that are consumed by predators and herbivores.

Fecundity. Rate at which an individual produces offspring, usually expressed only for females.

Field capacity. The amount of water that soil can hold against the pull of gravity.

Food chain. An abstract representation of the passage of energy through populations in the community.

Food chain efficiency. See Ecological efficiency.

Food web. An abstract representation of the various paths of energy flow through populations in the community.

Functional response. Change in the rate of exploitation of prey by an individual predator as a result of a change in prey density. See also Numerical response.

Generation time. The average age at which a female gives birth to her offspring.

Gross production. The total energy or nutrients assimilated by an organism, a population, or an entire community. See also Net production.

Gross production efficiency. The percentage of ingested food utilized for growth and reproduction by an organism.

Heterotroph. An organism that utilizes organic materials as a source of energy and nutrients. See also Autotroph.

Homeothermic. Able to maintain constant body temperature in the face of fluctuating environmental temperature; warm-blooded.

Hydrarch succession. Progression of terrestrial plant communities developing in an aquatic habitat such as a bog or swamp.

Hypolimnion. The cold, oxygen-poor part of a lake or other body of water that lies below the zone of rapid change in water temperature. See also Epilimnion.

Indirect competition. Exploitation of a resource by one individual that reduces the availability of that resource to others. See also Direct competition.

Interspecific competition. Competition between individuals of different species.

Intraspecific competition. Competition between individuals of the same species.

Ion. The dissociated parts of a molecule, each of which carries an electrical charge.

Laterization. Leaching of silica from soil, usually in warm, moist regions with an alkaline soil reaction.

Leaching. Removal of soluble compounds from leaf litter or soil by water.

Life form. Characteristic structure of a plant or animal.

Life table. A summary by age of the survivorship and fecundity of female individuals in a population.

Mesic. Referring to habitats with plentiful rainfall and well-drained soils.

Micelle. A complex soil particle resulting from the association of humus and clay particles, with negative electric charges at its surface.

Mutualism. Relationship between two species that benefits both parties.

Mycorrhizae. Close association of fungi and tree roots in the soil that facilitates the uptake of minerals by trees.

Natural selection. Change in the frequency of genetic traits in a population through differential survival and reproduction of individuals bearing those traits.

Negative feedback. Tendency of a system to counteract externally-imposed change and return to a stable state.

Net production. The total energy or nutrients accumulated by the organism by growth and reproduction, gross production minus respiration.

Net production efficiency. The percentage of assimilated food utilized for growth and reproduction by an organism.

Net reproductive rate. Number of offspring that females are expected to bear on average during their lifetimes.

Nitrification. Breakdown of nitrogen-containing organic compounds by microorganisms, yielding nitrates and nitrites.

Nitrogen fixation. Biological assimilation of atmospheric nitrogen to form organic nitrogen-containing compounds.

Numerical response. Change in the population size of a predatory species as a result of a change in the density of its prey. See also Functional response.

Oligotrophic. Referring to a body of water with low nutrient content and productivity.

Osmosis. Diffusion of substances in aqueous solution across the membrane of a cell.

Pattern–climax theory. The hypothesis that succession reaches a wide variety of non-discrete climax communities depending on local climate, soil, slope, grazing pressure, and so on.

Photosynthesis. Utilization of the energy of light to combine carbon dioxide and water into simple sugars.

Photosynthetic efficiency. Percentage of light energy assimilated by plants; based either on net production (net photosynthetic efficiency) or on gross production (gross photosynthetic efficiency).

Phytoplankton. See Plankton.

Plankton. Microscopic floating aquatic plants (phytoplankton) and animals (zooplankton).

Podsolization. Breakdown and removal of clay particles from the acidic soils of cold, moist regions.

Poikilothermic. Unable to regulate body temperature; cold-blooded.

Polyclimax theory. The hypothesis that succession leads to one of a variety of distinct climax communities, depending on local environmental conditions.

Potential evapotranspiration. See Evapotranspiration.

Primary productivity. Rate of assimilation (gross primary productivity) or accumulation (net primary productivity) of energy and nutrients by green plants and other autotrophs.

Primary succession. Sequence of communities developing in a newly exposed habitat devoid of life.

Rain shadow. Dry area on the leeward side of a mountain range.

Recruitment. Addition of new individuals to a population by reproduction.

Replacement series diagram. A diagram showing the outcome of competition between two species in experiments in which the initial ratio of the two species was varied.

Resource. A substance or object required by an organism for normal maintenance, growth, and reproduction. If the resource is scarce relative to demand, it is referred to as a limiting resource. Nonrenewable resources (such as space) occur in fixed amounts and can be fully utilized; renewable resources (such as food) are produced at a rate that may be partly determined by their utilization.

Respiration. Use of oxygen to break down organic compounds metabolically for the purpose of releasing chemical energy.

Search image. A behavioral selection mechanism that enables predators to increase searching efficiency for prey that are abundant and worth capturing.

Secondary succession. Progression of communities in habitats where the climax community has been disturbed or removed entirely.

Sere. A series of stages of community change in a particular area leading towards a stable state. See also Succession.

Soil.   The solid substrate of terrestrial communities resulting from the interaction of weather and biological activities with the underlying geological formation.

Soil horizon.   A distinctive zone of soil formed at a characteristic depth by weathering and organic contributions to the soil.

Species.   A group of actually or potentially interbreeding populations that are reproductively isolated from all other kinds of organisms.

Specific heat.   Amount of energy that must be added or removed to change the temperature of a substance by a specific amount. By definition, 1 calorie of energy is required to raise the temperature of 1 gram of water by 1 degree Celsius.

Stability.   Inherent capacity of any system to resist change.

Subclimax.   A stage of succession along a sere prevented from progressing to the climatic climax by fire, soil deficiencies, grazing, and similar factors.

Succession.   Replacement of populations in a habitat through a regular progression to a stable state.

Survivorship.   Proportion of newborn individuals that are alive at a given age.

Synecology.   The relationship of organisms and populations to biotic factors in the environment.

Taxon cycle.   Cycle of expansion and contraction of the geographical range and population density of a species or higher taxonomic category.

Thermocline.   The zone of water depth within which temperature changes rapidly between the upper warm water layer (epilimnion) and lower cold water layer (hypolimnion).

Transit time.   Average time that a substance or quantum of energy remains in the biological realm; ratio of biomass to productivity.

Transpiration.   Evaporation of water from leaves and other plant parts.

Transpiration efficiency.   The ratio of net primary production to transpiration of water by a plant, usually expressed by grams/kilogram of water.

Trophic level.   Position in the food chain determined by the number of energy-transfer steps to that level.

Trophic structure.   Organization of the community based on feeding relationships of populations.

Upwelling.   Vertical movement of water, usually near coasts and driven by onshore winds, that brings nutrients from the depths of the ocean to surface layers.

Wilting capacity. The minimum water content of the soil at which plants can obtain water.

Xerarch succession. Progression of terrestrial plant communities developing in habitats with well-drained soil.

Xeric. Referring to habitats in which plant production is limited by availability of water.

Zooplankton. See Plankton.

Selected Readings and
Text References

Chapter 1. Introduction

Colinvaux, P. A. *Introduction to Ecology*. Wiley, New York (1973).

Collier, B. D., G. W. Cox, A. W. Johnson and P. C. Miller. *Dynamic Ecology*. Prentice-Hall, Englewood Cliffs, New Jersey (1973).

Kormondy, E. J. *Concepts of Ecology*. Prentice-Hall, Englewood Cliffs, New Jersey (1969).

Krebs, C. J. *Ecology: The Experimental Analysis of Distribution and Abundance*. Harper & Row, New York (1972).

McNaughton, S. J. and L. L. Wolf. *General Ecology*. Holt, Rinehart, and Winston, New York (1973).

Odum, E. P. 1971. *Fundamentals of Ecology* (3rd ed.). Saunders, Philadelphia (1971).

Oosting, H. J. *The Study of Plant Communities* (2nd ed.). W. H. Freeman, San Francisco (1956).

Ricklefs, R. E. *Ecology*. Chiron, Portland, Oregon (1973).

Smith, R. L. *Ecology and Field Biology* (2nd ed.). Harper & Row, New York (1974).

Van Dyne, G. M. (ed.). *The Ecosystem Concept in Natural Resource Management*. Academic Press, New York (1969).

Warren, C. E. *Biology and Water Pollution Control*. Saunders, Philadelphia (1971).

Watt, K. E. F. *Ecology and Resource Management*. McGraw-Hill, New York (1968).

Whittaker, R. H. *Communities and Ecosystems* (2nd ed.). Macmillan, New York (1975).

Young, G. L. Human Ecology as an Interdisciplinary Concept: A Critical Inquiry. *Advances in Ecological Research* 8: 1–105 (1974).

Chapter 2. Life and the Physical Environment

Billings, W. D. Physiological Ecology. *Annual Review of Plant Physiology* 8: 375–392 (1957).

Hadley, N. F. Desert species and adaptation. *American Scientist* 60: 338–347 (1972).

Hutchinson, G. E. The biosphere. *Scientific American* 223: 44–58 (1970).

Redfield, A. C. The biological control of chemical factors in the environment. *American Scientist* 46: 205–221 (1958).

Smith, N. G. The advantage of being parasitized. *Nature* 219: 690–694 (1968).

Tansley, A. G. The use and abuse of vegetational concepts and terms. *Ecology* 16: 284–307 (1935).

Tevis, L., Jr. and I. M. Newell. Studies on the biology and seasonal cycle of the giant red velvet mite, *Dinothrombium pandorae* (Acari, Thrombidiidae). *Ecology* 43: 497–505 (1962).

Chapter 3.   Aquatic and Terrestrial Environments

Berkner, L. V. and L. C. Marshall. History of major atmospheric components. *Proceedings of the National Academy of Sciences* 53: 1215–1226 (1965).

Deevey, E. S. A re-examination of Thoreau's "Walden." *Quarterly Review of Biology* 17: 1–11 (1942).

Gates, D. M. The energy environment in which we live. *American Scientist* 51: 327–348 (1963).

Gates, D. M. Energy, plants and ecology. *Ecology* 46: 1–13 (1965).

Isaacs, J. D. The nature of oceanic life. *Scientific American* 221: 146–162 (1969).

Macan, T. T. *Freshwater Ecology*. Longmans, London (1963).

Moore, H. B. *Marine Ecology*. Wiley, New York (1958).

Ruttner, F. *Fundamentals of Limnology* (3rd ed.). University of Toronto Press, Toronto (1963).

Schmidt-Nielson, K. *How Animals Work*. Cambridge University Press, London (1972).

Schmidt-Nielson, K. and B. Schmidt-Nielson. The desert rat. *Scientific American* 189: 73–78 (1953).

Tait, R. V. *Elements of Marine Ecology*. Plenum, New York (1968).

Weisskopf, V. F. How light interacts with matter. *Scientific American* 219: 60–71 (1968).

Chapter 4.   Soil Formation

Brady, N. C. *Nature and Properties of Soils* (8th ed.). Macmillan, New York (1974).

Bunting, B. T. *The Geography of Soil* (rev. ed.). Aldine, Chicago (1967).

Crocker, R. L. Soil genesis and the pedogenic factors. *Quarterly Review of Biology* 27: 139–168 (1952).

Crocker, R. L. and J. Major. Soil development in relation to vegetation and surface age at Glacier Bay, Alaska. *Journal of Ecology* 43: 427–448 (1955).

Eyre, S. R. *Vegetation and Soils* (2nd ed.). Aldine, Chicago (1968).

Olson, J. S. Rates of succession and soil changes on southern Lake Michigan sand dunes. *Botanical Gazette* 119: 125–170 (1958).

Chapter 5.   Variation in the Environment

Geiger, R. *The Climate Near the Ground.* Harvard University Press, Cambridge, Massachusetts (1966).

Flohn, H. *Climate and Weather.* World University Library, McGraw-Hill, New York (1969).

Harper, J. L., J. T. Williams, and G. R. Sagar. The behavior of seeds in soil. *Journal of Ecology* 51: 273–286 (1965).

Hutchinson, G. E. *A Treatise on Limnology, Vol. 1: Geography, Physics, and Chemistry.* Wiley, New York (1957).

Lowry, W. P. *Weather and Life.* Academic Press, New York (1969).

Merriam, C. H. Laws of temperature control of the geographic distribution of terrestrial animals and plants. *National Geographic Magazine* 6: 229–238 (1894).

Thornthwaite, C. W. An approach to a rational classification of climate. *Geographical Review* 38: 55–94 (1948).

Chapter 6.   The Diversity of Biological Communities

Bourlière, F. and M. Hadley. The ecology of tropical savannas. *Annual Review of Ecology and Systematics* 1: 125–152 (1970).

Braun, E. L. *Deciduous Forests of Eastern North America.* McGraw-Hill — Blakiston, New York (1950).

Carpenter, J. R. The biome. *American Midland Naturalist.* 21: 75–91 (1939).

Dansereau, P. *Biogeography — An Ecological Perspective.* Ronald Press, New York (1957).

Eltringham, S. K. *Life in Mud and Sand.* Crane Russak, New York (1971).

Friedlander, C. P. *Heathland Ecology.* Harvard University Press, Cambridge, Massachusetts (1961).

Hardy, A. *The Open Sea: Its Natural History.* Houghton Mifflin, Boston (1971).

Holdridge, L. *Life Zone Ecology.* Tropical Science Center, San José, Costa Rica (1967).

Jaeger, E. C. *The North American Deserts.* Stanford University Press, Stanford, California (1957).

Odum, E. P. *Fundamentals of Ecology* (3rd ed.). Saunders, Philadelphia (1971).

Pearsall, W. H. *Mountains and Moorlands.* Collins, London (1960).

Popham, E. J. *Life in Fresh Water.* Harvard University Press, Cambridge, Massachusetts (1961).

Raunkiaer, C. *The Life Forms of Plants and Statistical Plant Geography.* Clarendon Press, Oxford (1934).

Richards, P. W. *The Tropical Rainforest.* Cambridge University Press, London (1952).

Schimwell, D. W. *Description and Classification of Vegetation.* University of Washington Press, Seattle (1971).

Shelford, V. E. *The Ecology of North America.* University of Illinois Press, Urbana (1963).

Stephenson, T. A. and A. Stephenson. *Life Between Tidemarks on Rocky Shores.* W. H. Freeman, San Francisco (1972).

Teal, J. and M. Teal. *Life and Death of a Salt Marsh*. Little Brown, Boston (1969).

Weaver, J. E. *Grasslands of the Great Plains*. Johnsen, Lincoln, Nebraska (1956).

Whittaker, R. H. *Communities and Ecosystems* (2nd ed.). Macmillan, New York (1975).

Chapter 7. Primary Production

Bray, J. R. and E. Gorham. Litter production in forests of the world. *Advances in Ecological Research* 2: 101–157 (1964).

Goldman, C. R. Aquatic primary production. *American Zoologist* 8: 31–42 (1968).

Jordan, C. F. A world pattern in plant energetics. *American Scientist* 59: 425–433 (1971).

Mann, K. H. Seaweeds: their productivity and strategy for growth. *Science* 182: 975–981 (1974).

McLaren, I. A. Primary production and nutrients in Ogac Lake, a landlocked fiord on Baffin Island. *Journal of the Fisheries Research Board of Canada* 26: 1562–1576 (1969).

Norman, A. G. Soil-plant relationships and plant nutrition. *American Journal of Botany* 44: 67–73 (1957).

Odum, H. T. Primary production in flowing waters. *Limnology and Oceanography* 1: 102–117 (1956).

Ovington, J. D., D. Heitkamp, and D. B. Lawrence. Plant biomass and productivity of prairie, savanna, oakwood, and maize field ecosystems in central Minnesota. *Ecology* 44: 52–63 (1963).

Transeau, E. N. The accumulation of energy by plants. *Ohio Journal of Science* 26: 1–10 (1926).

Whittaker, R. H. and G. Likens. The primary production of the biosphere. *Human Ecology* 1: 299–369 (1973).

Wilde, S. A. Mycorrhizae and tree nutrition. *BioScience* 18: 482–484 (1968).

Witter, S. H. Maximum production capacity of food crops. *BioScience* 24: 216 (1974).

Woodwell, G. M. The energy cycle of the biosphere. *Scientific American* 223: 64–74 (1970).

Chapter 8. Energy Flow in the Community

Englemann, M. D. Energetics, terrestrial field studies, and animal productivity. *Advances in Ecological Research* 3: 73–115 (1966).

Harley, J. L. Fungi in ecosystems. *Journal of Animal Ecology* 41: 1–16 (1972).

Kozlovsky, D. G. A critical evaluation of the trophic level concept. I. Ecological efficiencies. *Ecology* 49: 48–60 (1968).

Lindeman, R. L. The trophic-dynamic aspect of ecology. *Ecology* 23: 399–418 (1942).

Mann, K. H. The dynamics of aquatic ecosystems. *Advances in Ecological Research* 6: 1–81 (1969).

Menhinick, E. F. Structure, stability and energy flow in plants and arthropods in Sericea Lespedeza stand. *Ecological Monographs* 37: 255–272 (1967).

Odum, E. P. Relationships between structure and function in the ecosystem. *Japanese Journal of Ecology* 12: 108–118 (1962).

Odum, E. P. Energy flow in ecosystems: A historical review. *American Zoologist* 8: 11–18 (1968).

Odum, H. T. Trophic structure and productivity of Silver Springs, Florida. *Ecological Monographs* 27: 55–112 (1957).

Phillipson, J. *Ecological Energetics*. Edward Arnold, London (1966).

Ryther, J. H. Photosynthesis and fish production in the sea. *Science* 166: 72–76 (1969).

Stark, N. Nutrient cycling pathways and litter fungi. *BioScience* 22: 355–360 (1972).

Teal, J. M. Energy flow in the salt marsh ecosystem of Georgia. *Ecology* 43: 614–624 (1962).

Wiegert, R. G. Energy dynamics of the grasshopper populations in old-field and alfalfa field ecosystems. *Oikos* 16: 161–176 (1965).

Wiegert, R. G. and D. F. Owen. Trophic structure, available resources and population density in terrestrial vs. aquatic ecosystems. *Journal of Theoretical Biology* 30: 69–81 (1971).

Chapter 9.   Nutrient Cycling

Alexander, M. *Microbial Ecology*. Wiley, New York (1971).

Beeton, A. M. Eutrophication of the St. Lawrence Great Lakes. *Limnology and Oceanography* 10: 240–254 (1965).

Bormann, F. H. and G. E. Likens. Nutrient cycling. *Science* 155: 424–429 (1967).

Brady, N. C. *Nature and Properties of Soils*. (8th ed.). Macmillan, New York (1974).

Cloud, P. and A. Givor. The oxygen cycle. *Scientific American* 223: 111–123 (1970).

Delwiche, C. C. The nitrogen cycle. *Scientific American* 223: 137–146 (1970).

Likens, G. E., F. H. Bormann, N. M. Johnson, and R. S. Pierce. The calcium, magnesium, potassium, and sodium budgets for a small forested ecosystem. *Ecology* 48: 772–785 (1967).

Pomeroy, L. R. The strategy of mineral cycling. *Annual Review of Ecology and Systematics* 1: 171–190 (1970).

Schindler, D. W. Eutrophication and recovery in experimental lakes: implications for lake management. *Science* 184: 897–899 (1974).

Witkamp, M. Decomposition of leaf litter in relation to environment, microflora, and microbial respiration. *Ecology* 47: 194–201 (1966).

Chapter 10.   Environment and the Distribution of Organisms

Andrewartha, H. G. and L. C. Birch. *The Distribution and Abundance of Animals*. University of Chicago Press, Chicago (1954).

Billings, W. D. The environmental complex in relation to plant growth and distribution. *Quarterly Review of Biology* 27: 251–265 (1952).

Brett, J. R. Some principles in the thermal requirements of fishes. *Quarterly Review of Biology* 31: 75–87 (1956).

Cain, S. A. Life-forms and phytoclimate. *Botanical Reviews* 16: 1–32 (1950).

Clausen, J., D. D. Keck, and W. M. Hiesey. Experimental studies on the nature of species. III. Environmental responses of climatic races of *Achillea*. *Carnegie Institution of Washington Publications* 581: 1–129 (1948).

Good, R. *The Geography of the Flowering Plants*. Longmans, London (1964).

Harrison, A. T., E. Small, and H. A. Mooney. Drought relationships and distribution of two Mediterranean-climate California plant communities. *Ecology* 52: 869–875 (1971).

Hiesey, W. M. and H. W. Milner. Physiology of ecological races and species. *Annual Review of Plant Physiology* 16: 203–216 (1965).

Niering, W. A., R. H. Whittaker, and C. H. Lowe. The saguaro: a population in relation to environment. *Science* 142: 15–23 (1963).

Raunkaier, C. *The Life Form of Plants and Statistical Plant Geography*. Clarendon Press, Oxford (1934).

Turesson, G. The genotypic response of the plant species to the habitat. *Hereditas* 3: 211–350 (1922).

Waring, R. H. and J. Major. Some vegetation of the California coastal region in relation to gradients of moisture, nutrients, light, and temperature. *Ecological Monographs* 34: 167–215 (1964).

Wecker, S. C. Habitat selection. *Scientific American* 211: 109–116 (1964).

Chapter 11.   Homeostatic Responses of Organisms

Daubenmire, R. *Plants and Environment* (2nd ed.). Wiley, New York (1959).

Hochachka, P. W., and G. N. Somero. *Strategies of Biochemical Adaptation*. Saunders, Philadelphia (1973).

Langley, T. L. *Homeostasis*. Reinhold, New York (1965).

Irving, L. Adaptations to cold. *Scientific American* 214: 94–101 (1966).

Ricklefs, R. E., and F. R. Hainsworth. Temperature dependent behavior of the cactus wren. *Ecology* 49: 227–233 (1968).

Schmidt-Nielson, K. *How Animals Work*. Cambridge University Press, London (1972).

Schmidt-Nielson, K. *Animal Physiology. Adaptation and Environment*. Cambridge University Press, London (1975).

Chapter 12.   Evolutionary Responses

Antonovics, J. The effects of a heterogeneous environment on the genetics of natural populations. *American Scientist* 59: 593–599 (1971).

Brower, L. P. Ecological chemistry. *Scientific American* 220: 22–29 (1969).

Cott, H. B. *Adaptive Coloration in Animals*. Oxford University Press, London (1940).

Fenner, F. Evolution in action: myxomatosis in the Australian wild rabbit. Pp. 463–471 in A. Kramer (ed.). *Topics in the Study of Life. The Bio Source Book*. Harper & Row, New York (1971).

Janzen, D. H. Seed-eaters versus seed size, number, toxicity and dispersal. *Evolution* 23: 1–27 (1969).

Kettlewell, H. B. D. Darwin's missing evidence. *Scientific American* 200: 48–53 (1959).

Lack, D. Evolutionary ecology. *Journal of Animal Ecology* 34: 223–231 (1965).

Maynard-Smith, J. *The Theory of Evolution.* Penguin, Baltimore (1958).

Orians, G. H. Natural selection and ecological theory. *American Naturalist* 96: 257–263 (1962).

Robinson, M. H. Defenses against visually hunting predators. *Evolutionary Biology* 3: 225–259 (1969).

Whittaker, R. H., and P. O. Feeny. Allelochemics: chemical interactions between species. *Science* 171: 757–770 (1971).

Chapter 13.   Population Growth and Regulation

Calhoun, J. B. Population density and social pathology. *Scientific American* 206: 1399–1408 (1962).

Chitty, D. Population processes in the vole and their relevance to general theory. *Canadian Journal of Zoology.* 38: 99–113 (1960).

Christian, J. J., and D. E. Davis. Endocrines, behavior and population. *Science* 146: 1550–1560 (1964).

Davidson, J., and H. G. Andrewartha. The influence of rainfall, evaporation, and the atmospheric temperature on fluctuations in the size of a natural population of *Thrips imaginis* (Thysanoptera). *Journal of Animal Ecology* 17: 193–199 (1948).

Deevey, E. S., Jr. Life tables for natural populations of animals. *Quarterly Review of Biology* 22: 283–314 (1947).

Ehrlich, P. R. and L. C. Birch. The "balance of nature" and "population control." *American Naturalist* 101: 97–107 (1967).

Elton, C. *Voles, Mice and Lemmings. Problems in Population Dynamics.* Clarendon Press, Oxford (1942).

Frank, P. W., C. D. Boll, and R. W. Kelly. Vital statistics of laboratory cultures of *Daphnia pulex* De Geer as related to density. *Physiological Zoology* 30: 287–305 (1957).

Harcourt, D. G., and E. J. Leroux. Population regulation in insects and man. *American Scientist* 55: 400–415 (1967).

Krebs, C. J., and J. Myers. Population cycles in small mammals. *Advances in Ecological Research* 8: 267–399 (1974).

Lack, D. *The Natural Regulation of Animal Numbers.* Oxford University Press, London (1954).

Morris, R. F. The dynamics of epidemic spruce budworm populations. *Memoirs of the Entomological Society of Canada.* 31: 1–332 (1963).

Murdoch, W. W. Population regulation and population inertia. *Ecology* 51: 497–502 (1970).

Neilson, M. M., and R. F. Morris. The regulation of European spruce sawfly numbers in the maritime provinces of Canada from 1937 to 1963. *Canadian Entomologist* 96: 773–784 (1964).

Nicholson, A. J. An outline of the dynamics of animal populations. *Australian Journal of Zoology* 2: 9–65 (1954).

Nicholson, A. J. The self-adjustment of populations to change. *Cold Spring Harbor Symposium on Quantitative Biology* 22: 153–173 (1958).

Solomon, M. E. Analysis of processes involved in the natural control of insects. *Advances in Ecological Research* 2: 1–58 (1964).

Southwick, C. H. The population dynamics of confined house mice supplied with unlimited food. *Ecology* 36: 212–225 (1955).

Taber, R. D., and R. F. Dasmann. The dynamics of three natural populations of the deer *Odocoileus hemionus columbianus*. *Ecology* 38: 233–246 (1957).

Wynne-Edwards, V. C. *Animal Dispersion in Relation to Social Behavior*. Hafner, New York (1962).

Wynne-Edwards, V. C. Population control in animals. *Scientific American* 211: 68–74 (1964).

Chapter 14. Competition

Ayala, F. J. Competition between species. *American Scientist* 60: 348–357 (1972).

Birch, L. C. The meanings of competition. *American Naturalist* 91: 5–18 (1957).

Brown, W. L., Jr., and E. O. Wilson. Character displacement. *Systematic Zoology* 5: 49–64 (1956).

Connell, J. H. The influence of interspecific competition and other factors on the distribution of the barnacle *Chthamalus stellatus*. *Ecology* 42: 710–723 (1961).

DeBach, P. The competitive displacement and coexistence principles. *Annual Review of Entomology* 11: 183–212 (1966).

DeBach, P., and R. A. Sundby. Competitive displacement between ecological homologues. *Hilgardia* 34: 105–166 (1963).

Gause, G. F. *The Struggle for Existence*. Williams and Wilkins, Baltimore (1934).

Hardin, G. The competitive exclusion principle. *Science* 131: 1292–1297 (1960).

Harper, J. L. The evolution and ecology of closely related species living in the same area. *Evolution* 15: 209–227 (1961).

Harper, J. L. Approaches to the study of plant competition. *Symposium of the Society of Experimental Biologists* 15: 1–39 (1961).

Harper, J. L. A Darwinian approach to plant ecology. *Journal of Ecology* 55: 247–270 (1967).

MacArthur, R. H., H. Recher, and M. Cody. On the relation between habitat selection and species diversity. *American Naturalist* 100: 319–332 (1966).

Marshall, D. R., and S. K. Jain. Interference in pure and mixed populations of *Avena fatua* and *A. barbata*. *Journal of Ecology* 57: 251–270 (1969).

Miller, R. S. Pattern and process in competition. *Advances in Ecological Research* 4: 1–74 (1967).

Muller, C. H. The role of chemical inhibition (allelopathy) in vegetational composition. *Bulletin of the Torrey Botanical Club* 93: 332–351 (1966).

Paine, R. T. Trophic relationships of eight sympatric predatory gastropods. *Ecology* 44: 63–73 (1963).

Park, T. Beetles, competition, and populations. *Science* 138: 1369–1375 (1962).

Schultz, A. M., J. L. Launchbaugh, and H. H. Biswell. Relationship between grass diversity and brush seedling survival. *Ecology* 36: 226–238 (1955).

Chapter 15. Predation

Batzli, G. O., and F. A. Pitelka. Influence of meadow mouse populations on California grassland. *Ecology* 51: 1027–1039 (1970).

DeBach, P. (ed.). *Biological Control of Insect Pests and Weeds.* Chapman & Hall, London (1964).

Dodd, A. P. The biological control of prickly pear in Australia. *Monographiae Biologiae* 8: 567–577 (1959).

Errington, P. L. The phenomenon of predation. *American Scientist* 51: 180–192 (1963).

Gross, J. E. Optimum yield in deer and elk populations. *Transactions of the North American Wildlife Conference* 34: 372–386 (1969).

Harper, J. L. The role of predation in vegetational diversity. *Brookhaven Symposia on Biology* 22: 48–62 (1969).

Holling, C. S. The components of predation as revealed by a study of small mammal predation of the European pine sawfly. *Canadian Entomologist* 91: 293–320 (1959).

Holling, C. S. The functional response of predators to prey density and its role in mimicry and population regulation. *Memoirs of the Entomological Society of Canada* 48: 1–85 (1966).

Huffaker, C. B. Experimental studies on predation: dispersal factors and predator-prey oscillations. *Hilgardia* 27: 343–383 (1958).

Huffaker, C. B. Life against life — nature's pest control scheme. *Environmental Research* 3: 162–175 (1970).

Le Cren, E. D. and M. W. Holdgate (eds.). *Exploitation in Natural Animal Populations.* Wiley, New York (1962).

Paine, R. T., and R. Vadas. The effects of grazing by sea urchins, *Strongylocentrotus* spp., on benthic algal populations. *Limnology and Oceanography* 14: 710–719 (1969).

Pimentel, D. Population regulation and genetic feedback. *Science* 159: 1432–1437 (1968).

Ricker, W. E. Stock and recruitment. *Journal of the Fisheries Research Board of Canada* 11: 559–623 (1954).

Slobodkin, L. B. How to be a predator. *American Zoologist* 8: 43–51 (1968).

Utida, S. Population fluctuation, an experimental and theoretical approach. *Cold Spring Harbor Symposia on Quantitative Biology* 22: 139–151 (1957).

Chapter 16. Extinction

MacArthur, R. H., and E. O. Wilson. An equilibrium theory of insular zoogeography. *Evolution* 17: 373–387 (1963).

MacArthur, R. H., and E. O. Wilson. *The Theory of Island Biogeography.* Princeton University Press, Princeton, New Jersey (1967).

Myers, G. S. The endemic fish fauna of Lake Lanao, and the evolution of higher taxonomic categories. *Evolution* 14: 323–333 (1960).

Ricklefs, R. E., and G. W. Cox. Taxon cycles in the West Indian Avifauna. *American Naturalist* 106: 195–219 (1972).

Simpson, G. G. History of the fauna of Latin America. *American Scientist* 38: 361–389 (1950).

Simpson, G. G. *The Major Features of Evolution.* Columbia University Press, New York (1953).

Stehli, F. G., R. G. Douglas, and N. D. Newell. Generation and maintenance of gradients in taxonomic diversity. *Science* 164: 947–949 (1969).

Wilson, E. O. The nature of the taxon cycle in the Melanesian ant fauna. *American Naturalist* 95: 169–193 (1961).

Chapter 17. The Community as a Unit of Ecology

Borchert, J. R. The climate of the central North American grassland. *Annals of the Association of American Geographers* 40: 1–39 (1950).

Darlington, P. J., Jr. *Zoogeography: The Geographical Distribution of Animals.* Wiley, New York (1957).

Daubenmire, R. Vegetation: identification of typal communities. *Science* 151: 291–298 (1966).

Hairston, N. G., F. E. Smith, and L. B. Slobodkin. Community structure, population control, and competition. *American Naturalist* 94: 421–425 (1960).

Janzen, D. H. Coevolution of mutualism between ants and acacias in Central America. *Evolution* 20: 249–275 (1966).

Janzen, D. H. Herbivores and the number of tree species in tropical forests. *American Naturalist* 104: 501–528 (1970).

MacArthur, R. H. Patterns of species diversity. *Biological Reviews* 40: 510–533 (1965).

MacArthur, R. H. Patterns of communities in the tropics. *Biological Journal of the Linnean Society* 1: 19–30 (1969).

McIntosh, R. P. The continuum concept of vegetation. *Botanical Review* 33: 130–187 (1967).

McMillan, C. Ecotypes and ecosystem function. *BioScience* 19: 131–134 (1969).

Murdoch, W. W. "Community structure, population control, and competition" — A critique. *American Naturalist* 100: 219–226 (1966).

Paine, R. T. Food web complexity and species diversity. *American Naturalist* 100: 65–75 (1966).

Pianka, E. R. Latitudinal gradients in species diversity: a review of concepts. *American Naturalist* 100: 33–46 (1966).

Watt, A. S. The community and the individual. *Journal of Ecology* 52 (Supplement): 203–211 (1964).

Whittaker, R. H. Dominance and diversity in land plant communities. *Science* 147: 250–260 (1965).

Whittaker, R. H. Gradient analysis of vegetation. *Biological Reviews* 42: 207–264 (1967).

Zaret, T., and R. T. Paine. Species introduction in a tropical lake. *Science* 182: 449–455 (1973).

Chapter 18. Community Development

Clements, F. E. Plant succession: analysis of the development of vegetation. *Carnegie Institute of Washington Publications.* 242: 1–512 (1916).

Clements, F. E. Nature and structure of the climax. *Journal of Ecology* 24: 252–284 (1936).

Cooper, C. F. The ecology of fire. *Scientific American* 204: 150–160 (1961).

Curtis, J. T., and R. P. McIntosh. An upland forest continuum in the prairie-forest border region of Wisconsin. *Ecology* 32: 476–496 (1951).

Keever, C. Causes of succession on old fields of the Piedmont, North Carolina. *Ecological Monographs* 20: 230–250 (1950).

Knapp, R. (ed.). *Vegetation Dynamics*. Junk, The Hague (1974).

Olson, J. S. Rates of succession and soil changes on southern Lake Michigan sand dunes. *Botanical Gazette* 119: 125–170 (1958).

Phillips, J. Succession, development, the climax, and the complex organism: an analysis of concepts. *Journal of Ecology* 22: 554–571; 23: 210–246, 488–508 (1934–1935).

Watt, A. S. Pattern and process in the plant community. *Journal of Ecology* 35: 1–22 (1947).

Whittaker, R. H. A consideration of climax theory: the climax as a population and pattern. *Ecological Monographs* 23: 41–78 (1953).

Woodwell, G. M. Success, succession, and Adam Smith. *BioScience* 24: 81–87 (1974).

Chapter 19.   Community Stability

Daubenmire, R. Ecology of fire in grasslands. *Advances in Ecological Research* 5: 209–266 (1968).

Frank, P. W. Life histories and community stability. *Ecology* 49: 355–357 (1968).

Fritts, H. C. Growth-rings of trees: their correlation with climate. *Science* 154: 973–979 (1966).

Hairston, N. G., et al. The relationship between species diversity and stability: an experimental approach with protozoa and bacteria. *Ecology* 49: 1091–1101 (1968).

Kozlowski, T. T., and C. E. Ahlgren (eds.). *Fire and Ecosystems*. Academic Press, New York (1974).

Margalef, R. Diversity and stability: a practical proposal and a model of interdependence. *Brookhaven Symposia on Biology* 22: 25–37 (1969).

Murdoch, W. W. Switching in general predators: experiments on predator specificity and stability of prey populations. *Ecological Monographs* 39: 335–354 (1969).

Patrick, R. The structures of diatom communities under varying ecological conditions. *Annals of the New York Academy of Sciences* 108: 353–358 (1963).

Pimentel, D. Species diversity and insect population outbreaks. *Annals of the Entomological Society of America* 54: 76–86 (1961).

Preston, F. W. Diversity and stability in the biological world. *Brookhaven Symposia on Biology* 22: 1–12 (1969).

Watt, K. E. F. Comments on fluctuations of animal populations and measures of community stability. *Canadian Entomologist* 96: 1434–1442 (1964).

Watt, K. E. F. Community stability and the strategy of biological control. *Canadian Entomologist* 97: 887–895 (1965).

Woodwell, G. M., and H. H. Smith (eds.). Diversity and stability in Ecological Systems. *Brookhaven Symposia on Biology* 22 (1969).

Illustration Credits, Acknowledgments, and References*

Drawings and Photographs

Figure 2-1   Photographs courtesy of the U.S. Soil Conservation Service.
Figure 2-3   Photograph courtesy of Philip L. Boyd, Deep Canyon Research Center.

Figure 3-1   After a drawing of *Calocalanus pavo* in R. S. Wimpenny, *The Plankton of the Sea*. Faber and Faber, London (1966).
Figure 3-2   Photograph courtesy of the U.S. Bureau of Commercial Fisheries.
Figure 3-3   Scanning electron micrograph courtesy of M. V. Parthasarathy. Appreciation to John Neill and Roy Tedoff, of W. W. Norton & Company, for assistance.
Figure 3-4   Photograph courtesy of the U.S. Fish and Wildlife Service.

Figures 4-1,2   Photographs courtesy of the U.S. Soil Conservation Service.
Figure 4-3   After Table 3 in B. T. Bunting, *The Geography of Soil* (rev. ed.). Aldine, Chicago (1967).
Figure 4-4   After Table 131 in E. W. Russel, *Soil Conditions and Plant Growth* (9th ed.). Wiley, New York (1961).
Figure 4-5   After Figure 7 in S. R. Eyre, *Vegetation and Soils* (2nd ed.). Aldine, Chicago (1968).
Figure 4-6   After N. C. Brady, *Nature and Properties of Soils* (8th ed.). Macmillan, New York (1974).
Figures 4-7,8   Photographs courtesy of the U.S. Soil Conservation Service.

Figure 5-3   After E. B. Espenshade, Jr. (ed.), *Goode's World Atlas* (13th ed.). Rand-McNally, Chicago (1971).
Figure 5-4   After E. P. Odum, *Fundamentals of Ecology* (3rd ed.). Saunders, Philadelphia (1971) and A. C. Duxbury, *The Earth and Its Oceans*. Addison-Wesley, Reading, Massachusetts (1971).
Figure 5-7   Photograph courtesy of the U.S. Forest Service.

*Material not cited has been adapted from original figures and tables in R. E. Ricklefs, *Ecology*. Chiron, Portland, Oregon (1973).

Figure 5-8   Modified from E. J. Kormondy, *Concepts of Ecology.* Prentice-Hall, Englewood Cliffs, New Jersey (1969).

Figure 5-10   After C. W. Thornthwaite. *Geogr. Rev.* 38: 55–94 (1948).

Figure 5-13   Photographs courtesy of the U.S. Soil Conservation Service, U.S. Forest Service, W. John Smith, and R. H. Whittaker, from R. H. Whittaker and W. A. Niering. *Ecology* 46: 429–452 (1965).

Figure 5-15   Photograph courtesy of the U.S. Forest Service.

Figure 5-17   After N. F. Hadley. *Ecology* 51: 434–444 1970).

Figure 5-18   After J. L. Harper, J. T. Williams, and G. R. Sagar. *J. Ecol.* 53: 273–286 (1965).

Figure 6-1   After H. A. Fowells, *Silvics of Forest Trees of the United States. Agric. Handbook* No. 271, U.S. Department of Agriculture (1965).

Figure 6-2   After P. Dansereau. *Biogeography — An Ecological Perspective.* Ronald, New York (1957).

Figure 6-3   After C. Raunkiaer, *Plant Life Forms.* Clarendon, Oxford (1937).

Figure 6-4   After data in P. W. Richards, *The Tropical Rainforest.* Cambridge, London (1952); P. Dansereau, *Biogeography — An Ecological Perspective.* Ronald, New York (1957); R. Daubenmire, *Plant Communities.* Harper & Row, New York (1968).

Figure 6-5   After L. Holdridge, *Life Zone Ecology.* Tropical Science Center, San Jose, California (1967).

Figure 6-7   After R. H. Whittaker, *Communities and Ecosystems.* Macmillan, New York (1970).

Page 84   Photograph courtesy of W. J. Smith (top).

Page 85   Photographs courtesy of W. J. Smith (top) and the U.S. National Park Service.

Page 86   Photograph courtesy of W. J. Smith (bottom).

Page 87   Photograph courtesy of W. J. Smith.

Page 90   Photographs courtesy of the U.S. Fish and Wildlife Service (top) and the U.S. National Park Service.

Page 92   Photograph courtesy of the U.S. Soil Conservation Service.

Page 93   Photographs courtesy of the U.S. Forest Service.

Page 94   Photographs courtesy of the U.S. Forest Service.

Page 95   Photographs courtesy of James Lane, Archbold Biological Station (top), and the U.S. Forest Service.

Page 96   Photographs courtesy of the U.S. National Park Service (top), and the U.S. Department of Agriculture.

Page 97   Photographs courtesy of the U.S. Soil Conservation Service.

Page 99   Photograph courtesy of the U.S. Department of Agriculture (bottom).

Page 100   Photographs courtesy of the U.S. Forest Service.

Page 102   Photographs courtesy of the U.S. Forest Service.

Page 103   Photograph courtesy of the U.S. Forest Service.

Page 104   Photograph courtesy of the U.S. Soil Conservation Service (top).

Page 106   Photographs courtesy of F. B. Bowles.

Page 107   Photographs courtesy of the U.S. National Park Service (top), and F. B. Bowles.

Page 108   Photographs courtesy of the U.S. National Marine Fisheries Service (top left), P. J. Tzimoulis, American Littoral Society (top right), and W. J. Smith.

Page 109   Photographs courtesy of James Porter, From J. W. Porter. *Science* 186: 543–545 (1974). Copyright 1974 by the American Association for the Advancement of Science.

Figure 7-1   After Figure 3 in R. H. Whittaker and G. M. Woodwell. *J. Ecol.* 56: 1-25 (1968).

Figure 7-2   After I. A. McLaren. *J. Fish. Res. Bd. Can.* 26: 1562–1576 (1969).

Figure 7-3   After J. H. Ryther. *Limnol. Oceanogr.* 1: 61–70 (1956) and P. J. Kramer and J. P. Decker. *Plant Physiol.* 19: 350–358 (1944).

Figure 7-4   A. After R. Emerson and C. M. Lewis. *J. Gen. Physiol.* 25: 579–595 (1942). B. After R. A. Moss and W. E. Loomis. *Plant Physiol.* 27: 370–391 (1952). C. After C. A. Federer and C. B. Tanner. *Ecology* 47: 555–560 (1966). D. After F. T. Haxo and L. R. Blinks. *J. Gen. Physiol.* 33: 389–422 (1950).

Figure 7-5   After D. Gates. *Brookhaven Symp. Biol.* 22: 115–126 (1969).

Figure 7-7   Photograph courtesy of the U.S. Forest Service.

Figure 7-8   After R. H. Whittaker and G. E. Likens. *Human Ecol.* 1: 357–370 (1973).

Figure 8-6   Based on data in H. Welch, *Ecology* 49; 755–759 (1968) and data summarized by R. E. Ricklefs, *Ecology.* Chiron, Portland, Oregon (1973).

Figure 8-7   After C. A. Edward and G. W. Heath. Pp. 76–84 in J. Doeksen and J. Van Der Drift (eds.). *Soil Organisms.* North-Holland, Amsterdam (1963).

Figure 8-8   Photograph courtesy of the U.S. National Park Service.

Figure 8-9   After E. P. Odum. *Jap. J. Ecol.* 12: 108–118 (1962).

Figure 8-10   After R. C. Ball and F. F. Hooper. Pp. 217–228 in V. Schultz and A. W. Klement (eds.). *Radioecology.* Reinhold, New York (1963).

Figure 9-2   After G. Borgstrom, *Too Many.* Macmillan, New York (1969) and G. E. Hutchinson, *A Treatise on Limnology,* Vol. I. Wiley, New York (1957).

Figure 9-3   After I. Waldron and R. E. Ricklefs, *Environment and Population.* Holt, Rinehart, and Winston, New York (1973).

Figure 9-6   Photograph courtesy of the U.S. Soil Conservation Service.

Figure 9-7   Photograph courtesy of D. W. Schindler, from D. W. Schindler. *Science* 184: 897–899 (1974). Copyright 1974 by the American Association for the Advancement of Science.

Figure 9-9   Photograph courtesy of the U.S. Forest Service.

Figure 9-10   Photograph courtesy of the U.S. Forest Service.

Figure 9-11   G. E. Likens, F. H. Bormann, R. S. Pierce, and D. W. Fisher. Pp. 553–563 in *Productivity of Forest Ecosystems.* UNESCO (1971).

Figure 9-12   After data in G. E. Likens, F. H. Bormann, N. M. Johnson, and R. S. Pierce. *Ecology* 48: 772–785 (1967).

Figure 9-13   Photograph courtesy of the U.S. Forest Service.

Figures 10-1,2   After H. A. Fowells, *Silvics of Forest Trees of the United States. Agric. Handbook* No. 271, U.S. Department of Agriculture (1965).

Figure 10-3   After R. O. Erickson. *Ann. Missouri Bot. Garden* 32: 413–460 (1945).

Figures 10-4,5   After R. H. Waring and J. Major. *Ecol. Monogr.* 34: 167–215 (1964).

Figure 10-6   After E. W. Beals and J. B. Cope. *Ecology* 45: 777–792 (1964).

Figure 10-8   After F. S. Bodenheimer. *Bull. Soc. Entomol. Egypte* 1924: 149–157 (1925), and W. C. Allee, A. E. Emerson, O. Park, T. Park, and K. P. Schmidt, *Principles of Animal Ecology.* Saunders, Philadelphia (1949).

Figure 10-9   After R. F. Dasmann, *Wildlife Biology.* John Wiley, New York (1964).

Figure 10-11   Distributions after R. H. Whittaker and W. A. Niering. *Ecology* 46: 429–452 (1965).

Figure 10-12   After A. T. Harrison, E. Small, and H. A. Mooney. *Ecology* 52: 869–875 (1970).

Figure 10-13   After J. Clausen, D. D. Keck, and W. M. Hiesey. *Carnegie Inst. Wash. Publ.* 581: 1–129 (1948).

Figure 11-1   After B. Wallace and A. M. Srb, *Adaptation* (2nd ed.). Prentice-Hall, Englewood Cliffs, New Jersey (1964).

Figure 11-3   After data of H. Werntz, in C. L. Prosser and F. A. Brown, *Comparative Animal Physiology* (2nd ed.). Saunders, Philadelphia (1961).

Figure 11-5   After L. Irving. *Sci. Amer.* 214: 94–101 (1966).

Figure 11-6   Photograph courtesy of the U.S. Fish and Wildlife Service.

Figure 11-8   Data courtesy of George T. Austin.

Figure 11-9   After G. C. West. *Comp. Biochem. Physiol.* 42A: 867–876 (1972).

Figure 11-10   After F. E. J. Fry and J. S. Hart. *J. Fish. Res. Bd. Can.* 7: 169–175 (1948).

Figure 11-11   Photographs courtesy of the U.S. Forest Service.

Figure 11-12   Photograph courtesy of the U.S. Department of Agriculture.

Figure 12-1   After H. B. D. Kettlewell. *Heredity* 12: 51–72 (1958).

Figure 12-2   Photographs courtesy of, and from the experiments of, H. B. D. Kettlewell.

Figure 12-3   After J. Antonovics and A. D. Bradshaw. *Heredity* 25: 349–362 (1970) and J. Antonovics. *Amer. Scient.* 59: 593–599 (1971).

Figures 12-4,5   After F. Fenner and F. N. Ratcliffe, *Myxoamatosis.* Cambridge University Press, London (1965).

Figure 12-6   Photograph E courtesy of W. J. Smith.

Figure 13-2   Photograph courtesy of the American Museum of Natural History.

Figure 13-3   Based on data in N. Keyfitz and W. Flieger, *World Population. An Analysis of Vital Data.* University of Chicago Press, Chicago (1968).

Figure 13-4   Photograph courtesy of the American Museum of Natural History.

Figure 13-7   After P. W. Frank, C. D. Bell, and R. W. Kelly. *Physiol. Zool.* 30: 287–305 (1957).

Figure 13-8   Modified from R. Laughlin. *J. Anim. Ecol.* 34: 77–91 (1965).

Figure 13-9   Photograph courtesy of J. Ewing.

Figure 13-10   After J. Davidson. *Trans. R. Soc. South Australia* 62: 342–346 (1938).

Figure 13-11   After C. C. Davis. *Limnol. Oceanogr.* 9: 275–283 (1964).

Figure 13-12   After J. Davidson and H. G. Andrewartha. *J. Anim. Ecol.* 17: 193–199 (1948).

Figure 13-13   Photograph courtesy of the U.S. Forest Service.

Figure 13-14   After M. M. Neilson and R. F. Morris. *Canad. Entomol.* 96: 773–784 (1964).

Figure 13–15   After D. A. MacLulich. *Univ. Toronto Studies, Biol. Ser.* No. 43 (1937).

Figure 13–16   Photograph courtesy of the U.S. Forest Service.

Figures 13–17,18.   After A. J. Nicholson. *Cold Spring Harbor Symp. Quant. Biol.* 22: 153–173 (1958).

Figure 13–19   Photograph courtesy of the U.S. Bureau of Sport Fisheries and Wildlife.

Figure 13–20   After C. H. Southwick. *Ecology* 36: 212–225; 627–634 (1955).

Figure 13–21   Photograph courtesy of the U.S. Bureau of Sport Fisheries and Wildlife.

Figure 14–1   Photograph courtesy of the U.S. Bureau of Sport Fisheries and Wildlife.

Figure 14–2   After G. F. Gause, *The Struggle for Existence.* Williams and Wilkins, Baltimore (1934).

Figure 14–3   After G. A. Schad. *Nature* 198: 404–406 (1963).

Figure 14–4   After H. A. Bess, R. V. Bosch, and F. H. Haramoto. *Proc. Haw. Entomol. Soc.* 27: 367–378 (1961).

Figures 14–5,6   After P. DeBach and R. A. Sunby. *Hilgardia* 34: 105–166 (1963).

Figures 14–7,8   After J. L. Harper. *J. Ecol.* 55: 247–270 (1967).

Figures 14–9,10,11   After D. R. Marshall and S. K. Jain. *J. Ecol.* 57: 251–270 (1969).

Figure 14–12   After T. Park. *Physiol. Zool.* 27: 177–238 (1954).

Figures 14–13,14   After R. D. Wright and H. A. Mooney. *Amer. Midl. Nat.* 73: 257–284 (1965).

Figure 14–15   Photograph courtesy of the American Museum of Natural History.

Figure 14–16   After J. P. Schulz, *Ecological Studies on Rainforest in Northern Suriname.* North-Holland, Amsterdam (1960).

Figure 14–17   Photograph courtesy of the U.S. Forest Service.

Figures 14–18,19   After A. M. Schultz, J. L. Launchbaugh, and H. H. Biswell. *Ecol.* 36: 226–238 (1955).

Figure 14–20   After R. I. Yeaton and M. L. Cody. *Theoret. Pop. Biol.* 5: 42–58 (1974).

Figure 14–21   Photographs courtesy of C. H. Muller, from C. H. Muller. *Bull. Torrey Bot. Club* 93: 332–351 (1966).

Figure 14–23   After R. H. MacArthur. *Ecology* 39: 599–619 (1958).

Figure 14–24   After R. W. Storer. *Auk* 83: 423–436 (1966).

Figure 14–26   After D. Lack, *Darwin's Finches.* Cambridge University Press, London (1947).

Figure 15–1   Photograph courtesy of the U.S. Department of Agriculture.

Figure 15–2   After C. B. Huffaker and C. E. Kennett. *Hilgardia* 26: 191–222 (1956).

Figure 15–3   Photograph courtesy of W. H. Haseler, Department of Lands, Queenland, Australia.

Figure 15–4   After C. J. Alexopoulos and H. C. Bold, *Algae and Fungi.* Macmillan, New York (1967).

Figure 15–5   Photograph courtesy of the U.S. Department of Agriculture.

Figure 15–6   After S. Utida. *Cold Spring Harbor Symp. Quant. Biol.* 22: 139–151 (1957).

Figure 15–7   Photographs courtesy of C. B. Huffaker. From C. B. Huffaker. *Hilgardia* 27: 343–383 (1958).

Figures 15–8,9   After C. B. Huffaker. *Hilgardia* 27: 343–383 (1958).

Figures 15–11,13   After C. S. Holling. *Canad. Entomol.* 91: 293–320 (1959).

Figure 15–15   After J. A. Gulland. In E. D. LeCren and M. W. Holdgate (eds.), *The Natural Exploitation of Natural Populations.* Wiley, New York (1962).

Figure 15–16   Photographs courtesy of D. Pimentel. After D. Pimentel. *Science* 159: 1432–1437 (1968). Copyright 1968 by the American Association for the Advancement of Science; D. Pimentel, W. P. Nagel, and J. L. Madden. *Amer. Nat.* 97: 141–167 (1963).

Figure 15–17   After D. Pimentel. *Science* 159: 1432–1437 (1968).

Figure 15–18   After F. O. Batzli and F. A. Pitelka. *Ecol.* 51: 1027–1039 (1970).

Figure 15–19   Upper photograph courtesy of the U.S. Forest Service.

Figure 15–20   R. M. Belyea. *J. Forestry* 50: 729–738 (1952).

Figures 16–1,2   Based on data in G. G. Simpson, *The Major Features of Evolution.* Columbia University Press, New York (1953).

Figure 16–3   After G. G. Simpson. *Amer. Scient.* 38: 361–389 (1950).

Figures 16–4,5   After R. H. MacArthur and E. O. Wilson. *Evolution* 17: 373–387 (1963).

Figure 16–6   After R. H. MacArthur. *Biol. J. Linnean Soc.* 1: 19–30 (1969).

Figure 16–8   After R. E. Ricklefs and G. W. Cox. *Amer. Nat.* 106: 195–219 (1972).

Figure 17–3   After C. D. White. *Vegetation — Soil Chemistry Correlations in Serpentine Ecosystems.* Ph.D. Diss., University of Oregon (1971).

Figure 17–4   After H. A. Fowells, *Silvics of Forest Trees of the United States. Agric. Handbook* No. 271, U.S. Department of Agriculture (1965).

Figure 17–5   After R. H. Whittaker. *Ecol. Monogr.* 30: 279–338 (1960) and R. H. Whittaker and W. A. Niering. *Ecology* 46: 429–452 (1965).

Figure 17–6   After R. H. Whittaker. *Ecol. Monogr.* 26: 1–80 (1956).

Figure 17–7   After R. L. Dressler. *Evolution* 22: 202–210 (1968).

Figure 17–8   Photographs courtesy of D. H. Janzen. After D. H. Janzen. *Evolution* 20: 249–275 (1966).

Figure 17–9   After K. F. Laglar, J. E. Bardach, and R. R. Miller, *Icthyology.* Wiley, New York (1962).

Figure 17–10   After D. Lack, *Darwin's Finches.* Cambridge University Press, London (1947) and D. Amadon *Evolution* 1: 63–68 (1947).

Figure 17–11   After D. Lack, ibid.

Figure 17–12   After F. Bourlière. Pp. 279–292 in B. J. Meggars, E. S. Ayensu, and W. D. Duckworth, *Tropical Forest Ecosystems in Africa and South America: A Comparative Review.* Smithsonian Inst. Press, Washington (1973).

Figure 18–1   Bottom photograph courtesy of the U.S. Forest Service.

Figure 18–2   Photographs courtesy of the U.S. Soil Conservation Service.

Figure 18–3   Photograph courtesy of the U.S. Forest Service.

Figure 18–4   Photograph courtesy of D. Whitehead.

Figure 18-6   After H. J. Oosting, *The Study of Plant Communities* (2nd ed.). W. H. Freeman, San Francisco (1956).

Figure 18-7   After J. S. Beard. *Ecology* 36: 89–100 (1955) and R. H. Whittaker, *Communities and Ecosystems.* Macmillan, New York (1970).

Figure 18-8   After J. T. Curtis and R. P. McIntosh. *Ecology* 32: 476–496 (1951).

Figure 18-10   After R. L. Dix. *Ecology* 38: 663–665 (1957).

Figure 18-11   After W. A. Eggler. *Ecology* 19: 243–263 (1938).

Figure 18-12   After A. S. Watt. *J. Ecol.* 35: 1–22 (1947).

Figure 18-13   After J. P. Grime and D. W. Jeffrey. *J. Ecol.* 53: 621–642 (1965).

Figures 19-2,3   Photographs courtesy of the U.S. Forest Service.

Figures 19-4,5   After H. Fritts. Pp. 45–65 in *Ground Level Climatology,* AAAS, Washington (1967).

Figure 19-6   After F. G. Stehli, R. G. Douglas, and N. D. Newell. *Science* 164: 947–949 (1969).

Figures 19-8,9   Photographs courtesy of the U.S. Forest Service.

Figure 19-10   After R. Patrick. *Ann. N.Y. Acad. Sci.* 108: 353–358 (1963).

Figure 19-11   From data in H. Heatwole and R. Levins. *Ecology* 53: 531–534 (1972).

Tables

Table 3-3   Compiled from G. K. Reid, *Ecology of Inland Waters and Estuaries.* Reinhold, New York (1961) and M. S. Gordon, *Animal Function: Principles and Adaptations.* Macmillan, New York (1968).

Table 8-2   Compiled from various sources summarized in Table 41-1 in R. E. Ricklefs, *Ecology.* Chiron, Portland, Oregon. (1973).

Table 8-3   From data in R. H. Whittaker and G. E. Likens. *Human Ecology* 1: 357–370 (1973).

Table 8-4   After R. G. Wiegert and D. F. Owen. *J. Theoret. Biol.* 30: 69–81 (1971).

Table 8-5   After R. L. Lindeman. *Ecology* 23: 399–418 (1942).

Table 8-6   After R. L. Lindeman (loc. cit.) and H. T. Odum. *Ecol. Monogr.* 27: 55–112 (1957).

Table 9-1   Based partly on E. J. Kormondy, *Concepts of Ecology.* Prentice-Hall, Englewood Cliffs, New Jersey (1969).

Table 9-2   After M. Witcamp. *Ecology* 47: 194–201 (1966).

Table 9-3   After G. E. Likens, F. H. Bormann, N. M. Johnson, and R. S. Pierce. *Ecology* 48: 772–785 (1967), and P. Duvigneaud and S. Denaeyer-de Smet. Pp. 199–225 in D. E. Reichle (ed.), *Analysis of Temperate Forest Ecosystems.* Springer-Verlag, New York (1970).

Table 10-1   After T. Sargent and S. A. Hessel. *J. Lepidop. Soc.* 24: 105–117 (1970).

Table 10-2   After A. T. Harrison, E. Small, and H. A. Mooney. *Ecology* 52: 869–875 (1970); H. A. Mooney and E. L. Dunn. *Amer. Nat.* 104: 447–453 (1970); H. Hellmers, J. S. Horton, G. Juhren, and J. O'Keefe. *Ecology* 36: 666–678 (1955).

Table 13-1   After data of O. Murie in E. S. Deevey, Jr. *Quart. Rev. Biol.* 22: 283–314 (1947).

Table 13-3   Census data from Audubon Field Notes; production estimates from R. H. Whittaker and G. E. Likens. *Human Ecol.* 1: 357–370 (1973).

Table 13-4   From E. L. Chaetum and C. W. Severinghaus. *Trans. North Amer. Wildl. Conf.* 15: 170–189 (1950).

Table 13-5   From H. G. Andrewartha. *J. Council Sci. Indust. Res. Australia* 8: 281–288 (1935).

Table 13-6   After D. A. Mullen. *Univ. Calif. Publ. Zool.* 85: 1–24 (1968).

Table 14-1   Data from G. Cox and R. E. Ricklefs, unpublished.

Table 15-1   From F. A. Pitelka, P. O. Tomich, and G. W. Treichel. *Ecol. Monogr.* 25: 85–117 (1955).

Table 15-2   From various sources given in Table 37-5 of R. E. Ricklefs, *Ecology.* Chiron, Portland, Oregon (1973).

Table 16-1   From F. Harper, *Extinct and Vanishing Mammals of the Old World.* Amer. Comm. Intern. Wildl. Prot., New York (1945).

Tables 16-2,3   From R. E. Ricklefs and G. W. Cox. *Amer. Nat.* 106: 195–219 (1972).

Table 17-1   From D. H. Janzen. *Evolution* 20: 249–275 (1966).

Table 17-2   Unpublished data of G. W. Cox and R. E. Ricklefs.

Table 18-1   From W. A. Eggler. *Ecology* 19: 243–263 (1938).

Table 19-1   From Table 48-3 in R. E. Ricklefs, *Ecology.* Chiron, Portland, Oregon (1973).

Table 19-2   From K. E. F. Watt *Canad. Entomol.* 97: 887–895 (1965).

Table 19-3   From E. P. Odum, *Fundamentals of Ecology* (3rd ed.). Saunders, Philadelphia (1971).

Table 19-4   From L. E. Hurd, M. V. Mellinger, L. L. Wolf, and S. J. McNaughton. *Science* 173: 1134–1136 (1971).

Index

Absorption spectrum, 119
Acacia-ant mutualism, 365
Acclimation, 210
Acidity, 170
Adaptation and environment, 192
Adaptive radiation, 372
Adaptive zone, 372
Age distribution of forests, 392
Age structure, of population, 238
Agriculture, productivity of, 125
Algal blooms, 171
Allelopathy, 290
Ammonia, in nitrogen cycle, 164
Ammonification, 166
Anaerobic conditions, 28
Aquatic environment, 23
  seasonality, 56
  variation in, 67
Aquatic habitats, 99
  availability of oxygen, 27
Aquatic herbivores, 330
Aquatic production, 117
Assimilation efficiency, 137
Association, definition of, 351
Atmosphere, properties of, 24
Autecology, 2
Autotroph, 11
  definition of, 130
Azotobacter, 167

Bacteria
  role in leaf decomposition, 167
  role in litter decomposition, 144
  role in nitrogen cycle, 166
  sulfur, 111

Barnacles, 282
Behavior and population regulation, 261
Biological control, 271, 318
  of mites, 301
  of prickly pear cactus, 302
Biological optimum, 187
Biomass accumulation ratio, 145
Birds
  adaptive radiations on island, 372
  ecological release on islands, 288
  extinction of, 334
  migration of, 213
  population density of, 249
  rapid evolution in, 225
  temperature regulation in, 204
Bogs, 65
Bomb calorimeter, 113

Calcification, 46
Caliche, 46
Calorie, definition of, 110
Capillary attraction, 42
Carotenoids, 119
Cation exchange, 172
Calcium budget, in forest, 177
Carbon cycle, 162
Carbon dioxide, 115
Carrying capacity
  artificial manipulation, 249
  of environment, 247
Celcius temperature scale, 54
Chaparral, 393
  influence of fire, 213
Chaparral vegetation, 92
  protein content of, 191
  water relations, 192

Character displacement, 296
Chlorophyll, 119
  use in measuring productivity, 116
Clausen, J., 196
Clay, composition of, 40
Clay-humus complex, 41
Clear-cutting, effect on minerals, 177
Clements, F., 377, 385
Climate
  global pattern, 50
  influence on competition, 272
  integrated description, 59
  maritime, 54
  seasonality, 54
Climate climax, 387
Climax, establishment of, 390
Climax community, 385
Climograph, 60
Coadaptation, 362
Coexistence of species, 267, 292
Community
  closed, 356
  definition of, 350
  influence of predation in, 352
  open, 356
Community boundaries, 356
Community energetics, 149
Community simplification, 415
Community stability, influence of diversity
    on, 410
Community structure, influence of
    competition on, 371
Compartment model, 156
Compensation point, 26
Competition, 266
  avoidance of, 292
  chemical aspects of, 290
  and community structure, 371
  direct and indirect, 289
  grass-shrub, 285
  in insect parasites, 271
  and role in extinction, 342
  and specialization, 269
  and succession, 390
Competitive ability, 347
Competitive exclusion principle, 268
Constancy, definition of, 402
Consumer, definition of, 130
Continuum, 359
Continuum index, 387
Convergent evolution, 373
Coral reefs, 108
Counter-adaptation, 348
Countercurrent heat exchange, 205

Cowbird, giant, 17
Cowles, Henry C., 48
Crocker, R. L. and J. Major, 49
Crops, productivity of, 125
Cycles, population, 257
Cyclic climax, 395

DDT, 6
Dansereau, P., 75
Darwin, Charles, 244
Deciduous plants, 213
Deer populations, density and
    reproduction, 249
Defoliation of trees, 330
Density-dependence, 253
Density-independence, 254
Denitrification, 166
Desert animals, temperature regulation,
    207
Desert vegetation, 91
  leaf form, 192
Deserts, occurrence of, 53
Detritus pathways, 134, 142
Detritivore, definition of, 130
Developmental responses, 200
Diapause, 217
Direct competition, 289
Dispersal, 310
Dispersion, population, 235
Distribution and community structure, 368
Disturbance and community stability, 413
Diversification of species, 269
Diversity
  alpha and beta, 371
  influence of predation on, 353
  influence on stability, 410
  maintenance of, 335
  of species, 342
Doldrums, 52
Dormancy, 216
Doubling times, 245
Dune succession, 378
Dust Bowl, 10

Ecocline, 383
Ecological efficiency, 133
Ecological overlap, 292
Ecological release, 283
Ecological replacement of species, 279
Ecology
  applied, 4
  definition of, 2

Ecosystem concept, 11
Ecotone, 356
Ecotypes, 194
Edge species, 359
Efficiency
  ecological, 133, 137
  photosynthetic, 121
  transpiration, 121
Energetic efficiency, 136
Energy, in water cycle, 160
Energy flow
  community, 149
  rate of, 144
Energy flux, 13
Environment
  geological influences, 65
  influence on competition, 272
  influence on population size, 253
  suitability of, 189
  topographic influences, 63
  variability in, 403
Environmental gradient, 188
Environmental heterogeneity, 310
Epilimnion, 58
Equilibrium
  biological, 9
  definition of, 400
  predator-prey, 312
  of species diversity, 342
Euphotic zone, 26
Eutrophication, 170
Eutrophy, 171
Evaporation
  and rainfall, 51
  role in water cycle, 160
Evapotranspiration, 61
Evolution
  conservation of, 341
  convergent, 373
  of insecticide resistance, 221
  of predator-prey equilibrium, 322
Evolutionary divergence, 295
Evolutionary responses, 219
Exploitation efficiency, 137, 142
Exploitation of populations, 320
Exposure, influence on vegetation, 63
Extinction, 333
  causes of, 341
  of island birds, 347
  of South American mammals, 337

Fahrenheit temperature scale, 54
Fall bloom, 59

Fall overturn, 59
Fecundity, definition of, 240
Field capacity of soil, 43
Fire, influence on chaparral, 213
Fire climax, 393
Fish
  diversity of in lakes, 345
  geographical distribution of, 368
  influence of temperature on, 211
Fisheries, management, 322
Floristic analysis, 75
Flour beetles, competition between, 279
Food, storage of, 212
Food chain, 128
Food source, alternative, 312
Food specialization, 294
Food web, 131
Forest, seasonal, 58
Forests
  age distribution of trees, 392
  cation exchange in, 173
  conifer, 103
  temperate, 94
Fossil record of diversity, 335
Functional response, 313
Fungi
  mycorrhizal associations, 282
  role in leaf decomposition, 167
  role in litter decomposition, 144

Galapagos Islands, 297
Gause, G. F., 268, 308
Generation time, calculation of, 242
Genetic traits, selection of, 219
Geographical distributions, 180
Geologic time scale, 336
Geometric growth, 244
Glucose, 110
Gradient analysis, 358
Grasslands, 96
Granite, weathering of, 39
Gross production, 112
Gross production efficiency, 137
Growth potential of populations, 244
Growth rings, 406

Habitat complexity, 312
Hairston, N. G. *et al.*, 354
Harper, J. L., 70
Harvest method, 113
Heat, units of, 110
Herbivores, 328

Heterotroph, 11
  definition of, 130
Hibernation, 217
Holdridge, L. R., 78
Holling, C. S., 313
Homeostasis, 198
  and ecosystem function, 217
  role in community stability, 406
Homeothermy, cost of, 204
Hubbard Brook Experimental Forest, 173
Huffaker, C. B., 308
Human population, effect of plague on, 326
Human survivorship curve, 238
Humidity, influence on populations, 189
Hydrarch succession, 380
Hydrological cycle, 159
Hypolimnion, 58

Ice, properties of, 57
Immigration rate, to islands, 342
Immunity, 328
Indirect competition, 289
Industrial melanism, 221
Insect population cycles, 259, 306, 413
Insecticides, 221, 301
Insects
  competition among parasites, 271
  developmental responses in, 200
  diapause, 217
  evolution of insecticide resistance, 221
  influence on vegetation, 330
  population dynamics, 253
Interspecific competition, 266
Intertidal habitats, 353
Intertidal organisms, density of, 249
Intertidal snails, 295
Intertidal zone, 67
  competition for space, 282
Intraspecific competition, 266
Ions, 31
Islands
  adaptive radiation on, 372
  diversity of birds on, 371
  diversity of species on, 342
  ecological release on, 288
  extinctions on, 334
  taxon cycles on, 346

Janzen, D., 353, 365

Kangaroo rat, 208
  adaptation to desert, 34

Keever, K., 390
Kelp beds, 304
Kettlewell, H. B. D., 221
Kilocalorie, definition of, 110
Kuchler, A. W., 72

Lakes
  community structure in, 354
  entrophication of, 171
Laterite, 45
Laterization, 44
Leaching, 39
  of minerals from leaf litter, 143
Leaf form, 192
Lemmings
  population cycles, 259
  predators on, 316
Lethal conditions, 188
Lichens, 364
Life, properties of, 8
Life forms, 75
Life table, 240
Life zones, 64
Light, 25
Light and dark bottle method, 115
Light intensity, influence on
    photosynthesis, 118
Likens, G., 124
Limiting resources, 268
Lindeman, R., 150
Lithosol, 36
Loam, 43
Lynx–hare cycles, 257

Magnesium budget, in forest, 177
Mammal fauna of South America, 337
Mammals
  effect of herbivores on vegetation, 329
  extinctions of, 334
  hibernation, 217
Marine environment, 108
Maritime climate, 54
Mark-release-recapture method, 223
Mean generation time, 242
Melanism, 221
Merriam, C. H., 64
Metals, evolution of tolerance to, 226
Methane, 163
Micelle, 41
Microenvironment, 69
Microhabitat, 69
Midge control program, 6

Migration, 213
Mimicry, 17
Mineral cycles, 156
Minerals
  availability in water, 30
  biological function, 30
Mite, giant red velvet, 14
Mites, predator-prey cycles of, 309
Moisture, influence on plant distribution 181
Murdoch, W., 354
Mutualism, 364
Mycorrhizae, 364
Mycorrhizal associations, 282
Myxomatosis, 228

Natural selection, 219
Negative feedback, 199
Net annual aboveground productivity, 114
Net production, 112
Net production efficiency, 137
Net reproductive rate, 242
Nicholson, A. J., 259
Nitrates and production, 117
Nitrification, 166
Nitrobacter, 166
Nitrogen, occurrence of, 165
Nitrogen cycle, 164
Nitrogen excretion, 34
Nitrogen fixation, 166
Nitrogen-fixing bacteria, 364
Nitrosomonas, 166
Numerical response, 315
Nutrient cycles, 14, 156

Ocean currents, 52
Odum, E. P., 146
Odum, H. T., 152
Old-field succession, 389
Oligotrophy, 171
Olson, Jerry, 48
Optimum, biological, 187
Oropendola, chestnut-headed, 17
Oxygen
  availability of, 26
  dissolved, 187
  influence on production, 123
  occurrence of, 158
Oxygen cycle, 161

Paine, R. T., 353
Paramo vegetation, 89

Parasite, brood, 17
Parasite-host interactions, 324
Parasites, competition between, 290
Passenger pigeon, extinction of, 333
Pattern-climax theory, 392
Peppered moth, melanism in, 221
Perturbation, effect on community, 413
Pesticides, influence on community, 5
Phosphates and production, 117
Phosphorus, radioactive tracer, 146
Phosphorus cycle, 169
Photosynthesis, 9, 110
  chemical equation for, 110
  influence of desiccation on, 192
  influence of environment, 118
Photosynthetic efficiency, 121
Physical environment
  influence of organisms on, 9
  influence on organisms, 16
Phytoplankton, 24
Pimentel, D., 322, 415
Plague, 326
Plant forms, 75
Plants
  chemical defenses, 231
  competition among, 274
  developmental responses in, 200
  ecotypes in, 195
  evidence for competition, 283
  evolution of metal tolerance, 226
Podsolization, 43
Pollination, adaptations for, 365
Pollution
  influence on kelp beds, 305.
  selection for industrial melanism, 223
Polyclimax, 392
Population cycles, 257, 259
Population density, 248
Population dispersion, 235
Population fluctuations, 251
Population growth, 408
Population regulation, 245
  behavioral aspects, 261
  growth potential, 244
  impact of predators, 300
Potassium budget, in forest, 177
Prairie, 394
Prairie vegetation, 96
Precipitation, See also Rainfall
  map of United States, 74
Predation
  effect on prey population, 300
  influence on community structure, 352
  influence on species diversity, 353

Predator efficiency, 318
Predator-prey cycles, 257, 306
Predator-prey equilibrium, 316
    evolution of, 322
Predator-prey stability, 311
Predator prudency, 320
Predators, effectiveness of, 301
Predictability, definition of, 402
Prickly pear cactus, 302
Primary production, 110
    aquatic, 124
    measurement of, 112
    terrestrial, 124
Primary succession, 380
Produce, definition of, 130
Production, gross and net, 112
Protective coloration, 231
Protein in food plants, 191
Protozoa, 308
Pseudomonas, 167
Pyramid of productivity, 131

Rabbits, 228
Rain shadow, 54
Rainfall, See also Precipitation
    influence on plant distribution, 181
    latitudinal trend, 51
    variability in, 403
Raunkiaer, C., 75
Recruitment, 316
Replacement series diagram, 277
Resources partitioning, 292, 295
Resources
    competition for, 267
    renewable, 248
Respiration, 112
    role in nutrient cycles, 158
Root nodules, 168

Sage, direct competition, 290
Salinity, 68
    influence on oysters, 187
Salt balance, 33, 202
Sand dunes, vegetation of, 48
Sawfly, spruce, 254
Search image, 315
Seasonality, 54, 190
Secondary succession, 380
Seed germination, 70
Self-thinning curve, 275
Sere, 378

Serpentine barrens, 65, 358
Sex ratio, 238
Shade tolerance and succession, 398
Smith, Neal, 17
Snails, 295
Social behavior, 261
Social pathology, 262
Soil
    effect on plant competition, 280
    influence on plant distribution, 184
    map of United States, 73
Soil formation, 36
    role of vegetation, 47
Soil horizons, 37
Soil temperature, 69
Soil water, 42
Soils
    alkaline, 46
    acid, 43
    tropical, 44
Solar constant, 126
Solar equator, 52
Specialization, 269
Species diversity, 342
Species number, 353
Specific heat, 29
Spring turnover, 57
Stability
    definition of, 402
    individual and population, 405
    influence of diversity on, 410
    predator-prey interactions, 311
Stable equilibrium, 400
Steady state, 343
Stratification, thermal, 58
Subclimax, 387
Succession, 377
    causes of, 388
    changes in vegetation structure, 398
Successional species, 398
Sugar maple, 181
Sulfur bacteria, 111
Survivorship, definition of, 240
Survivorship curve
    calculation of, 240
    of human population, 238
Symbiosis, 168
Synecology, 3

Tannins, 231
Tansley, A. G., 11
Taxon cycles, 346

Temperature
  influence on photosynthesis, 119
  influence on plant distribution, 181
  influence on populations, 189
  lethal, 211
  map of United States, 73
Temperature regulation, 204
Temperature scales, 54
Terrestrial environment, 23
Territories, 236
Thermal conductance, 29
Thermal properties of water and air, 29
Thermocline, 57
Thornthwaite, C. W., 61
Thrips, 252
Tidal cycles, 68
Time lag, 259
  in predator-prey cycles, 307
Topography and environment, 63
Torpidity, 205
Transeau, E. N., 138
Transient climax, 395
Transit time, 144
Transpiration, 61, 159, 192
Transpiration efficiency, 121
Tree rings, 285
Trophic level, 128, 147
  number in community, 153
  regulation of size, 354
Trophic structure, 13
Tropical vegetation, 86
Tropics, rainfall in, 51
Trypanosomiasis, 325
Tundra habitats, 104
Türesson, G., 196
Turnover in lakes, 57

Upwelling, 56
Urea, 34
Uric acid, 34

Vector, disease, 328
Vegetation
  continuum index, 387
  decomposition of litter, 143
  gradient analysis of, 358
  protein content of, 191
  role in mineral cycling, 177
  map of United States, 74
Vegetation types, 72
Virus, myxoma, 228
Viscosity, of air and water, 24

Water
  properties of, 23
  retention in soil, 42
Water cycle, 159
Water loss, 34
Watt, K., 412
Weathering, 39
Whittaker, R. H., 81, 124, 361, 393
Wilting capacity, 43
Wind patterns, 52

Xerarch succession, 380

Yield, maximum population, 320

Zinc tolerance, 226